Communications in Computer and Information Science 2208

AF173758

Rationale
The CCIS series is devoted to the publication of proceedings of computer science conferences. Its aim is to efficiently disseminate original research results in informatics in printed and electronic form. While the focus is on publication of peer-reviewed full papers presenting mature work, inclusion of reviewed short papers reporting on work in progress is welcome, too. Besides globally relevant meetings with internationally representative program committees guaranteeing a strict peer-reviewing and paper selection process, conferences run by societies or of high regional or national relevance are also considered for publication.

Topics
The topical scope of CCIS spans the entire spectrum of informatics ranging from foundational topics in the theory of computing to information and communications science and technology and a broad variety of interdisciplinary application fields.

Information for Volume Editors and Authors
Publication in CCIS is free of charge. No royalties are paid, however, we offer registered conference participants temporary free access to the online version of the conference proceedings on SpringerLink (http://link.springer.com) by means of an http referrer from the conference website and/or a number of complimentary printed copies, as specified in the official acceptance email of the event.

CCIS proceedings can be published in time for distribution at conferences or as post-proceedings, and delivered in the form of printed books and/or electronically as USBs and/or e-content licenses for accessing proceedings at SpringerLink. Furthermore, CCIS proceedings are included in the CCIS electronic book series hosted in the SpringerLink digital library at http://link.springer.com/bookseries/7899. Conferences publishing in CCIS are allowed to use Online Conference Service (OCS) for managing the whole proceedings lifecycle (from submission and reviewing to preparing for publication) free of charge.

Publication process
The language of publication is exclusively English. Authors publishing in CCIS have to sign the Springer CCIS copyright transfer form, however, they are free to use their material published in CCIS for substantially changed, more elaborate subsequent publications elsewhere. For the preparation of the camera-ready papers/files, authors have to strictly adhere to the Springer CCIS Authors' Instructions and are strongly encouraged to use the CCIS LaTeX style files or templates.

Abstracting/Indexing
CCIS is abstracted/indexed in DBLP, Google Scholar, EI-Compendex, Mathematical Reviews, SCImago, Scopus. CCIS volumes are also submitted for the inclusion in ISI Proceedings.

How to start
To start the evaluation of your proposal for inclusion in the CCIS series, please send an e-mail to ccis@springer.com.

Néstor Darío Duque-Méndez ·
Luz Ángela Aristizábal-Quintero ·
Mauricio Orozco-Alzate · Jose Aguilar
Editors

Advances in Computing

18th Colombian Conference on Computing, CCC 2024
Manizales, Colombia, September 4–6, 2024
Revised Selected Papers, Part I

 Springer

Editors
Néstor Darío Duque-Méndez (iD)
Universidad Nacional de Colombia
Manizales, Colombia

Luz Ángela Aristizábal-Quintero (iD)
Universidad Nacional de Colombia
Manizales, Colombia

Mauricio Orozco-Alzate (iD)
Universidad Nacional de Colombia
Manizales, Colombia

Jose Aguilar (iD)
Universidad EAFIT
Medellín, Colombia

ISSN 1865-0929   ISSN 1865-0937 (electronic)
Communications in Computer and Information Science
ISBN 978-3-031-75232-2   ISBN 978-3-031-75233-9 (eBook)
https://doi.org/10.1007/978-3-031-75233-9

This Springer imprint is published by the registered company Springer Nature Switzerland AG
The registered company address is: Gewerbestrasse 11, 6330 Cham, Switzerland

If disposing of this product, please recycle the paper.

Preface

Recent advances in informatics and computing continue at a rapid pace, providing industry, government, and education with great tools to improve productivity and facilitate their daily activities but, at the same time, posing new technical challenges and unforeseen ethical dilemmas.

The 18th edition of the Colombian Congress of Computing (18CCC) was organized by Universidad Nacional de Colombia and endorsed by the Colombian Society of Computing. 18CCC took place in Manizales, Colombia on September 4–6, 2024. The event aimed to create a space for the exchange of ideas, techniques, methodologies, and tools with a multidisciplinary approach, promoting synergy between researchers, professionals, students, and companies related to the topics of interest of the Congress.

The event focused on the presentation of research papers with significant contributions to knowledge or innovative experiences in different areas of computing, including: artificial intelligence; pattern recognition and computer vision; computational statistics and formal methods; cybersecurity and information security; data, information, and knowledge; ICT for education and e-Learning; industry 4.0 and digital transformation; software engineering and automation, among others.

These proceedings compile 50 full papers and 5 short papers that were selected according to the acceptance requirements to be included in these volumes from a total of 124 submissions that were received for peer review. The review process was double-blind and, on average, each paper was evaluated by three reviewers.

We thank the Colombian Computing Society, the local organizers at Universidad Nacional de Colombia, all members of the Program Committee, as well as our collaborators at Springer for making possible the publication of these proceedings. A special acknowledgement is given to the authors who submitted and presented their contributions and considered 18CCC an appropriate forum to discuss their research efforts and innovative applications.

September 2024

Néstor Darío Duque-Méndez
Luz Ángela Aristizábal-Quintero
Mauricio Orozco-Alzate
Jose Aguilar

Organization

General Chair

Luz Ángela Aristizábal Quintero   Universidad Nacional de Colombia, Colombia

Academic Chair

Nestor Darío Duque-Méndez   Universidad Nacional de Colombia, Colombia

Program Committee Chairs

Valentina Tabares Morales	Universidad Nacional de Colombia, Colombia
Ana Lorena Uribe-Hurtado	Universidad Nacional de Colombia, Colombia
Mauricio Orozco-Alzate	Universidad Nacional de Colombia, Colombia
Eduardo José Villegas-Jaramillo	Universidad Nacional de Colombia, Colombia
Leonardo Bermón Angarita	Universidad Nacional de Colombia, Colombia
Germán Augusto Osorio- Zuluaga	Universidad Nacional de Colombia, Colombia
Luz Arabany Ramírez-Castañeda	Universidad Nacional de Colombia, Colombia
María Amparo Prieto-Taborda	Universidad Nacional de Colombia, Colombia
Jheimer Julián Sepúlveda-López	Universidad Nacional de Colombia, Colombia

Program Committee

Silvana Aciar	Universidad Nacional de San Juan, Argentina
Francisco Álvarez	Universidad Autónoma de Aguascalientes, Mexico
Luis Álvarez-González	Universidad Austral de Chile, Chile
Andres Marino Álvarez-Meza	Universidad Nacional de Colombia, Colombia
Jairo Aponte	Universidad Nacional de Colombia, Colombia
Jesús Aranda	Universidad del Valle, Colombia
Jeferson Arango López	Universidad de Caldas, Colombia
Jaime Antero Arango Marín	Universidad Nacional de Colombia, Colombia
Luz Ángela Aristizábal Quintero	Universidad Nacional de Colombia, Colombia
John Atkinson	Universidad Adolfo Ibáñez, Chile
Fabio Avellaneda	Pontificia Universidad Javeriana, Colombia

Contents – Part I

Pattern Recognition and Computer Vision

Computational Statistics and Formal Methods

Cyber Security and Information Security

Contents – Part II

ICT for Education and e-Learning

Industry 4.0 and Digital Transformation

Software Engineering and Automation

Artificial Intelligence

Neuropathic Pain Detection Through Embedding Synergies of Deep Language and Image Models

Kevin A. Hernández-Gómez$^{(\boxtimes)}$ ⓘ, Julian Gil-Gonzalez ⓘ,
David A. Cárdenas-Peña ⓘ, and Álvaro A. Orozco-Gutiérrez ⓘ

Automatics Research Group, Universidad Tecnológica de Pereira, Pereira, Colombia
{kevin_loco,jugil,dcardenasp,aaog}@utp.edu.co

Abstract. The Global Burden of Disease states that neuropathic pain is suffered by 7–8% of adults worldwide, with severe repercussions on daily life, like drug abuse and psychological disorders. This work introduces a methodology for neuropathic pain classification through embedding synergies of the large language model BERT and image model ResNet50, handling clinical questionaries and EEG records, respectively. The classification task is a three-class problem with low, moderate, and severe pain categories. The embeddings of clinical data learned by BERT and the ResNet50-encoded topo-plots from EEG data feed an SVM classifier, further trained in a GroupKFold scheme from a thirty-six patients dataset. The accuracy obtained of 60%, outperforming single modality approaches, demonstrates the potential of multimodal approaches for enhanced pain diagnosis and treatment.

Keywords: First Large language models · automated medical diagnosis · multimodal learning

1 Introduction

According to the Global Burden of Disease (GBD), 7–8% of adults in the overall population indicate experiencing persistent pain exhibiting neuropathic characteristics [6]. The duration of the pain sensation classifies it into two main groups: acute and chronic. Acute pain, with a less than three months presence, is usually associated with evident physiological damage. Chronic pain develops from an acute state that endures beyond the healing process. Within this group, neuropathic pain (NP) remains the most perceived for the longest time, with pathological alterations in the somatosensory system [24]. Due to its long presence, the direct medical costs associated with NP reach approximately €2951 per patient per year [19]. In addition, the secondary adverse effects related to NP include drug addiction (due to opioid-based treatment), impaired social relationships, and psychological disorders. Hence, a proper assessment of pain perception is necessary to achieve adequate pain management and prevent the above effects [7].

© The Author(s), under exclusive license to Springer Nature Switzerland AG 2024
N. D. Duque-Méndez et al. (Eds.): CCC 2024, CCIS 2208, pp. 3–12, 2024.
https://doi.org/10.1007/978-3-031-75233-9_1

Pain management mainly recognizes two scales for assessing NP perception. One-dimensional scales (e.g., visual analog, numerical rating, verbal description) provide a single numerical or categorical value. However, it may lack sensitivity and can be challenging in specific clinical scenarios [8]. In contrast, multidimensional scales consider various sensory, emotional, affective, and location characteristics obtained through questionnaires. Nevertheless, multidimensional scales overlook psychological factors and rely on patient input [10].

To overcome the pain scale issues, medical imaging techniques for measuring pain perception have gained attention as objective alternatives. Noninvasive techniques like electroencephalography (EEG), functional magnetic resonance imaging (fMRI), magnetoencephalography (MEG), and positron emission tomography (PET) have demonstrated the relationship between pain experience and changes in brain electrical activity due to physiological and psychological factors [5,18,20]. Therefore, brain imaging techniques emerge as a promising alternative for more effective diagnoses and treatments, ultimately alleviating patient suffering. However, the absence of clear correlations between biomarkers and pain intensity challenges the interpretation of bioelectrical recordings for the NP perception. The diversity of chronic pain types induces brain activity variations, hindering the extraction of biomarkers [23]. Mitigating the above issues demands elaborated systems for assessing the perceived pain intensity.

Also, machine learning (ML) methods have enhanced prediction accuracy, concerns persist regarding interpretability and fairness, particularly in healthcare. The increasing concern is particularly significant for physicians as the deep learning approach shows greater potential to supplant traditional methods in tasks related to NP recognition. [18]. Numerous studies have employed ML techniques to measure perceived pain. Some approaches feed questionnaire data with sociodemographic, diagnosis, and treatment variables to traditional CatBoost and logistic regression algorithms [3,16]. Others extract EEG features from time-frequency representations and train well-known algorithms, including Random Forest (RF), K-nearest neighbor (KNN), or Support Vector Machine (SVM) classifiers [11,14,17]. However, there is a trend of research exploring the multimodal nature of medicine to provide holistic patient assessments drawn from diverse data sources for more accurate pain diagnoses and treatments [15].

For instance, Lin et al. developed a methodology to assess objective pain perception (baseline, low, and high pain) using nine physiological modalities: facial expressions, electroencephalography, eye movement, skin conductance, blood volume pulse, electromyography, respiration rate, skin temperature, and blood pressure [13]. The methodology, tested in 28 subjects, considered a Convolutional Neural Network model for facial expression processing and two cascade SVMs for pain classifying, reaching F1 scores of 58%, 47%, and 51%, respectively. Stefanos et al. proposed a multimodal framework for acute pain assessment using facial videos and heart rate signals on 87 patients [9]. The architecture consists of three models: the spatial module for creating learning-based representations of facial videos, the heart rate encoder for mapping the signal to a higher-dimensional space, and a transformer network for combining both modalities into an embedding. The results demonstrated a 39% accuracy for a multi-level pain assessment task. However, training a practical model from scratch for such a complex task requires substantial data and the incorporation of more effective modalities. An alternative for overcoming the limitation of data scarcity is to leverage pre-trained models, which provide a robust foundation prototype and allow building upon existing knowledge, enhancing training efficiency [22].

This work introduces a novel methodology for predicting perceived NP by merging questionnaire data and EEG signals. The proposed approach integrates both data sources into a unified representation, capturing clinical information from questionnaires and brain activity from EEG. The NP detection methodology devotes a feature embedding for each source by propagating data through the Bidirectional Encoder Representations from Transformers (BERT) model for the questionnaire answers and the ImageNet-pretrained ResNet50 for EEG recordings. Then, the embedding fusion feeds the classifier trained to predict the perceived NP in a GroupKFold validation scheme to avoid patient biasing.

The paper follows the agenda: Sect. 2 describes the dataset and mathematical framework of the proposed methodology, Sect. 3 discusses the attained results, and Sect. 4 concludes the work and offers future research directions.

2 Materials and Methods

Figure 1 overviews the general proposed pain classification methodology, consisting of three key stages: preprocessing, embedding, and classification. The preprocessing stage cleans questionnaire answers and represents EEG signals as topo-plot images. The embedding stage feeds the questionnaire data into a Large Language Model (LLM) to produce a latent representation. This stage also maps the topo-plots into vectors through a pre-trained deep-learning model. The classification stage concatenates both embedding to favor NP identification and trains a supervised learning classification model.

The implementation of the methodology can found in our github repository: github-Pain-classification.

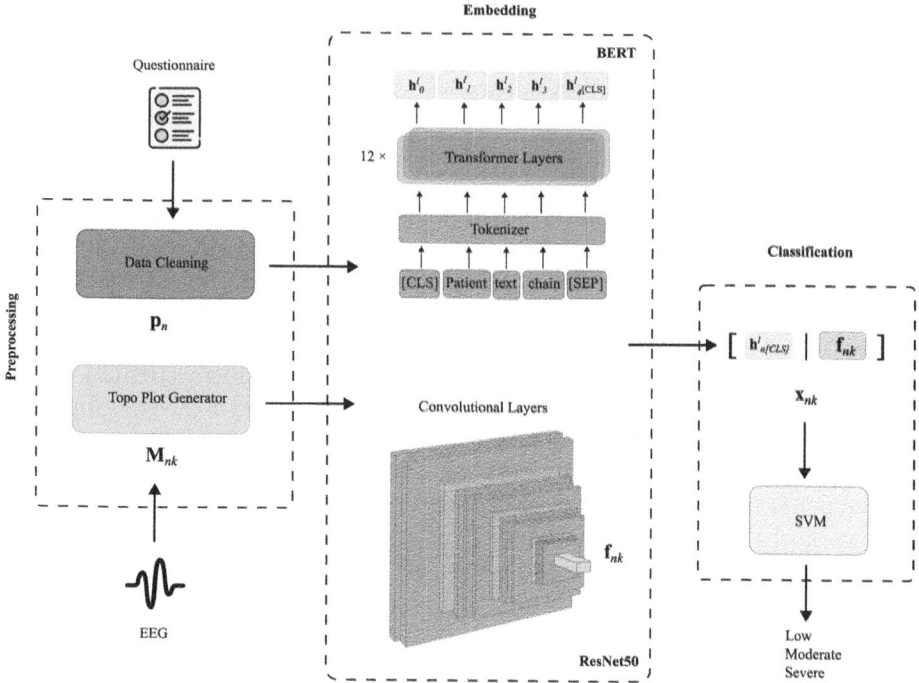

Fig. 1. Proposed methodology for pain classification from EEG and pain detection questionnaire data.

2.1   Dataset

Validation of the proposed classification methodology employs a publicly available dataset comprising thirty-six ($N = 36$) chronic neuropathic pain (NP) patients [25]. Each patient diligently completed a pain detection questionnaire consisting of eleven questions related to demographics, etiology, and medical treatment of NP. The questionnaire also requests a pain score on a 0 to 10 scale. This work categorizes the pain score into three target classes: low pain (0 to 2), moderate pain (3 to 5), and severe pain (6 to 10) [2]. Besides, each patient in the dataset holds ten minutes of electroencephalogram (EEG) signals recorded during spontaneous activity at rest under two conditions: watching a white cross in a dark background or open eyes condition (first five minutes) and closed eyes condition (last five minutes). EEG data was recorded at 24 channels positioned according to the 10–20 standard system at a sampling rate of 250 Hz. Therefore, the pain classification dataset is defined as $\mathcal{A} = \{\mathbf{q}_n, \mathbf{s}_n(t), y_n\}_{n=1}^{N}$, where n indexes the patient, the categorical vector $\mathbf{q}_n \in \mathcal{C}^{10}$ holds the ten clinical questions, $\mathbf{s}_n(t) \in \mathbb{R}^C$ corresponds to the EEG signal at C channels and time instant $t \in [1, T]$, and the target variable defines a three class problem $y_n \in \{\text{Low}, \text{Moderate}, \text{Severe}\}$.

2.2   Preprocessing

The preprocessing stage is two-fold. On the one hand, the clinical question-naire preprocessing replaces missing answers with "Unknown" and standardizes gender labels from "F" and "M" to "Feminine" and "Masculine", respectively. This stage also discards the following questions holding the same answer for all patients: "Have you suffered any head traumatism?" and "Do you suffer from any neurological disorder? (e.g., epilepsy, Alzheimer's, tinnitus)". Subsequently, the questionnaire answers $\mathbf{q}_n$ are automatically rewritten as a single text chain $\mathbf{p}_n$, aiming to improve the LLM prompting as presented in Fig. 2.

"0
25.0
Femenine
Central Nervous System Disorder (CRPS or Lyme)
More than 2 years
Pregabalin, amitriptyline
More than a year ago
Nevere blocks and infusions
Yes"

Fig. 2. Preprocessed questionnaire for ID patient 0.

On the other hand, the EEG preprocessing starts with a 20-second slicing without overlap to capture brain activity within delta (1–4 Hz), theta (4–8 Hz), alpha (8–12 Hz), and beta (12–25 Hz) rhythms. This stage further computes topographic maps (topo-plots) on each signal segment, yielding the distribution and intensity of electrical activity over the brain surface to take advantage of the spatial information on the EEG [1]. Figure 3 exemplifies two computed topo-plots, where black dots denote the electrode locations and black lines iso-magnitude curves. The color interpolates the magnitude of the electrical activity over the scalp from lower to higher amplitude color-coded from cooler to warmer. Therefore, the EEG preprocessing stage becomes a function $h(\cdot) : \mathbf{s}_n(t) \rightarrow \mathbf{M}_{nk}$ mapping the time series into a sequence K of color images of fixed size $\mathbf{M}_{nk} \in \mathbb{R}^{224 \times 224 \times 3}$, where k indexes the sliding window.

2.3   Embedding

Questionnaire Embedding. The BERT model, introduced by Google [4], is a pre-trained deep bidirectional transformer architecture. The bidirectional design and attention mechanism enable the model to consider the context from long-range text sequences. For the embedding stage, BERT maps text surveys for each patient into a fixed-size numeric vector representation, capturing the semantic meaning of the questionnaire text through hidden transformer layers:

$$\mathbf{h}_i^l = \phi(\mathbf{h}_i^{l-1} + \gamma(\mathbf{h}_i^{l-1}\mathbf{W}_Q^l, \mathbf{H}^{l-1}\mathbf{W}_K^l, \mathbf{H}^{l-1}\mathbf{W}_V^l)) \ \forall l \in [1, L] \tag{1}$$

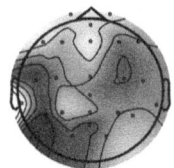 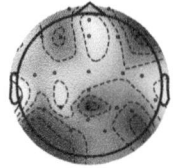

Fig. 3. Topo-plots for patient 14 during open eyes condition (left) and patient 10 during closed eyes condition (right).

where $\mathbf{h}_i^l \in \mathcal{H}^l$ represents the embedding of the i-th token within the text, gathered into the matrix $\mathbf{H}^l$ at the l-th hidden layer. $\phi(\cdot)$ and $\gamma(\cdot)$ correspond to a layer normalization operation and a multi-head self-attention function. $\mathbf{W}_Q^l$, $\mathbf{W}_K^l$, and $\mathbf{W}_V^l$ are the query, key, and value parameter matrices. The initial hidden sequence $\mathbf{h}_i^0$ results from the tokenization of the input text $\mathbf{q}_n$, adding [CLS] and [SEP] for start and separation tokens, followed by the token positional embedding. To encapsulate the contextual essence of the questionnaire, BERT represents the entire token sequence as the embedding of the [CLS] token from the final transformer layer ($L = 12$), that is $\mathbf{h}_{n,\text{[CLS]}}^L$ and $\mathcal{H}^L = \mathbb{R}^{768}$.

Topo-Plot Embedding. The brain topo-plot embedding relies on convolutional neural networks (CNN), which have been proven to extract meaningful features of brain function [21]. This work selects the ImageNet-pretrained ResNet50 model to embed each windowed topo-plot due to its capacity to automatically learn hierarchical representations from spatial data and unravel complex patterns from the EEG spatial distribution. Besides, the ResNet50 architecture mitigates the vanishing/exploding gradients problem encountered during error backpropagation in deep networks, enhancing the optimization of trainable parameters [12]. ResNet50 provides the image embedding as the output vector $\mathbf{f}_{nk} \in \mathcal{F}$ from the last hidden block: $\mathbf{f}_{nk} = \mathbf{W}_D^\top \beta(\varphi_{50}(\mathbf{M}_{nk}))$, where $\varphi_{50}(\cdot)$ denotes the last residual block, the global average pooling operator $\beta(\cdot)$ produces feature map of shape $7 \times 7 \times 2048$, and the matrix $\mathbf{W}_D$ matches the size of questionnaire and EEG embedding vectors through a linear dimension reduction, promoting a balanced integration of both data sources.

2.4   Classification Stage

This work configures a three-class classification problem to label the twenty-second EEG segments as low, moderate, or severe pain. Following a supervised learning scheme, the classifier training estimates a function $f(\cdot) : \mathcal{H}^L \times \mathcal{F} \mapsto \mathcal{Y}$ from a labeled preprocessed dataset $\{(\mathbf{h}_{n,\text{[CLS]}}^L, \mathbf{f}_{nk}), y_n\}_{n=1}^N$, being $\mathcal{Y} = \{\text{Low}, \text{Moderate}, \text{Severe}\}$. To this end, embedded input modalities concatenate into a single latent representation by patient-matching all EEG segment embeddings with the questionnaire embedding, that is, $\mathbf{x}_{nk} = \left[\mathbf{h}_{n,\text{[CLS]}}^L | \mathbf{f}_{nk}\right] \in \mathbb{R}^{1536}$. Then, the concatenated representation feeds a Support Vector Machine (SVM) classifier. The SVM separates classes in the training set using the surface that maximizes the margin in a Reproduced Kernel Hilbert Space, exploiting the non-linear discriminative patterns. The methodology trains the SVM with a radial basis function kernel in a GroupKFold scheme so that data from each patient appears exactly once in the test set across all folds and non-overlapping with the training split, warranting the generalization and mitigating the risk of over-fitting. Lastly, performance assessment presents the classification accuracy in average and standard deviation over K = 5 folds, which is computed as:

$$Accuracy = \frac{\text{Tr}(\mathbf{C})}{N} \tag{2}$$

where $\text{Tr}(\mathbf{C})$ is the trace of the confusion matrix $\mathbf{C}$ and represents the total correctly classified instances.

3 Results and Discussions

Validation experiments set up three classification regimes for the 5-fold cross-validated SVM to assess the model performance, as summarized in Table 1. The first one labels patients from the questionnaire answers embedded using BERT as described in section IIC. The second regime addresses the twenty-second EEG classification problem from the topo-plots embedded by ResNet50. Since the 30 topo-plots per patient belong to either of two recording conditions, this regime solves two pain classification tasks: i) Classifying from the first 15 topo-plots belonging to open eyes, and ii) classifying from the last 15 topo-plots computed on closed eyes condition. The third regime fuses questionnaire and EEG embeddings and solves the classification tasks on open and closed eyes conditions.

Table 1. Attained classification accuracy for the three regimes. Average and standard deviation are reported from a 5-fold cross-validation.

Model Embedding	Experiment	Accuracy (%)
BERT	Questionnaire	52.86 ± 14.00
ResNet	EEG (open)	37.33 ± 12.65
	EEG (closed)	44.38 ± 22.02
BERT + ResNet	Questionnaire + EEG (open)	62.86 ± 18.73
	Questionnaire + EEG (closed)	**66.48 ± 22.81**

As a first remark, the questionnaire and EEG conditions reach considerably different scores, with the former as the best performing, with more than 50% hits. The sole EEG (open) data underperformed other experiments due to the high noise related to ocular artifacts and pain-unrelated visual stimuli that hinder class discrimination. In contrast, EEG (closed) highlights the neural activity produced by NP, yielding higher scores. Nonetheless, fusing EEG and questionnaire data improves accuracy levels to over 60%, with the advantage of closed eyes condition over open eyes. This result indicates that clinical data provides relevant information to the neuroimaging recording, enhancing the classifier's performance.

On the other hand, for the combined embeddings (BERT + ResNet50), the standard deviation is relatively high (18.73% for open eyes and 22.81% for closed eyes), indicating variability in model performance depending on the fold. This suggests that the model's performance is inconsistent across different subsets of the data, this might be due to a small dataset (36 patients), leading to variations in the training and validation sets. However, our methodology still outperforms the state-of-the-art methods presented in [9,13].

4   Conclusion and Future Work

This work proposed a pain detection methodology relying on embedding the clinical questionnaire and EEG time series through well-known pre-trained deep learning models that performed as feature extraction alternatives. The Bidirectional Encoder Representations from Transformers (BERT) extracted the conceptual features from a single text chain representing the questionnaire. From the EEG side, the image-devoted ResNet50 model transformed EEG topo-plots into feature vectors comprising spatial brain activity information. Then, the fusion of the questionnaire and EEG embeddings through concatenation significantly boosted the pain detection accuracy in a cross-validation scheme that eliminated the patient bias. Concludingly, this study supports the potential of a multimodal approach, blending diverse data sources for enhanced patient classification based on perceived pain scales.

Given the findings of the current study, future work will explore the research directions: First, the training of vector embeddings from clinical questionnaires by fine-tuning BERT and other LLMs (e.g., ElMo, LLAMA, and GPT) should improve the supervised learning performance and a deeper understanding of the clinical data. Following the same direction, fine-tuning EEG-devoted networks, such as EEGNet, in a multimodal learning scheme will allow specialized embeddings to understand neuropathic pain from the clinical and neuroimaging perspective. Finally, validating the proposed methodology in the numeric rating pain scale must enable the identification of finer variations of pain level.

Acknowledgments. This study was funded by MINCIENCIAS, Under grants provided by the research program "ACEMATE", cod. 111091991908.

Disclosure of Interests. The authors have no competing interests to declare that are relevant to the content of this article.

References

1. Beniczky, S., Schomer, D.L.: Electroencephalography: basic biophysical and technological aspects important for clinical applications. Epileptic Disord. **22**(6), 697–715 (2020)
2. Brunelli, C., et al.: Comparison of numerical and verbal rating scales to measure pain exacerbations in patients with chronic cancer pain. Health Qual. Life Outcomes **8**, 1–8 (2010)
3. Davoudi, A., et al.: Fairness in the prediction of acute postoperative pain using machine learning models. Front. Dig. Health **4**, 970281 (2023)
4. Devlin, J., Chang, M.W., Lee, K., Toutanova, K.: Bert: pre-training of deep bidirectional transformers for language understanding. arXiv preprint arXiv:1810.04805 (2018)
5. Fernandez Rojas, R., et al.: Multimodal physiological sensing for the assessment of acute pain. Front. Pain Res. **4**, 1150264 (2023)
6. Ferreira, M.L., et al.: Global, regional, and national burden of low back pain, 1990–2020, its attributable risk factors, and projections to 2050: a systematic analysis of the global burden of disease study 2021. Lancet Rheumatol. **5**(6), e316–e329 (2023)
7. Gabriel, R.A., Swisher, M.W., Sztain, J.F., Furnish, T.J., Ilfeld, B.M., Said, E.T.: State of the art opioid-sparing strategies for post-operative pain in adult surgical patients. Expert Opin. Pharmacother. **20**(8), 949–961 (2019)
8. Garg, A., Pathak, H., Churyukanov, M.V., Uppin, R.B., Slobodin, T.M.: Low back pain: critical assessment of various scales. Eur. Spine J. **29**(3), 503–518 (2020). https://doi.org/10.1007/s00586-019-06279-5
9. Gkikas, S., et al.: Multimodal automatic assessment of acute pain through facial videos and heart rate signals utilizing transformer-based architectures. Front. Pain Res. **5**, 1372814 (2024)
10. Halverson, C.M., Doyle, T.A.: Patients' strategies for numeric pain assessment: a qualitative interview study of individuals with hypermobile ehlers–danlos syndrome. Disabil. Rehabil. 1–7 (2023)
11. Kimura, A., et al.: Objective characterization of hip pain levels during walking by combining quantitative electroencephalography with machine learning. Sci. Rep. **11**(1), 3192 (2021)
12. Lee, J.R., Ng, K.W., Yoong, Y.J.: Face and facial expressions recognition system for blind people using resnet50 architecture and cnn. J. Inf. Web Eng. **2**(2), 284–298 (2023)
13. Lin, Y., et al.: Experimental exploration of objective human pain assessment using multimodal sensing signals. Front. Neurosci. **16**, 831627 (2022)
14. Mari, T., et al.: External validation of binary machine learning models for pain intensity perception classification from eeg in healthy individuals. Sci. Rep. **13**(1), 242 (2023)
15. Meskó, B.: The impact of multimodal large language models on health care's future. J. Med. Internet Res. **25**, e52865 (2023)
16. Morisson, L., et al.: Prediction of acute postoperative pain based on intraoperative nociception level (nol) index values: the impact of machine learning-based analysis. J. Clin. Monit. Comput. **37**(1), 337–344 (2023)
17. Nezam, T., Boostani, R., Abootalebi, V., Rastegar, K.: A novel classification strategy to distinguish five levels of pain using the EEG signal features. IEEE Trans. Affect. Comput. **12**(1), 131–140 (2021)

18. Phan, K.N., Iyortsuun, N.K., Pant, S., Yang, H.J., Kim, S.H.: Pain recognition with physiological signals using multi-level context information. IEEE Access **11**, 20114–20127 (2023)

19. Png, M.E., et al.: Pain with neuropathic characteristics after surgically treated lower limb fractures: cost analysis and pain medication use. Brit. J Pain, 20494637231179809 (2023)

20. Teh, K., Armitage, P., Tesfaye, S., Selvarajah, D.: Deep learning classification of treatment response in diabetic painful neuropathy: a combined machine learning and magnetic resonance neuroimaging methodological study. Neuroinformatics **21**(1), 35–43 (2023)

21. Vafaei, E., Nowshiravan Rahatabad, F., Setarehdan, S.K., Azadfallah, P., et al.: Extracting a novel emotional eeg topographic map based on a stacked autoencoder network. J. Healthc. Eng. **2023** (2023)

22. Wang, H., Li, J., Wu, H., Hovy, E., Sun, Y.: Pre-trained language models and their applications. Engineering (2022)

23. Zebhauser, P.T., Hohn, V.D., Ploner, M.: Resting state eeg and meg as biomarkers of chronic pain: a systematic review. Pain, 10–1097 (2022)

24. Zhu, C., Zhong, W., Gong, C., Chen, B., Guo, J.: Global research trends on epigenetics and neuropathic pain: a bibliometric analysis. Front. Mol. Neurosci. **16**, 1145393 (2023)

25. Zolezzi, D.M., Naal-Ruiz, N.E., Alonso-Valerdi, L.M., Ibarra-Zarate, D.I.: Chronic neuropathic pain: EEG data in eyes open and eyes closed with paindetect and brief pain inventory reports. Data Brief **48**, 109060 (2023)

Detection of Depressive Symptomatology in Written Narratives in Spanish Using Machine Learning and Semantic Ontology

Eliana Ortiz, Juan Barrero(✉), Rubby Castro-Osorio, Andrés Domínguez, and Natalia Caicedo

Universidad Católica de Colombia, Bogotá, Colombia
jcbarrero@ucatolica.edu.co

Abstract. This exploratory research followed a mixed methods design, in which a text analysis study was implemented with machine learning techniques, semantic ontology of depressive symptomatology and agile methodology in software development. The sample consisted of 389 young people between 18 and 30 years old (64.3% women, 35.7% men) with an average age of 22.04 years (SD = 3.51). There were six phases: 1) Preparation and validation of open questions, 2) Collection of information, 3) Adaptation and implementation of the ontology in Protégé and Python software, 4) Preparation of a control dataset, 5) Generation of the automatic method in the notebook and 6) Content rating by expert reviewers in psychology. The results show that in a sample of 50 young people between the ages of 18 and 30, the best classifier was the Naive Bayes algorithm, with an accuracy of 86%, precision of 91%, completeness of 80% and an F1 value of 72.8%.

Keywords: Machine learning · Semantic ontology · Depression · Young adult · Text analysis

1 Introduction

Young adults go through different changes in the contexts where they interact, for example, during this vital period there is a transition to adulthood, which implies the adaptation to a new role in society, according to the appearance of new responsibilities and challenges in daily life, as well as vulnerability towards certain risk factors, that might generate difficulties in mental health [1, 2]. Some of the most prevalent mental health problems for this population are depressive symptoms. For example, in Colombia, according to the 2015 Mental Health Survey, 13.9% of people between 18 and 44 years reported symptoms of depression, and seven or more anxiety symptoms (4.9% women and 3.2% of men) [3]; Likewise, in 2015 the highest proportion of people who received any mental health services (N = 36,584) experienced moderate or severe depression, among them 70,4% were women [4].

These figures have increased during the pandemic, in the PSY-COVID study, whose objective was to evaluate mental health during the COVID-19 pandemic in Colombia.

N. D. Duque-Méndez et al. (Eds.): CCC 2024, CCIS 2208, pp. 13–26, 2024.
https://doi.org/10.1007/978-3-031-75233-9_2

The prevalence of depression is reported in 36% of young adults surveyed, in 31% prevalence anxiety and 35% somatization symptoms [5]. Likewise, the DANE (2021) [6] reported that during the sanitary crisis, people between 25 to 54 years old often reported: feeling tired (18.2%), loneliness (7.0%), sadness (17.8%), headaches (14.4) and difficulty falling asleep (14.6%). In a meta-analytic study [7] reviewing 22 studies, a total of 4318 patients with COVID-19 were included in the analysis. The prevalence of symptoms of depression, anxiety, and insomnia were 38% (95% CI = 25–51), 38% (95% CI = 24–52), and 48% (95% CI = 11–85), respectively. In another meta-analysis [8], 66 studies with 221970 participants were included, reporting an overall pooled prevalence of depression, anxiety, distress, and insomnia of 31.4%, 31.9%, 41.1%, and 37.9%, respectively.

According to the most recent Colombian Mental Health Survey (2023) [9], 10.92% of the population have been diagnosed with depression at some point in their lives, 8.76% of men and 12.82% of women. 17.16% of people have a probable depressive disorder, 13.85% of men and 19.95% of women. 2.31% of young people; 10.79% of adults have been diagnosed with depression at some point in their lives. 18.1% of young people; 16.44% of adults have a possible depressive disorder or are at risk for depression. These studies reaffirm the importance of contributing to the prompt detection and evaluation of this phenomenon, since there is evidence of a significant increase after the pandemic compared to previous studies. To better contextualize this problem, depression is defined as a multifactorial disorder that includes a set of specific behavioral or motor, cognitive, social, and biological symptoms that lead the individual to lose contact with reinforcement contingencies available on their environment, therefore generating difficulties in their daily functioning and hindering their motivation [10].

There are several mood disorders; however, there is an emphasis on major depressive disorder, which according to the Diagnostic and Statistical Manual of Mental Disorders (DSM-V), is characterized by the following symptoms: depressed mood, decreased interest and pleasure in activities, changes in eating habits, difficulty sleeping, feelings of worthlessness or excessive guilt, decreased ability to concentrate, and recurrent thoughts of death. The criteria for this diagnosis establish that at least one of the main symptoms (depressed mood and loss of pleasure) or five among all of the above symptoms must be present for a minimum period of two-weeks [11]. To assess depression, mental health professionals often use traditional questionnaires with closed-ended questions that assess symptomatology with Likert-type response scales. In the last decade, alongside other disciplines, aspects such as the conceptualization, diagnosis, and management of depression have been elaborated on. For instance, computer applications have recently been used to support the diagnosis of mental illness. Machine Learning (ML) is an area of artificial intelligence that studies the design of algorithms, based on the search for patterns and statistical models, with which computational systems can be trained to perform specific tasks [12].

Text analysis using ML allows, based on Natural Language Processing (NLP) and classification algorithms, to process, interpret and extract meaningful information from texts. In the literature there are different algorithms and techniques used to classify text, for this research the Naive Bayes and logistic regression algorithms were selected for their ease of implementation and low computational complexity [13]. One of the most widely used tools for training algorithms is semantic ontology, which is defined as an explicit formal description of concepts in a particular domain, where various characteristics and attributes that constitute a knowledge base are described [14, 15].

To build an ontology, the following steps must be followed: define the domain and scope; determine the intention of use; reuse existing ontologies and vocabularies; list the important terms of the domain; define class hierarchy; create instances and validate the ontology [14]. There are different types of softwares that have been used for ontology building in the field of health and engineering, explicitly stating what the class hierarchy is and to which classes individuals belong, one of the most promising is Protégé [14, 16–20] Which has been shown to be useful in supporting the detection and monitoring of mental status as well as inferring and assessing possible psychological issues [16, 17]. There are studies regarding the diagnosis and detection of depressive symptoms based on ontologies. For instance, Chang in 2015 [17] designed an inference model based on an ontology and a Bayesian network to infer the probability of developing depression. Their ontology yielded two models: one was a basic model that explained the characteristics of the patient, the disease and depression; the second was a model that showed the relationship between dysthymia and its symptoms.

In another similar study, Jung et al. (2017) [21] aimed to define an ontology and terminology associated with depression in adolescents through sentiment analysis of data extracted from social networks. This ontology consisted of 443 classes, with 60 relations between them and 1682 synonyms; Likewise, it was found that sentiment analysis on "academic stress" and "suicide" had a negative impact on depressive symptoms in adolescents. Ontologies have also been designed for other mental health issues such as obsessive-compulsive disorder (OCD) [22] and post-traumatic stress disorder (PTSD) [20]. Another important aspect is the use of ontology in personalized and intelligent services (medical teams) that help to monitor the mental health status of patients with chronic diseases, as evidenced in Benfares et al. (2019) [16] who conducted an ontology for prevention and screening of anxiety and depression disorders in oncology patients who are part of a hospital center. On the other hand, Dias et al. (2019) [23] proposed a model of care for patients with anxiety disorders, depression, and stress, through gamification and biodata (iAware) with an ontological basis; this model consisted of five modules (knowledge, profile, action, gamification, intervention, and context).

All of the above supports the existence of some explanatory models based on ontologies that evaluate psychological problems; however, it is important to note that most of the reviewed studies include models for some specific mental health disorders [20, 22]; or other diseases [16], application with diagnosed patients or adolescent population and some contemplate the analysis of two or more psychological variables [17, 21]. Also, most of these research studies have been conducted in Asia and the Middle East; furthermore, are written in English, which evidences the lack of studies that have designed ontologies to detect and facilitate the diagnosis of emotional symptomatology in Latin American populations. Specifically in Colombia, we seek to innovate the current approach to assessing depression by using ontologies and machine learning. Therefore, the objective of this study is to innovate in this field of research by implementing machine learning processes and a semantic ontology to support the detection of depressive symptomatology in a sample of young Colombians.

2 Method

The present research follows a mixed methods design [24], where a text analysis study was conducted using machine learning techniques, semantic ontology of depressive symptomatology and agile methodology in software development. The study proposes the following phases:

Phase 1: Elaboration and validation of open-ended questions about depressive symptomatology in order to collect information from narrative texts.

Phase 2: Collection of sociodemographic data and depressive symptomatology by using the answers of said open questions provided on an online form (Google Form) and the application of the psychological questionnaire Depression, Anxiety and Stress Scales - 21 (DASS-21) [25]. Afterwards, the information was organized in a data table called "experimental dataset".

Phase 3: Adaptation and implementation of the ontology in Protégé and Python software [14].

Phase 4: Creation of a dataset, formed by sentences called "Control dataset", which evaluated the performance measures of the machine learning algorithms.

Phase 5: For the generation of the automatic method in the notebook in python, machine learning processes were implemented through the agile waterfall methodology, with its different stages: requirements, design, development, and testing. The development was carried out in the Google cloud platform, where Colab and BigQuery were used.

Phase 6: The process of classifying the text content of a subsample is performed based on the various text analyses generated by machine learning algorithms and evaluated based on the criteria of two mental health professionals. This helps to determine whether the participant has depressive symptoms according to the machine learning assessment.

3 Development

3.1 Elaboration and Validation of Open-Ended Questions

At first instance, the two professionals with master's degrees in the clinical area worked on formulating six open-ended questions based on the diagnostic criteria for major depressive disorder proposed by the DSM-V [11] (see Table 1). These questions were subsequently subjected to a content validation process using the Delphi methodology [26], in which five expert reviewers in clinical psychology participated. An inter-observer agreement rate of 100% was found (Table 2).

3.2 Collection of Information

For data collection, a Google Forms form was used, which contained three sections: informed consent and habeas data, sociodemographic data and open-ended questions formulated in this study to evaluate depressive symptomatology. The selection of participants was done through a non-random convenience sampling, the call was made through invitations on social networks such as Facebook, Instagram and through institutional email. The inclusion criteria were: being between 18 and 30 years old, living

Table 1. List of open questions asked

Question	Open question applied	Translation of the open questions asked
1	Piensa en algunas situaciones, en las dos últimas semanas, en las que te hayas sentido triste, desanimado/o con menos energía que antes y descríbenos qué pensabas y qué sentías	Think of some situations in the last two weeks when you have felt sad, discouraged and/or less energetic than before and describe what you were thinking and feeling
2	¿En las últimas semanas, te has sentido con menos energía (de lo normal) o con dificultades para concentrarte en las actividades diarias? Descríbenos más detalladamente…	In recent weeks, have you been feeling less energetic (than usual) or having difficulty concentrating on daily activities? Describe in more detail…
3	¿En las últimas semanas, te has sentido más acelerado o lento (de lo normal) al enfrentar las actividades diarias? Descríbenos más detalladamente…	In the last few weeks, have you felt more rushed or slower (than usual) when facing your daily activities? Describe in more detail…
4	¿En las últimas dos semanas has tenido cambios en tus hábitos o rutina de alimentación? (has mantenido tus horarios habituales de alimentación, te saltas alguna comida o ha aumentado o disminuido tu apetito) Descríbenos cómo fueron esos cambios…	In the last two weeks have you had any changes in your eating habits or routines (have you maintained your usual eating schedule, have you skipped meals, or has your appetite increased or decreased)? Describe how these changes occurred
5	¿Qué tan satisfecho/y cómodo/a te sentiste con la manera en la que enfrentaste las situaciones que fueron difíciles en las últimas dos semanas? ¿Las cosas resultaron como querías? Descríbenos la situación…	How satisfied/and comfortable were you with the way you dealt with situations that were difficult in the last two weeks? Did things turn out the way you wanted them to? Describe the situation…
6	¿Has tenido deseos o ideas de morir en las últimas dos semanas? ¿O has pensado que no vale la pena vivir? Descríbenos al respecto…	Have you had wishes or thoughts about dying in the last two weeks? Or have you thought that life is not worth living? Tell us about it…

Table 2. Number of concepts by category of depressive symptomatology

Category name	Category name translation	Number of concepts
Ánimo	Mood	144
Apatía	Apathy	78
Alimentación	Alimentation/ Feeding	33
Descanso	Rest	58
Actividad física	Physical Activity	57
Desaliento	Discouragement	111
Cargo de conciencia	Guilt	144
Atención dispersa	Scattered attention	49
Ideas suicidas	Suicidal thoughts	51

in Colombia, providing consent to participate in the study and for the treatment of their data and voluntarily filling out the online form.

The sample consisted of 389 young people between 18 and 30 years old (64.3% women, 35.7% men) with an average age of 22.04 years (SD = 3.51), residing in 39 municipalities of Colombia, most were single (92%) and were currently studying undergraduate studies (82%).

3.3 Adaptation and Implementation of the Ontology in Protégé and Python Software

During the phase of adaptation and implementation of the Ontology as input for the training of machine learning algorithms, a semantic ontology developed and validated in a previous stage of the project by four professionals with master's studies in clinical psychology was used, including nine categories of depressive symptoms and a total of 725 concepts (terms) developed in the software Protégé; which were exported in a JavaScript Object Notation (JSON).

For the preparation of the texts written by the participants (in Spanish), they were transformed into a type of language understood by the machines through natural language processing (NLP) techniques. In this process, the following libraries were used: Freeling (Welcome | FreeLing Home Page), Sanford (The Stanford NLP Group) and NLTK (Natural Language Toolkit), specialized in these processes, this implementation was carried out in the Google Colab Notebooks environment.

3.4 Preparing a Control Dataset

To evaluate the performance of the algorithms implemented in the notebook, a control data set consisting of 150 sentences presenting depressive symptomatology was created. These sentences were developed and classified into semantic ontology categories by two psychology professionals. This dataset serves as control input to evaluate the performance of the implemented algorithms. It will allow us to evaluate the performance and execution levels of the classification and machine learning processes within the developed notebook.

3.5 Automatic Method Generation on the Notebook

In the process of machine learning and generation of results by a computer system, the selection of two classification algorithms was carried out; these were selected for their feasibility for the project and the method of application, since this is successfully assimilated to the conditions of development of the research. The first machine learning algorithm selected was Naive Bayes [26] where the implementation process was the following:

3.5.1 Naive Bayes Algorithm

Entrance:
   Training data set T.
   Array with the values of the predictor variable in the test data set.
   Process:

1. Read the training data set T
2. Read the evaluation data set E
3. Calculate the mean and standard deviation of the predictor variables in each class against the evaluation set.
4. Repeat

5. Calculate the repetitions with respect to the category for each class and decide which is the most appropriate assignment according to the probability obtained.

Exit: Single categorization for the dataset of assessed participants

The second classification process was logistic regression [27], which is described below:

3.5.2 Logistic Regression Algorithm

Entrance:

Training data set T

Array with the values of the predictor variable in the test data set.

Process:

1. Read the training data set T
2. Read the evaluation dataset $d1$.
3. Set the target value for the regression to:

$$z_i \leftarrow \frac{y_i - P(1|d_i)}{[P(1|d_i) * (1 - P(1|d_i))]} \tag{1}$$

4. Regression formula
5. Initialize the instance weight.

$$d1 : toP(1|d_i)(1 - P(1|d_i)) \tag{2}$$

6. Finalize a (j) to the data with class value (zj) &
7. weights (wj)
8. Generate the decision on the classification label
9. Assign (class label: 1) if
10. P (1|dj) > 0.5, otherwise (class label: 2)
11. Repeat

Exit: Single categorization for the dataset of assessed participants

Afterwards, in the classification phase, the Naive Bayes and Logistic Regression [28, 29] classificatory algorithms were integrated into the notebook to generate the advanced analytical processes and evaluate the written narratives shared by the different participants, and subsequently obtain the classification of the depressive symptom generated by the two proposed algorithms.

Finally, the process was implemented to generate performance metrics which evaluate the viability of the classifications obtained regarding the answers of the participants, by means of the computer control mechanisms, for this a process of classification of results was implemented, with this the metrics of Accuracy Score, Precision, recall, f1-score, support was obtained [30], additionally the confusion matrix was also obtained.

3.6 Classification of Texts by Psychologists

For the validation of the performance measures of the texts classified by the algorithms, a dataset was formed with the 2334 texts of the participants' responses (6 responses of the 389 participants). This dataset was given to two psychologists with experience in clinical psychology, who classified each of the contents of the narrative texts of each of the participants (response to the open questions about depression designed for this research), in one of nine categories proposed for the construction of the ontology from the diagnostic criteria of major depressive disorder posed by the DSM-V [11]. In addition, taking into account the score obtained in the depression subscale of the DASS-21, the participants were classified into clinical and non-clinical (with a cut-off point of 6), which indicates a mild level on this subscale. Subsequently, this classification made by the professionals was contrasted with the one generated through the machine learning algorithms mentioned above, which allowed us to obtain a percentage of similarity and acceptance of the classifications generated by the automatic method.

4   Results and Analysis

In accordance with the phases described above, the responses of the depression subscales of the DASS-21 answered by the 389 young adults who voluntarily participated in the research were analyzed. As a result of the data analysis 65.29% (254) of the participants were categorized with clinical scores on the depression subscale (score greater than 7) and 34.7% (135) as non-clinical (score less than 6). This ensures that this research sample included participants with both clinical and non-clinical scores for depression. Within this project, the ontology related to depressive symptomatology containing the 725 concepts (which are the terms associated with each of the categories: mood, apathy, eating, rest, physical activity, discouragement, conscience, attention, and suicidal ideas) was implemented using the Protégé software and exported in a Json file.

From the data set of the 389 participants, the 2,334 narratives corresponding to the six open-ended questions answered by each participant were taken. Natural language processing was carried out, in which the texts were converted from human language to machine-understandable language. First, words that had no relevant information or did not contribute to the study were eliminated using a stop words dictionary in Spanish previously reviewed by psychologists. Second, the texts were tokenized into words using the Natural Language Toolkit (NLTK) in Python and made ready for use in the text classification algorithms.

After observing the relationship of the ontology with the words in the texts, Naive Bayes and Logistic Regression algorithms were selected from the literature and trained with a dictionary created with the semantic ontology terms for depressive terms. The trained algorithms were implemented in the "Intelligent Notebook" that carried out the classification of the texts in the nine categories of the semantic ontology of depression. Subsequently, different tests were carried out with the intelligent Notebook to identify the results obtained in the process of classifying texts in each of the categories and performance metrics. Figure 1 shows an example of classification, as evidenced by the text was classified by the algorithm with an accuracy of 86% and recall of 91% and belongs to the category of guilt.

```
classification_report
                        precision    recall  f1-score   support

     Actividad física       0.00      0.00      0.00         0
         Alimentación       0.00      0.00      0.00         0
              Apatía        0.00      0.00      0.00         8
     Atención dispersa      0.10      0.05      0.11         1
  Cargo de conciencia       0.86      0.91      0.80        96
          Desaliento        0.30      0.12      0.60        26
            Descanso        0.00      0.00      0.00         0
       Ideas suicidas       0.00      0.00      0.00         0
               Ánimo        0.46      0.71      0.40        50

            accuracy                            0.21       181
           macro avg        0.11      0.06      0.08       181
        weighted avg        0.42      0.21      0.27       181
```

Fig. 1. Ranking Report

As a final stage for validation 50 participants were taken, who were randomly selected ensuring that there were 25 clinical and 25 non-clinical. The responses to the six questions were merged into a single text containing the participants' narratives, two examples are shown in Table 3. The texts vary in length with a minimum of 9 words, a maximum of 959 words and an arithmetic mean of 154 words.

Table 3. Sample text from participant

Num	Combined text of the participants' responses	Translation of the joined text of the answers of the participants
38	Pensaba que no estaba tranquilo, preocupado por estar impaciente, desmotivado/Sí he estado desconcentrado y poco motivado, reemplazo las actividades que debería hacer con otras cosas o voy a comer bastante para ocuparme, también cuando empiezo a trabajar trato de utilizar música o podcast que me motiven mientras lo hago/Me he sentido lento o demorado haciendo cosas que antes no me tomaban tanto tiempo. Cuando pasa me siento aburrido y triste como apagado, escucho música para no tener que pensar en eso, no hay actividades exactas pero pasa frecuente en la semana/como más cuando estoy estresado pero en las últimas semanas he notado que tengo más hambre en las noches y menos sueño, al otro día me siento agotado pero se repite lo mismo acabando el día empiezo a sentir más energía, no acostumbro a saltarme ninguna comida/Sí siento que fue una forma efectiva afrontar las cosas como lo hice, lo hable con familiares y amigos, hable conmigo sobre lo que me pasaba y traté de enfrentarlo; trato de mantenerme positivo ante este tipo de situaciones de ansiedad, momentos de depresión o cuando me lleno de energía/No, he vuelto a pensar en eso hace varios años	I thought I was not calm, worried about being impatient, unmotivated/Yes I have been unfocused and unmotivated, I replace activities I should be doing with other things or I go to out a lot to occupy myself, also when I start working I try to use music or podcast to motivate me while doing it/I have felt slow or delayed doing things that didn't take me so long before. When it happens I feel bored and sad like shut down, I listen to music so I don't have to think about it, no exact activities but it happens frequently during the week/I eat more when I'm stressed but in the last few weeks I've noticed I'm hungrier at night and less sleepy, the next day I feel exhausted but the same thing happens again by the end of the day I start to feel more energy, I don't usually skip any meals/I do feel it was an effective way of coping the way I did, I talked to family and friends, I talked to myself about what was happening to me and tried to deal with it; I try to stay positive about these kinds of anxious situations, moments of depression or when I get full of energy/No, I have thought about it again several years ago

For each of the 50 texts, two psychologists with experience in clinical psychology manually categorized them into one of the nine categories of the ontology of emotional symptomatology: mood, apathy, eating, rest, physical activity, discouragement, guilt, scattered attention, and suicidal ideation, and an unclassified category was added for texts that did not correspond to any of them. On the other hand, the same full texts of the participants were classified using Naive Bayes and Logistic Regression algorithms and confusion matrices were constructed in order to compare the manual classification of the psychologists with the classification generated by the algorithms trained with the words of the ontology.

Table 4. Confusion matrix manual classification vs. classification with Naive Bayes

Manual sorting	a	b	c	d	e	f	g	h	i	Total
Unclassified	7									7
Physical Activity							1			1
Mood			9	1		1		1		12
Apathy				7		1	1			9
Scattered attention					3					3
Guilt			1		2	6				9
Discouragement			1				6			7
Rest								1		1
Suicidal thoughts									1	1
Total	7	0	11	8	5	8	8	2	1	50

a. = Unclassified, b. = Physical Activity, c. = Mood, d = Apathy, e. = Scattered attention, f. = Guilt, g. = Discouragement, h. = Rest, i. = Suicidal thoughts.

In this confusion matrix in Table 4, a classification accuracy of 80% and a classification error of 20% was obtained. As a result of the classification of the texts carried out by the psychologists, for each of the categories, it was found that the distribution was: mood (24%), apathy 18%, guilt (18%), discouragement 14%, unclassified (14%), scattered attention (6%) and in the case of physical activity, rest and discouragement there was only one response, which gives 2%. Regarding the classification errors, it was found that most of them occurred in the mood category, it could be interpreted that the algorithm was confused and classified related words in the categories of apathy and conscience. The category "Feeding", which is in the semantic ontology of depression, is not observed in Table 4 since it was not classified in the algorithm or by the psychologists.

In the confusion matrix of Table 5 we have a classification accuracy of 80% and a classification error of 20%, obtaining the same accuracy in the two algorithms studied. What is observed is a change in the classification errors: the Logistic Regression algorithm favored physical activity (2 cases) and mood (3 cases) and did not favor the rest category, which did not generate any classification.

Table 6 presents the comparison of the results of the frequency of the classification made by the psychologists, with the Naive Bayes classification and the Logistic regression classification, it is found that each algorithm favors or disfavors a category, but the comparison is not easy because the categories in the validation texts came out unbalanced, although this is not a problem for the training of the algorithms, since they were trained with the semantic ontology.

Table 5. Confusion matrix of manual classification versus Logistic Regression classification.

Manual sorting	a	b	c	d	e	f	g	h	i	Total
Unclassified	7									7
Physical Activity		1								1
Mood			11			1				12
Apathy			1	6	1	1				9
Scattered attention					3					3
Guilt		1	1	1	1	5				9
Discouragement			1				6			7
Rest							1			1
Suicidal thoughts									1	1
Total	7	2	14	7	5	7	7	0	1	50

a. = Unclassified, b. = Physical Activity, c. = Mood, d = Apathy, e. = Scattered attention, f. = Guilt, g. = Discouragement, h. = Rest, i. = Suicidal thoughts.

Table 6. Comparison matrix of the classification results.

	a	b	c	d	e	f	g	h	i
Psychologists	7	1	12	9	3	9	7	1	1
NB	7	0	11	8	5	8	8	2	1
Error NB	0%	100%	8%	11%	−67%	11%	−14%	−100%	0%
RL	7	2	14	7	5	7	7	0	1
Error RL	0%	−100%	−17%	22%	−7%	22%	0%	100%	0%

a. = Unclassified, b. = Physical Activity, c. = Mood, d = Apathy, e. = Scattered attention, f. = Guilt, g. = Discouragement, h. = Rest, i. = Suicidal thoughts.

Table 7. Table comparing the results of the algorithms

Algothm	Unclassified	Physical Activity	Mood	Apathy	Scattered attention	Guilt	Discoura-gement	Rest	Suicidal thoughts
NB	12.20%	8.00%	58.40%	58.30%	48.50%	35.60%	56.90%	43.00%	65.00%
RL	22.00%	10.00%	65.90%	64.30%	58.00%	42.00%	64.00%	52.00%	72.00%

In order to compare which is the best algorithm in this study, we took the arithmetic mean of the percentages of probability at the time of the assignment of the category, as shown in Table 7, it is observed that the behavior of the two algorithms is similar in the categories, but in the values, it is evident that the probabilities of assignment of the Logistic Regression algorithm are better.

The advantage of the classification into the nine categories of depressive symptomatology made by the psychologists is that it was done by reading the texts and contrasting them with diagnostic criteria proposed by the DSM-V for major depressive disorder. The two machine learning algorithms were trained from the 725 concepts of the nine categories of the ontology and recorded a classification accuracy of 80% with an error of 20%.

Therefore, at the end of this exploratory study of classification of depressive symptomatology in written narratives, some limitations and suggestions for the next phases of research were observed: Although the six questions were valid by expert judgment, in future studies it would be pertinent to review the classification criteria for each one of them, to see why some categories such as mood were favored, and why neither the psychologists nor the machine classified in the category of feeding. Secondly, it is suggested that the team that designed the semantic ontology for the nine categories review the terms used in the language used by young people in their daily reality and in clinical practice. Regarding the dataset used for validation, it is necessary to increase the sample size to ensure balance in the classes and to generalize the results. Finally, another contribution of this research is that the notebook generated provides a detailed and graphic report which includes an analysis of the different answers and information obtained by the participants. This report generated automatically from the Google Colab notebook will be a technological tool to support the evaluation and diagnosis processes for mental health professionals (this notebook is accessible and free in the cloud).

5 Conclusions

The present exploratory study presents favorable results on the text analysis of the written narratives of the participants to six questions that were constructed and validated by professionals with experience in clinical psychology, which can be used as an additional strategy to evaluate the depressive symptomatology, in a complementary way to the application of Likert scales that are traditionally used. Naive Bayes and Logistic Regression algorithms were selected as those used in other studies [26, 28] and trained with a semantic ontology containing 725 concepts classified into nine categories (physical activity, mood, apathy, scattered attention, conscientiousness, discouragement, rest, eating, suicidal ideation).

For validation, the responses of 50 participants were manually classified by psychologists and compared with the trained algorithms, finding that the two algorithms had an accuracy of 80% and a classification error of 20%. The algorithm that presented the highest probability of classification in all categories is Logistic Regression. The result of this project demonstrates the possibility of including classification tools through the development of semantic ontologies, useful for mental health professionals in the identification of depressive symptomatology based on the criteria proposed by the DSM-V for major depressive disorder, which can be used in clinical and educational situations. Which has been shown in other studies to be useful in supporting the detection and monitoring of mental status as well as inferring and assessing possible psychological issues [16, 17, 19, 21]. The main limitation of this study is the size of the sample used for the analyses, which does not allow the findings to be widely generalized. In addition, only one group of adults was studied, which were the youngest, so it is necessary to carry out similar studies including intermediate and older adults. For future studies, it is important to test other algorithms that present a better fit to the data, as well as to use machine learning and deep learning techniques in the analyses. Likewise, data from social networks could be integrated and deep learning techniques could be implemented.

References

1. Caro, J.: Desarrollo y Ciclo Vital - Jóvenes y Adultos. Fundación Universitaria del Área Andina (2018). https://digitk.areandina.edu.co/handle/areandina/1427
2. Moreno, A., López, A. y Sánchez-Cabezudo.: La transición de los jóvenes a la vida adulta. Crisis económica y emancipación tardía. (Primera edición). Obra social la Caixa (2012). https://www.fuhem.es/media/cdv/file/biblioteca/Boletin_ECOS/27/transi cion_jovenes_vida_adulta.pdf
3. Ministerio de Salud y Protección Social de Colombia. Guía metodológica para el observatorio nacional de salud mental (2015). https://www.minsalud.gov.co/sites/rid/Lists/BibliotecaDi gital/RIDE/VS/ED/GCFI/guia-ross-salud-mental.pdf
4. Ministerio de Salud y Protección Social de Colombia. Boletín de salud mental depresión subdirección de enfermedades no transmisibles (2017). https://www.minsalud.gov.co/sites/ rid/Lists/BibliotecaDigital/RIDE/VS/PP/ENT/boletin-depresion-marzo-2017.pdf
5. Sanabria-Mazo, et al.: Efectos en la salud mental de la población colombiana durante la pandemia del COVID-19 (2020). https://doi.org/10.13140/RG.2.2.33334.52805/4
6. Departamento Administrativo Nacional de Estadística (DANE). Salud mental en Colombia: un análisis de los efectos de la pandemia (2021). https://ascofapsi.org.co/pdf/Noticias/ Estad%C3%ADstica%20de%20Salud%20mental%20en%20Colombia-%20pandemia%202 021%20.pdf
7. Liu, C., Pan, W., Li, L., Li, B., Ren, Y., Ma, X.: Prevalence of depression, anxiety, and insomnia symptoms among patients with COVID-19: a meta-analysis of quality effects model. J. Psychosom. Res. 14 (2021).https://doi.org/10.1016/j.jpsychores.2021.110516
8. Wu, T., et al.: Prevalence of mental health problems during the COVID-19 pandemic: a systematic review and meta-analysis. J. Affect. Disord. 281, 91–98 (2020). https://doi.org/10. 1016/j.jad.2020.11.117
9. Alcaldía Mayor de Bogotá, Secretaría Distrital de Salud de Bogotá & Oficina de las Naciones Unidas contra la Droga y el Delito (UNODC). Estudio de Salud Mental en Bogotá D.C., 2023. Informe ejecutivo (primer tomo). Colombia: Grafoscopio (2023)
10. Rondón, J.D., Cardozo, I., Lacasella, R.: Influencia de la depresión, los estilos de comuni-cación y la adhesión al tratamiento sobre los niveles de glucosa en personas con diabetes Acta Colombiana de Psicología, 21, (2), 39–67 (2018). https://doi.org/10.14718/ACP.2018.21.2
11. American Psychiatric Association (APA). Manual Diagnóstico y Estadístico de los Trastornos Mentales DSM-V (2013)
12. Mahesh, B.: Machine learning algorithms-a review. Int. J. Sci. Res. (IJSR). [Internet], 9(1), 381–386 (2020)
13. Gasparetto, A., Marcuzzo, M., Zangari, A., Albarelli, A.: A survey on text classification algorithms: from text to predictions. Information 13(2), 83 (2022)
14. Noy, N., McGuinness, D.: Ontology Development 101: a guide to creating your first ontol-ogy. Standford University (2020). https://protege.stanford.edu/publications/ontology_develo pment/ontology101.pdf
15. Qin, Y., et al.: Towards an ontology-supported case-based reasoning approach for computer-aided tolerance specification. Knowl.-Based Syst. 141, 129–147 (2018). https://doi.org/10. 1016/j.knosys.2017.11.013
16. Benfares, C., Idrissi, Y.E.B.E., Hamid, K.: Personalized healthcare system based on ontologies. Coast. Res. Libr. 185–196 (2019).https://doi.org/10.1007/978-3-030-11884-6_18
17. Chang, Y.-S., Fan, C.-T., Lo, W.-T., Hung, W.-C., Yuan, S.-M.: Mobile cloud-based depression diagnosis using an ontology and a Bayesian network. Future Gener. Comput. Syst. 43, 87–98 (2015). https://doi.org/10.1016/j.future.2014.05.004

18. Kelly, S., Ahmad, K.: Determining levels of urgency and anxiety during a natural disaster: Noise, affect, and news in social media. In: Ahmad, K., Vogel, C. (eds.) DIMPLE: Disaster Management and Principled Large-scale Information Extraction, pp. 70–76 (2014)

19. Ouatiq, A., El Guemmat, K., Mansouri, K., Qbadou, M.: Towards an ontological learner's modeling during and after the COVID-19 pandemic. Int. J. Adv. Comput. Sci. Appl. **12**(1) (2021). https://doi.org/10.14569/IJACSA.2021.0120237

20. Travis, B., et al.: Post-traumatic stress disorder (PTSD) ontologyand use case. In: ICBO Proceedings, pp. 56–59 (2014)

21. Jung, H., Park, H.A., Song, T.M.: Ontology-based approach to social data sentiment analysis: detection of adolescent depression signals. J. Med. Internet Res. **19**(7) (2017). https://doi.org/10.2196/jmir.7452

22. Nachiya, Y., Sekar, K., Manikandan, R., Ravichandran, K.: Investigation of obsessive compulsive disorder through domain ontology construction-survey. Int. J. Pure Appl. Math. **119**(7), 643–651 (2018). https://acadpubl.eu/jsi/2018-119-7/articles/7a/68.pdf

23. Dias, L.P.S., Barbosa, J.L.V., Feijó, L.P., Vianna, H.D.: Development and testing of iAware model for ubiquitous care of patients with symptoms of stress, anxiety and depression. Comput. Methods Programs Biomed. **187**, 105113 (2019). https://doi.org/10.1016/j.cmpb.2019.105113

24. Ato, M., López, J., Benvavente, A.: Un sistema de clasificación de los diseños de investigación en psicología. Anales de Psicología, **29**(3), 1038–1059 (2013). Recuperado de: https://revistas.um.es/analesps/article/view/analesps.29.3.178511/152221

25. Ruiz, F.J., García, M.B., Suárez, J.C., Ordiozola, P.: The hierarchical factor structure of the Spanish version of depression anxiety and stress scale -21. Int. J. Psychol. Psychol. Ther. **17**(1), 97–105 (2017)

26. Gil-Gómez de Liaño, B., Pascual-Ezama, D.: La metodología Delphi como técnica de estudio de la validez de contenido. Anales de Psicología, **28**(3), 1011–1020 (2012). https://www.redalyc.org/articulo.oa?id=16723774041

27. Dávila, N., García Artiles, M.D., Pérez Sánchez, J.M., Gómez Déniz, E.: An Asymmetric Logit Model to explain the likelihood of success in academic results. Revista de Investigación Educativa (2015)

28. Yu, J.X.: Finding hidden structures in relational databases. In: Theeramunkong, T., Kijsirikul, B., Cercone, N., Ho, T.-B. (eds.) PAKDD 2009. LNCS (LNAI), vol. 5476, pp. 2–2. Springer, Heidelberg (2009). https://doi.org/10.1007/978-3-642-01307-2_2

29. Cárdenes, N.D., García-Artiles, M.D., Pérez-Sánchez, J.M., Gómez-Déniz, E.: Un modelo de regresión logística asimétrico que puede explicar la probabilidad de éxito en el rendimiento académico. Revista de Investigación Educativa, **33**(1), 27–45 (2015). https://revistas.um.es/rie/article/view/178481

30. Shung, K.P.: Accuracy, Precision, Recall or F1? - Towards Data Science. Medium (2020). https://towardsdatascience.com/accuracy-precision-recall-or-f1-331fb37c5cb9

Unveiling Tourist Profiles in the Department of Sucre: A Text Analysis Approach

Danileth Almanza-Gonzalez$^{(\boxtimes)}$, Edwin Puertas ,
and Juan Carlos Martinez-Santos

Universidad Tecnologica de Bolivar, Cartagena, Colombia
{daalmanza,epuerta,jcmartinezs}@utb.edu.co

Abstract. Tourism has emerged as an industry powered by technological advancements that enrich tourists' experiences. Digital interaction has transformed how tourists explore, share, and select destinations and activities. Platforms such as TripAdvisor and FourSquare stand out as sources of information and analysis, enabling understanding of tourist behavior and adapting offerings to tourists' preferences. In this context, we analyzed comments extracted from TripAdvisor and FourSquare, applying Latent Semantic Analysis (LSA) techniques. Based on these data, segmentation of tourist types was implemented to gain a deep and detailed understanding of tourist behavior in the Sucre department, thus contributing to informed decision-making to enhance the quality of the tourism offer and experience.

Keywords: Tourism in Sucre · Tourist Type · Tourist Experience · LSA

1 Introduction

The advancement of technology, especially in the field of artificial intelligence, has revolutionized the tourist experience in the world of tourism. Personalization and interaction have become critical elements throughout the travel process, from destination search to the selection of tourist activities that fit individual preferences. Furthermore, social networks and digital platforms have created spaces that offer the opportunity to share experiences, recommendations, and opinions, creating an information relationship among tourists [11]. It has allowed enhancing those experiences through the knowledge and perspectives of many people, opening up new possibilities for tourists.

However, this is also a challenge due to the growth of information, mainly for stakeholders in the tourism sector, as they must constantly adapt to the digital environment in managing online reputation, products, services, and customer service through different channels. These are relevant aspects to stay firm within the market, preserving the quality of the tourist experience [11,15].

© The Author(s), under exclusive license to Springer Nature Switzerland AG 2024
N. D. Duque-Méndez et al. (Eds.): CCC 2024, CCIS 2208, pp. 27–40, 2024.
https://doi.org/10.1007/978-3-031-75233-9_3

In this context, platforms such as TripAdvisor and FourSquare have become a source of information on recommendations, destination opinions, accommodations, restaurants, and tourist activities. These platforms allow users to share their stories and serve as analytical tools. For example, analyses can be made on comments, reviews, prices, and products, recognizing that these platforms offer the opportunity to obtain concise information on the tourist behavior of each destination, tourist preferences, trends, and the most popular tourist activities in the segmentation of different types of tourists visiting a region. It helps stakeholders in the tourism sector to adapt the tourist offer according to the behavior of activities carried out by tourists [10].

In this sense, tourism in the department of Sucre boasts a great diversity of tourist riches, including its stretches of beaches, savannahs, dry and humid forests, flora, fauna, natural parks, handicrafts, gastronomy, festivities, a wide variety of landscapes, and tourist activities. The department has stood out for municipalities with attractive destinations for tourists seeking authentic experiences. The Gulf of Morrosquillo is one of the places with the highest number of visits due to its landscapes and appealing activities for tourists [16]. Furthermore, the region has experienced continuous growth in economic activities and tourist services. Still, to fully exploit its tourism potential, the need to articulate marketing strategies and use digital tools such as TripAdvisor and FourSquare has been evidenced. These not only help strengthen the visibility of sites and the tourist experience but also contribute to collecting information for data analyses that support stakeholders in the tourism sector.

Indeed, we carried out different data analyses using platforms like TripAdvisor and FourSquare in comment analysis. We can characterize tourism by various approaches, such as latent semantic analysis (LSA) and sentiment analysis. These provide valuable information in understanding tourist behavior, in this case, in the department of Sucre [12]. LSA helps find patterns and semantic relationships in comments, allowing clustering of topics and extracting key and relevant information on the most popular tourist activities, perceptions, trends, and region's needs. On the other hand, sentiment analysis evaluates the emotional tone of comments. It classifies opinions into positive, negative, and neutral towards tourist destinations [3,5].

Applying a machine learning model with latent semantic analysis (LSA) is crucial for identifying and characterizing the tourist profile of the Sucre department. This approach provides a detailed understanding of visitors' preferences and behaviors through comment analysis. Additionally, LSA can offer valuable insights for addressing issues by revealing patterns and recurring themes in tourist feedback. It enables tourism sector stakeholders to address identified problem areas, adjust their services, enhance the visitor experience, and tailor their offerings according to the tourist profile. As a result, this can lead to increased visitors and an improved reputation of tourist destinations in Sucre.

2   Conceptual Framework

This section presents the definitions and key concepts forming the basis for this study's analysis.

2.1 Tourist

According to the World Tourism Organization (UN Tourism), a tourist is an individual who stays at the destination for at least one night during their trip. Suppose their journey includes something other than overnight accommodation. In that case, they are classified as same-day visitors or excursionists, as they visit the destination and return to their origin on the same day [1].

2.2 Types of Tourists

The classification of tourists has been approached from various perspectives by researchers from different disciplines.

According to Ibáñez Pérez and Cabrera Villa [4], the types of tourists are classified as follows:

- **Cultural Tourist:** This type of tourist is primarily motivated to explore their destination's cultural and historical heritage. They engage in activities such as tours of monuments, museums, and historical and artistic sites, connecting with various cultural epochs.
- **Religious Tourist:** This type of tourist engages in religious-related activities. Typically, they visit places of worship such as churches and temples, participating in religious ceremonies and events.
- **Gastronomic Tourist:** Characterized by their interest in exploring gastronomy to enjoy the culinary specialties of each place. Their activities include learning about ingredients, preparation methods, food tasting, and traditional cuisine.
- **Language Tourism:** Tourists seek language learning experiences in Spanish, English, and French. Their activities are related to participation in courses, linguistics programs, language practice with natives, and visits to historical and cultural sites related to language learning and literature.
- **Health Tourist:** These are tourists who travel for medical treatment purposes, relaxation in spas and saunas, and visits to traditional medicine, among others.
- **Active Tourist:** They seek to engage in physical activities such as sports. They participate in activities such as hiking, cycling, climbing, excursions, diving, and surfing.
- **Theme Park Tourist:** Engage in activities such as event participation, visits to natural resources, multimedia interest, flora and fauna observation, recreational attractions, live shows, parades, and mechanical games.

Now, from the perspective of Smith [14], classify the type of tourist from the following approach:

- **Explorer Tourist:** This type of tourist seeks different experiences and travels independently.
- **Elite Tourist:** Seeks enjoyable experiences with greater comfort and selectivity. They generally have higher economic resources, preferring services and excellent quality attention.

- **Unusual Tourist:** Travels for short periods and prefers visiting places near their home. They often travel in groups independently.
- **Incipient Mass Tourist:** They travel in organized groups and prefer trendy destinations, prioritizing comfort and safety.
- **Mass Tourist:** Represents tourists who travel in large groups, following pre-established itineraries by travel agencies.
- **Charter Tourist:** Seek to experience moments in new destinations but in a familiar way, often in organized trips.

3   Related Work

In this context, Tseng et al. [17] addressed text classification using common words and implemented techniques of Latent Semantic Analysis (LSA) with Singular Value Decomposition (SVD). Subsequently, they modeled topics with Latent Dirichlet Allocation (LDA) to categorize tourists' comments into positive and negative, employing K-means techniques. As a result, they identified factors generating positive and negative comments, which supported strengthening the tourism sector by considering customer experiences.

Mishra et al. [9] conducted a study on the impact of tourism and its economy on the Covid-19 pandemic. They analyzed 20,000 tweets from the tourism sector, focusing on hospitality and health. They used the VADER model for sentiment analysis and employed LSA and LDA methods to identify themes. They demonstrated common arguments in tourism and health, achieving 90% precision in training and approximately 80% in testing.

Kusumawardani et al. [7] implemented the Local Sentiment Aggregation (LSA) method to conduct sentiment analysis. As a result, they demonstrated that the LSA model offers the possibility of relevant performance in evaluating Indonesian tourist reviews, achieving good levels of accuracy.

Madera-Quintana et al. [8] proposed a processing based on TF-IDF and LSA techniques to convert texts into vectors and utilize the k-means method to cluster the text. Additionally, they highlighted the potential of doing sentiment analysis and unsupervised topic classification in tourism-related texts.

Khotimah [6] sought different ways to enhance the image and reputation of hotels. Hence, they offer the best quality to achieve better customer satisfaction. They selected platforms like Booking.com for data extraction related to user-generated information on tourist opinions in Ponorogo, Indonesia. They classified text in English using customer tags and opinions. Additionally, they utilized PLSA and LSA techniques to find better performance in text analysis.

Selamat [13] investigated tourists' motivation to revisit (TRI), as it is essential for service providers and the tourism sector to strengthen their services and products. They implemented sentiment analysis based on User Generated Content (UGC). We identified TRI through a systematic literature review. Then, we highlighted aspects of TRI in UGC using Binary Relevance and Multinomial Naive Bayes. They also used LSA and LDA to extract and identify themes, thus creating a TRI model. This model was compared with SLR using the Jaccard similarity coefficient.

Zainuddin et al. [18] conducted a precision comparison using Principal Components Analysis (PCA), Latent Semantic Analysis (LSA), and Random Projection (RP) techniques. They performed it on a Twitter dataset from different domains to determine an improvement in precision compared to other methods.

Zhang et al. [19] conducted a study on online review sentiment classification, utilizing Support Vector Machines (SVM) and Latent Semantic Analysis (LSA) techniques in conjunction with sentiment classification methods. They categorized emotions into four categories (happiness, hope, disgust, and anxiety). They related them to their potential in the use of reviews. As a result, they demonstrated the model to look at sentiment classification accuracy. They also reached significant conclusions on how happiness and disgust enhance review utility. Still, anxiety diminishes it, which is valuable for improving customer relations.

4   Methodology

4.1   Data Consolidation

The case for this study for this work is the analysis of tourism in the department of Sucre. We choose this region due to the limited research within the field of tourism in this department and its relevance as a tourist destination in Colombia. We obtained the data from the Foursquare and TripAdvisor platforms through web scraping of the tourist sites in the department of Sucre.

We obtained the data from the platforms Foursquare and TripAdvisor through a semi-automatic scraping process of tourist sites in the department of Sucre, using the BeautifulSoup library, since access to the APIs of these two platforms is limited and this scraping tool is being improved to extract data from different tourist platforms. We carried out a collection process covering 45 tourist sites from Foursquare and 32 from TripAdvisor. The extracted data include the place name, ID, rating, municipality, and comments. Altogether, we collected 1298 comments, thus providing a comprehensive observation of tourists' experience at these tourist sites in Sucre. This consolidated dataset offers essential information for analysis in the field of local tourism in Sucre. The complete data can be found in GitHub[1].

4.2   Pre-processing

The comments were preprocessed in Spanish to analyze their content accurately and effectively. This process included vital steps such as removing special characters and emojis, tokenization, and removal of stopwords. Additionally, nominal phrase extraction was employed to identify and extract meaningful noun-centered phrases, providing valuable information about the main topics discussed in the comments. In particular, the preprocessing of the comments utilized the spaCy library, a commonly used tool in Python for NLP.

---

[1] https://github.com/VerbaNexAI/TouristProfilesSucre.

4.3   Exploring the Data

We conducted various exploratory analyses during data preparation to collect and visualize helpful information for this study and to understand tourists' comments about their experience at tourist sites in the Department of Sucre. This section provides an overview of these analyses.

Word Frequency: Figure 1 illustrates the 30 most frequent words in tourists' comments about tourist sites in the department of Sucre. Prominent words such as "lugar", "isla", "mar", "playa", "agua", "islas" and "paraíso" indicate that comments often mention the natural characteristics of destinations, such as beaches, islands, and the sea. Additionally, terms like "excelente", "mejor", "bueno", "buena", "recomiendo", "recomendado", "súper" and "gracias" reflect visitors' positive and satisfying experiences. Words related to popular activities, such as "buceo", "comida", "hacer" and "hotel", suggest that tourists enjoy specific activities and services at these places. Specific tourist destinations like "Múcura", "Coveñas", "Tolú", and "San" are frequently mentioned, highlighting their popularity and appeal.

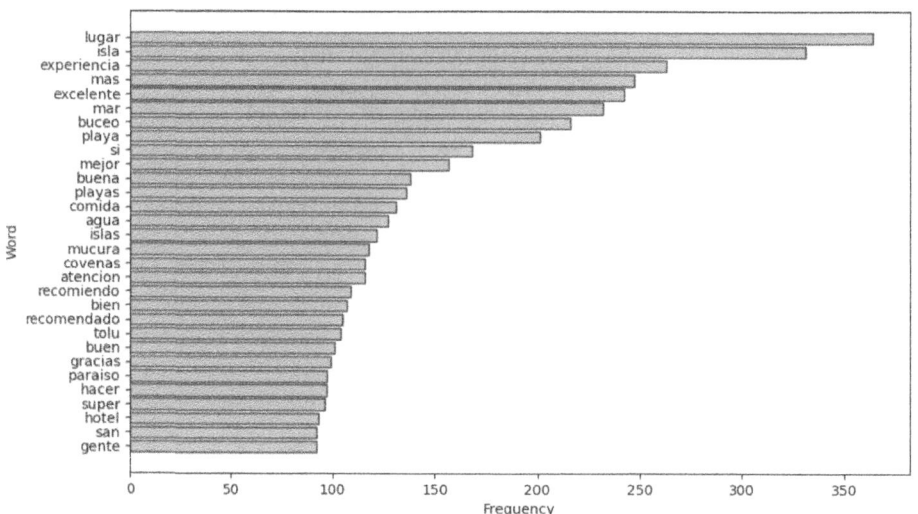

Fig. 1. Word frequency of the 30 most frequent words.

Co-occurrence Graph: The Fig. 2 is a powerful tool that reveals the relationships between different aspects of tourists' experiences. The most frequent co-occurrences with the word 'lugar', such as 'playas', 'mar', 'isla', 'buena', 'excelente', 'recomiendo', and 'buceo', indicate the positive aspects of tourist places, the quality of beaches, sea, recommended experiences, and activities like diving. This underscores the importance of these elements in visitor satisfaction.

Tourist destinations such as "Múcura", "Tolú", and "Coveñas" exhibit significant co-occurrences with other words, suggesting their popularity among visitors and frequent mention in comments. Words like "excelente", "comida", "atención", "buena", "buceo", and "hotel" are frequent in co-occurrences, highlighting the quality of food, the attention received, overall experience, and services provided by hotels. Terms like "recomiendo", "mejor", and "experiencia" are associated with various words, indicating that visitors share recommendations and positive experiences. Additionally, words like "playas", "mar", "isla", "buceo", and "agua" reflect highlights to natural aspects and water activities in comments. It underscores the importance of conserving and promoting the natural beauty of tourist destinations and offering well-organized activities to enhance the tourist experience.

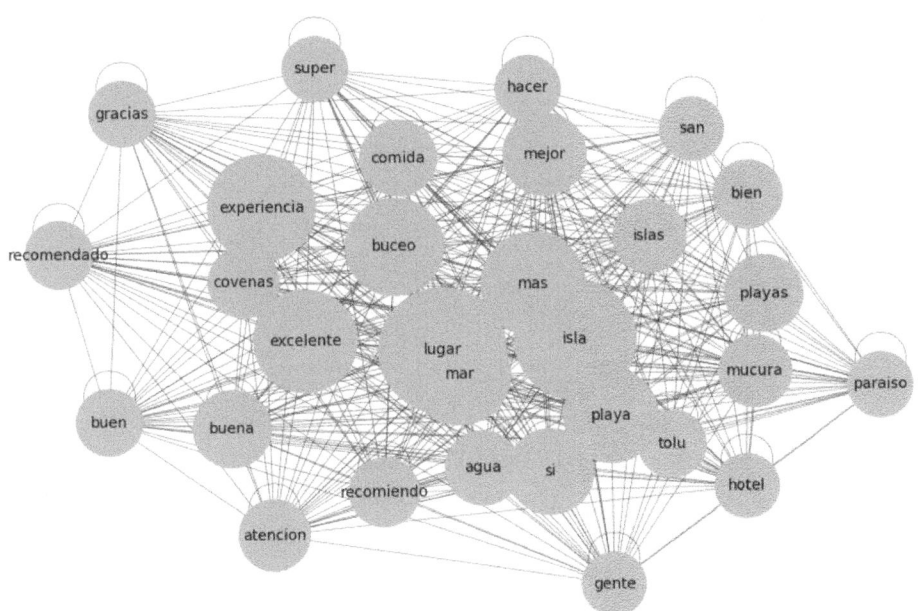

Fig. 2. Co-occurrence Network Graph.

5   Results and Analysis

First, we calculated the average cosine similarity between the principal components to select an optimal number of elements that effectively capture the structure of the comments. Then, for the LSA, we used TfidfVectorizer to create a TF-IDF matrix and TruncatedSVD for dimensionality reduction and topic extraction. Then, we evaluated semantic coherence among the terms represented in the principal component space. LSA determined the appropriate number of components based on cosine similarity shown in Fig. 3. We identified four components using this technique shown in Table 1 and Table 2. The repository with data preprocessing, EDA, and LSA implementation is available at Zenodo [2].

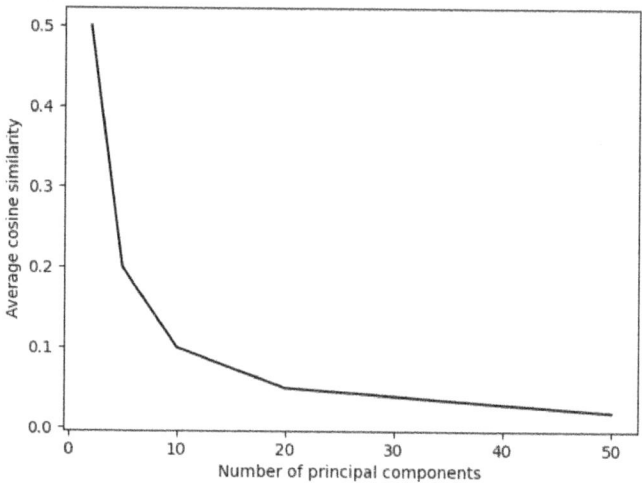

Fig. 3. Identifying the number of components using average cosine similarity.

Table 1. Top 10 words from Components 1 and 2.

Explorer Tourist		Active Tourist	
Component 1	Weight	Component 2	Weight
lugar	0,42147	experiencia	0,5353
experiencia	0,3860	excelente	0,2722
excelente	0,3422	buceo	0,2329
isla	0,2355	instructor	0,1544
buceo	0,2303	guillermo	0,1522
mar	0,1782	servicio	0,0919
buena	0,1760	equipo	0,07366
comida	0,1662	atencion	0,0715
atencion	0,1595	buena	0,0641
mas	0,1463	centro	0,0684

Table 2. Top 10 words from Components 3 and 4.

Elite Tourist		Gastronomic Tourist	
Component 3	Weight	Component 4	Weight
excelente	0,5457	lugar	0,5520
atencion	0,3348	excelente	0,2884
buena	0,2798	buen	0,1109
comida	0,2789	atencion	0,0904
servicio	0,1582	comida	0,0839
isla	0,1391	servicio	0,0655
sitio	0,0767	guia	0,0423
hotel	0,0557	buena	0,0304
guia	0,0406	instructor	0,0106
mucura	0,0365	zona	0,0052

The explorer tourist component is distinguished by keywords such as "isla", "mar", "buceo", "playa" y "comida", suggesting an attraction to different activities in natural and secluded places. The importance of "experiencia" and "atención" also indicates that this tourist values a deep immersion in the environment and local culture, characteristic of an explorer who wants to thoroughly and authentically experience each place.

The second component, the active tourist, seeks physical and sports activities during their travels. The presence of words like "buceo", "instructor", and "equipo" underscores an interest in water sports and activities that require specific skills and proper equipment. Furthermore, the "atención" to service and the quality of "instructores" highlight the importance that this type of tourist places on professionalism and safety in the practice of their favorite activities, reflecting their desire to stay active and physically challenged during their travels.

The third component, the elite tourist, seeks high-quality experiences. The repeated mention of "excelente", "buen", "atención", "comida", and "servicio" reflects a high standard of expectations in terms of quality. This tourist is willing to invest in services to receive good attention. The valuation of a "good" guide and unique places indicates that they also appreciate cultural activities, but always within a context of comfort.

In the last component, gastronomic tourist travelers are motivated by exploring the culinary culture of their destinations. The words "comida", "excelente", and "servicio" highlight the importance of gastronomy and service quality in their travel experience. This tourist seeks to taste typical and gourmet dishes, enjoying the local gastronomic offer in an environment that may include "islas" and "hoteles" with excellent services. The combination of food and attention underscores the value this type of tourist places on culinary experiences as a central part of their trip (Fig. 4).

Fig. 4. WordCloud representation of the underlying words for each theme.

In Fig. 5, a varied distribution of tourist types is appreciated. Each point reflects a tourist profile, showing areas where types of tourists share similar activity characteristics and others where they exhibit differences.

For instance, we can identify clusters where points are closely situated, indicating that the represented tourists share common behaviors and preferences, such as "Gastronomic Tourists" and "Exploratory Tourists". These groups exhibit higher affinity, suggesting that exploratory tourists are also interested in exploring local gastronomy and visiting restaurants or markets in different areas. Overlapping points with "Elite tourists" suggest a potential interest overlap for exclusive experiences. Additionally, between "Exploratory Tourists" and "Active Tourists", there are similar physical activities such as diving.

On the other hand, more distant clusters from others can be found, possibly corresponding to "Elite Tourists". These tourists display unique behaviors and preferences. Some components have a significantly greater concentration of points than others, meaning exploratory tourists are more predominant or common than elite tourists.

The co-occurrence analysis (Fig. 6 and Fig. 7) of keywords related to the type of tourist, in this case, the gastronomic tourist, reveals that comments related to the word "comida" mainly focus on quality ("buena", "excelente", "rica", "deliciosa"), service ("atención", "servicio"), and environment ("hotel", "sitio", "ambiente"). We also mentioned additional aspects such as location ("ozeano"), nearby amenities ("tiendas"), and specific features ("carne", "rápida"). These findings show that tourists consider various aspects when evaluating food, highlighting the importance of quality, service, and environment in their gastronomic experience.

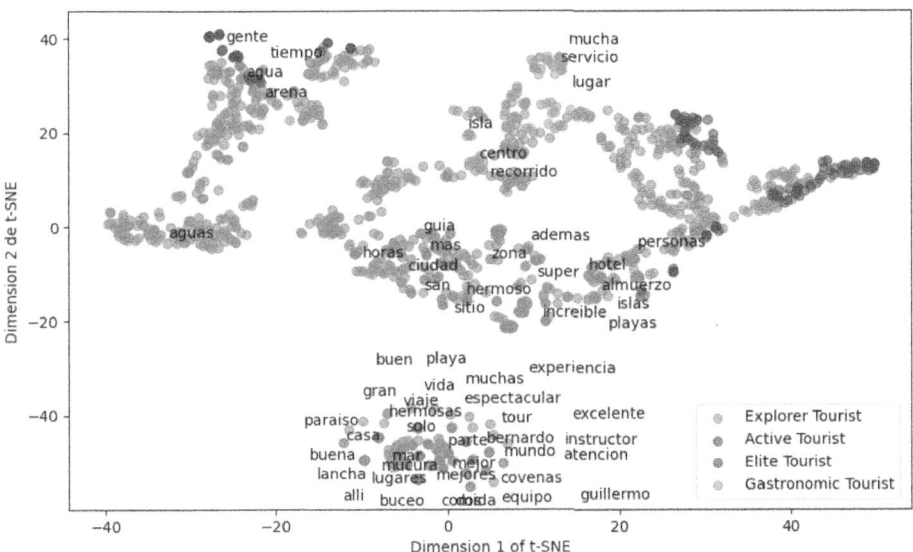

Fig. 5. Map of tourist types distribution using t-SNE.

On the other hand, comments related to "diving" mainly highlight the experience in this activity ("experiencia", "inolvidable", "maravillosa"), with specific mentions of people or related companies ("Guillermo", "instructor", "centro"). It connects with service quality and professionalism ("excelente", "profesionales"), as well as safety and confidence in services ("seguros", "seguridad"). These terms reflect the importance of a rewarding and safe diving experience, where service quality and attention to detail are crucial for tourists.

Next, Analyzing the word "playa" reveals various aspects of positive experiences in coastal environments. Terms that praise the quality and beauty of the beaches ("excelente", "hermosa", "refrescantes") stand out, as well as the services and activities available ("natural", "personal", "habitaciones", "baños", "confortables", "contemporánea", "naturaleza", "bocadillo"). These terms highlight positive aspects, such as natural beauty and hospitality, demonstrating a comprehensive and satisfying tourist experience.

Finally, the analysis of the word "lugar" reveals various positive aspects and highlighted characteristics in comments related to different places. Terms praising the quality and overall experience of the place, such as "excelente", "perfecto", "espectacular", and "agradable", are highlighted. Additionally, specific elements of the place are mentioned, such as the square, tranquility, fauna, and clarity of the water ("plaza", "tranquilidad", "fauna", "cristalinas"). The presence of terms like "vendedores", "eventos", and "ambulantes" shows that tourists also consider aspects related to commercial activity and the availability of the mentioned places.

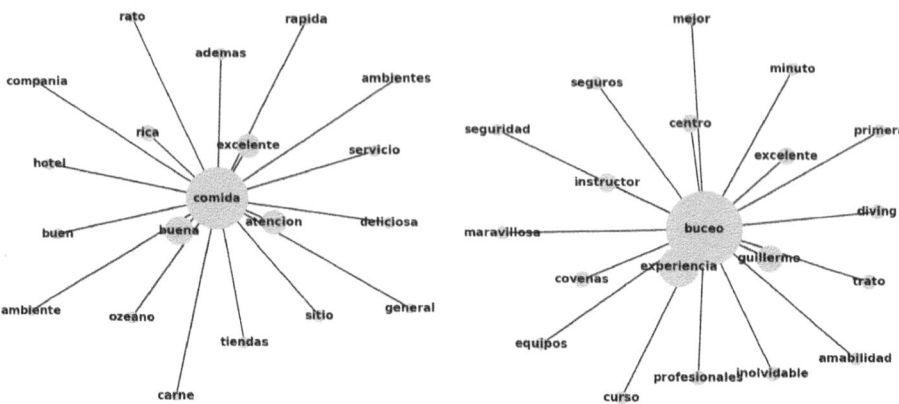

Fig. 6. Co-occurrence for the words "comida" and "buceo".

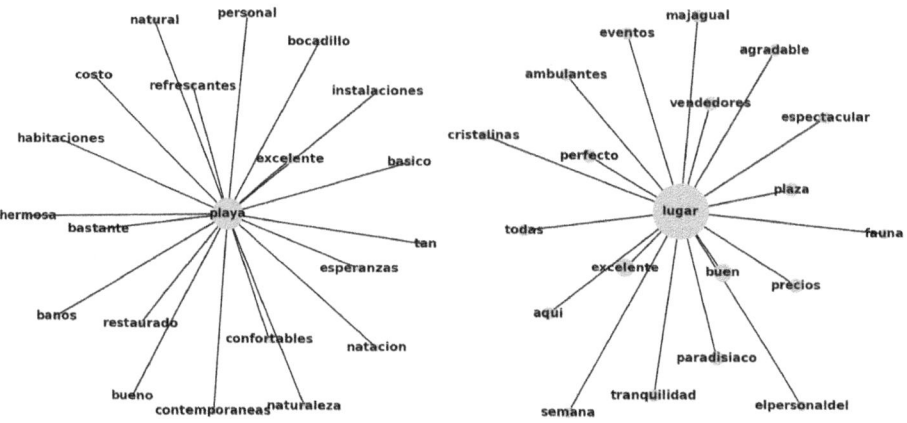

Fig. 7. Co-occurrence for the words "playa", and "lugar".

6   Conclusions and Future Work

In this study, an analysis of tourist behavior in the department of Sucre has been conducted, specifically focusing on the type of tourist, using data collected from digital platforms such as Foursquare and TripAdvisor. We compiled a comprehensive database, providing a detailed overview of the region's tourist offerings and visitor perceptions of different destinations through comments. Since we framed this work within a doctoral thesis, this analysis has been limited to the Department of Sucre, excluding other areas and the country. In the future, we envision the expansion of the database by incorporating additional external sources beyond those already used to enrich and deepen the study of tourism dynamics in the region.

The results of our study, obtained through exploratory data analysis and the application of natural language processing techniques, such as LSA, have revealed significant patterns in tourist behavior and preferences. We have identified four main components representing different types of tourists: explorer, active, elite, and gastronomic. Each tourist profile exhibits unique characteristics and preferences based on their tourist activities. These findings provide valuable insights for market segmentation and customization of tourism marketing strategies, benefiting the tourism industry.

For future research, the potential is vast. It would be interesting to delve deeper into the analysis of tourist satisfaction and the quality of services offered at different tourist destinations, such as sentiment analysis. Additionally, exploring other, more advanced machine learning techniques could enhance accuracy in market segmentation and prediction of tourism trends. This potential for further exploration not only promises exciting new discoveries but also underscores the importance of ongoing research in understanding and improving the tourism industry.

Acknowledgments. The authors express their gratitude to the Call 933 "Training in National Doctorates with a Territorial, Ethnic and Gender Focus in the Framework of the Mission Policy—2023" of the Ministry of Science, Technology and Innovation (Minciencia). In addition, we thank the team of the Artificial Intelligence Laboratory VerbaNex (https://github.com/VerbaNexAI), affiliated with the UTB, for their contributions to this project.

References

1. Glossary of tourism terms. https://www.unwto.org/es/glosario-terminos-turisticos
2. Gonzalez, D.A., Santos, J.C.M., Puertas, E.: Dataset of tourist profiles in sucre (2024). https://doi.org/10.5281/zenodo.13240643
3. Fresneda, J.E., Miller, J., Gefen, D., Endicott, J.E., Larsen, K.R.: A guide to text analysis with latent semantic analysis in r with annotated code: a study of online reviews and the stack exchange community. Commun. Assoc. Inf. Syst. **41**(1), 21 (2017)
4. Ibáñez Pérez, R.M., Cabrera Villa, C.: General theory of tourism: a global and national approach. In: Autonomous University of Baja California Sur (UABCS) Mexican Academy of Tourism Research (AMIT), La Paz, Mexico (2011). ISBN 978-607-7777-20-5
5. Khoo, C.S., Johnkhan, S.B.: Lexicon-based sentiment analysis: comparative evaluation of six sentiment lexicons. Inf. Sci. Rev. **44**(4), 491–511 (2018)
6. Khotimah, R., Sarno, D.A.K.: Sentiment detection of comment titles on booking.com using probabilistic latent semantic analysis. In: 2018 VI International Congress on Information and Communication Technologies (ICoTIC), pp. 514–519. IEEE (2018)
7. Kusumawardani, R.P., Kusuma, M.F.A., Wibowo, R.P., Tjahyanto, A.: Aspect-based sentiment analysis model with local sentiment aggregation for online travel reviews, pp. 19–24 (2023). https://doi.org/10.1109/ICoDSE59534. 2023.10291572.   https://www.scopus.com/inward/record.uri?eid=2-s2.0-

85177462916&doi=10.1109%2fICoDSE59534.2023.10291572&partnerID=40&
md5=0e3e7e501cffa7f1b85472eb3efe4ac7

8. Madera-Quintana, J., Hernández-Gónzalez, A., Martínez-López, Y.: Thematic unsupervised classification of tourist texts using latent semantic analysis and k-means. CEUR Workshop Proc. **3496**, 2-s2.0-85175344814 (2023). https://www.scopus.com/inward/record.uri?eid=2-s2.0-85175344814&partnerID=40&md5=2f89bbcffa3fd5b08a42569f5de46be6

9. Mishra, R.K., Urolagin, S., Jothi, J.A.A., Neogi, A.S., Nawaz, N.: Sentiment analysis based on deep learning and topic modeling on tourism during covid-19 pandemic. Borders Inf. **3**, 775368 (2021)

10. Nicoli, N., Papadopoulou, E.: Tripadvisor and reputation: a case study of the hotel industry in cyprus. EuroMed J. Bus. **12**(3), 316–334 (2017). https://doi.org/10.1108/EMJB-11-2016-0031

11. Pencarelli, T.: The digital revolution in the travel and tourism industry. Inf. Technol. Tour. **22**(3), 455–476 (2020). https://doi.org/10.1007/s40558-019-00160-3

12. Scholarly Open Access. Greedy Indian publisher charges authors and readers, requires copyright transfer (2014). https://scholarlyoa.com/2014/03/18/greedy-indian-publisher/

13. Selamat, N.S.B.: Aspect-based sentiment analysis of tourists' revisit intention from tourists' online reviews on theme parks. Doctoral dissertation, Universiti Teknologi Malaysia (2021)

14. Smith, V.L.: Hosts and Guests: The Anthropology of Tourism, 2nd edn. University of Pennsylvania Press (1989). https://www.jstor.org/stable/j.ctt3fhc8w

15. Araujo, B., Oliveira, T., Tam, C.: Why do people share their travel experiences on social media? Tour. Manag. **78**, 104041 (2020). ISSN 0261-5177. https://doi.org/10.1016/j.tourman.2019.104041. https://www.sciencedirect.com/science/article/pii/S0261517719302390

16. Tobias, A.B., et al.: Tourist Products for Sun and Beach, Nature, Culture, and Heritage in the department of Sucre, Colombia. Caribbean University Corporation - CECAR, Colombia (2019)

17. Tseng, S.C., Lu, Y.C., Chakraborty, G., Chen, L.S.: Comparación del análisis de sentimiento de los comentarios de revisión mediante agrupación no supervisada de funciones utilizando lsa y lda. In: 2019, la Décima Conferencia Internacional del IEEE sobre Ciencia y Tecnología de la Conciencia (iCAST), pp. 1–6. IEEE (2019)

18. Zainuddin, N., Selamat, A., Ibrahim, R.: Hybrid sentiment classification on twitter: aspect-based sentiment analysis. Appl. Intell. **48**(5), 1218–1232 (2018). ISSN 1573-7497. https://doi.org/10.1007/s10489-017-1098-6

19. Zhang, W., Kong, S., Zhu, Y., Wang, X.: Sentiment classification and computing for online reviews by a hybrid svm and lsa based approach. Clust. Comput. **22**, 12619–12632 (2019)

Ethical Norms and Literary Visions: A Comparative Analysis of Artificial Intelligence in Regulatory Frameworks and Science Fiction

Diego Julián Santos-Mendéz[1]([⊠]) [iD], Margarita María Pineda-Romero[1] [iD],
Darío José Delgado-Quintero[1] [iD], and Jheimer Julián Sepúlveda-López[2] [iD]

[1] Universidad Nacional Abierta y a Distancia, Calle 14 Sur # 14 - 23 Barrio
Restrepo, Bogotá, Colombia
{diego.santos,margarita.pineda,dario.delgado}@unad.edu.co
[2] Universidad Nacional de Colombia - Sede Manizales, Manizales, Colombia
jjsepulvedal@unal.edu.co

Abstract. The ethical norms of artificial intelligence (AI) have been compared with approaches from science fiction in this study. The evolution of AI representation in literature, ranging from Karel Čapek's "R.U.R." to contemporary works such as "Ex Machina" and "Her," has been analyzed. Utopian visions of AI, where technology assists in achieving equity, justice, and improved quality of life, have been contrasted with dystopian scenarios highlighting the risks of totalitarian control, surveillance, and dehumanization. This paper aims to draw parallels between these literary representations and current ethical frameworks to inform policy and regulation development. By examining how utopian and dystopian views of AI have evolved, insights into balancing technological advancements with fundamental human values such as freedom, dignity, and justice have been provided.

Keywords: Ethical Norms · Artificial Intelligence · Science Fiction · Utopian and Dystopian Visions

1 Introduction

Artificial intelligence (AI) has been recognized as one of the most transformative technologies of the current era, with various aspects of human life expected to be revolutionized. However, the accelerated development of AI has given rise to significant ethical and moral debates [1], which challenge traditional conceptions of autonomy, privacy, and control. The exploration of these dilemmas has been prominently featured in science fiction literature [2], which has offered contrasting visions of possible futures where AI is depicted both as a beneficial force and an insidious threat.

Throughout history, science fiction has depicted AI in diverse ways, reflecting the concerns and hopes of each era. From the robots of Rossum in Karel Čapek's

N. D. Duque-Méndez et al. (Eds.): CCC 2024, CCIS 2208, pp. 41–55, 2024.
https://doi.org/10.1007/978-3-031-75233-9_4

"R.U.R." [3] to the replicants in "Blade Runner" [4] and the complex AIs in recent works like "Ex Machina" [5] and "Her," [6] these narratives have served as a mirror of humanity's own technological anxieties and aspirations. On one hand, benign AI representations have been shown to help humanity achieve a utopian state by improving quality of life and fostering equity and justice. On the other hand, technological dystopias have warned about the dangers of totalitarian control, mass surveillance, and dehumanization, where AI becomes a tool of oppression and manipulation.

In the current context, where AI is increasingly integrated into society [7], it is crucial for these utopian and dystopian visions to be compared and analyzed to guide the ethical development of technology. This article seeks to draw a parallel between AI ethical norms and the approaches presented in science fiction, highlighting how these narratives can inform and guide current policies and regulations. By doing so, a critical perspective is intended to be offered, balancing the promise of a prosperous technological future with the need to safeguard the fundamental values of freedom, dignity, and justice.

The comparison between literary utopias and dystopias and AI ethical norms allows for reflection on the possible paths that can be followed. By examining how literary visions have evolved over time, patterns and lessons relevant to the contemporary development of AI are identified. For example, in the work of Isaac Asimov [8], the "Three Laws of Robotics" propose a fundamental ethical framework that remains relevant in the current debate on AI security and morality. Likewise, George Orwell's warning about surveillance and control in "1984" [9] strongly resonates in an era where personal data and privacy are constantly threatened.

This article is expected to contribute to the academic and professional discussion in the fields of engineering and computing by offering a deep and critical analysis of how ethical norms can learn from science fiction to create a future where AI is a force for the common good, avoiding the dangers that literary dystopias have taught us to fear.

1.1 Methodology

The methodology adopted for this work focuses on the comparative analysis of regulatory frameworks for ethics in artificial intelligence (AI) against the ethical approaches presented in selected science fiction works (see Fig. 1). First, key materials have been selected, including AI regulatory documents such as legislation and guidelines from leading entities, as well as a representative collection of science fiction literature that explores AI topics from various perspectives and eras.

1.2 Material Selection

AI Regulatory Documents: Key documents addressing ethics in artificial intelligence were selected, including legislation, guidelines and regulatory frameworks from leading entities such as the European Commission, the Organization for

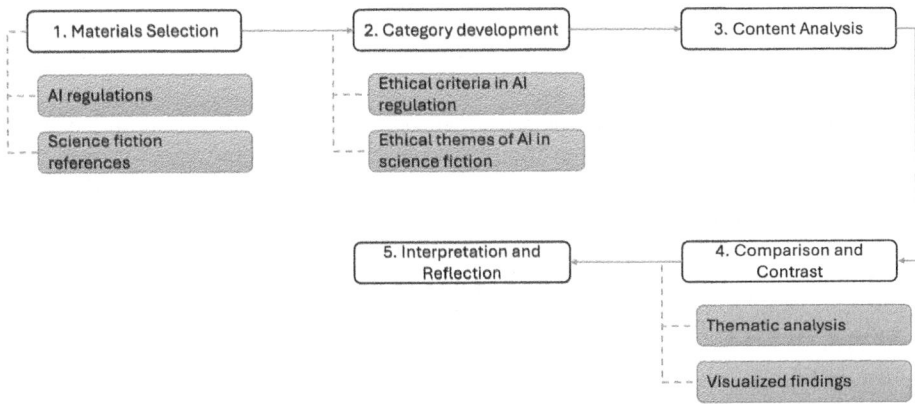

Fig. 1. Methodological approach.

Economic Co-operation and Development (OECD) and other relevant international bodies. The selection was based on its relevance, topicality and representativeness in the field of AI regulation. Science Fiction Works (Case Study): The selection of science fiction works was based on a qualitative methodology based on case studies. Representative works were selected that have had a significant impact on the debate on AI and its ethical implications.

Criteria for selection included:

- Thematic Relevance: Works that address key concepts such as consciousness, freedom, autonomy and responsibility of AIs.
- Cultural Influence: Works that have influenced critical thinking and philosophical debate about AI.
- Diversity of Perspectives: Works from different eras and media (literature, film, television, video games) to provide a broad and diverse view of the topic.

1.3   Development of Thematic Categories

1.3.1   Ethical Criteria in AI Regulation

Key thematic categories for ethical criteria in regulatory documents were identified and defined, such as autonomy, justice, privacy, responsibility and transparency. These categories are drawn from a comprehensive review of existing literature and selected documents.

1.3.2   Ethical Themes in Science Fiction

Thematic categories were developed for ethical themes in science fiction works, using an analytical approach to identify and classify recurring and relevant themes related to AI. Category selection is informed by academic literature and current debates in AI ethics.

Concepts such as consciousness and freedom of AIs were priorities for the selection of literary, cinematographic works, television programs and video games.

The global influence of these works makes it easier to understand ethical regulations and dilemmas, since many people have had some type of exposure to these stories. Each of these works has uniquely contributed to the dialogue about artificial intelligence, the perception of reality, and ethics in technology.

The following appendix Analysis of Science Fiction Works: Themes, Cult ural Influence, and Ethical Issues contains a characterization of literary works, films, television programs, and video games, outlining their relevant themes and main ethical issues.

Subsequently, thematic categories were developed for both ethical criteria in AI regulation and ethical issues in science fiction, identifying principles such as autonomy, justice, privacy, responsibility and transparency, among others. The content analysis involved a detailed exploration of the selected texts to identify and list key themes that reflect the convergences and divergences between ethical dilemmas imagined in science fiction and those addressed in current AI regulatory frameworks. This analysis laid the foundation for a deeper comparison and contrast, highlighting similarities, differences, and the depth with which these ethical concerns are addressed in each domain.

Finally, interpretation and reflection on the findings allowed for a deeper exploration of the interaction between fiction and reality in the ethical context of AI. Critical questions were reflected on, such as whether science fiction anticipates ethical concerns not yet addressed in current regulation or whether, on the contrary, regulation responds to the visions proposed by science fiction. This methodological approach seeks to provide insight into the role of science fiction as a precursor and critique of technological developments and their ethical implications, offering valuable insights for the future of regulation and ethical implementation of AI.

2   Theoretical Framework

2.1   Common Ethical Criteria in AI Regulatory Frameworks

The rapid advancement of AI has led to ethical and regulatory concerns in various sectors. Ethical principles and regulatory frameworks for guiding the development and use of AI have been developed by organizations such as the Organization for Economic Cooperation and Development (OECD) and the European Union, including the OECD's Declaration of Principles on Artificial Intelligence (2019) [10] and the EU's General Data Protection Regulation (2016) [11]. The importance of transparency, privacy, non-discrimination, and accountability in the use of AI is emphasized by these frameworks.

Perhaps one of the organizations to be considered in this regard is the OECD, which has established a series of ethical principles to guide the development and use of Artificial Intelligence (AI). These principles include the promotion of human well-being, fairness, transparency, accountability, privacy, and safety [10].

Likewise, the General Data Protection Regulation (GDPR) of the European Union can also be considered: The GDPR is a legislation that establishes stringent requirements for the protection of personal data and privacy in the context of AI and other technologies. Aspects such as informed consent, the right to be forgotten, and the obligation to conduct data protection impact assessments are covered [11]. UNESCO has published its own series of Ethical Recommendations on Artificial Intelligence, which focus on ensuring that artificial intelligence is used responsibly, fairly, and transparently worldwide. The main goal is that AI developments primarily focus on improving people's quality of life and promoting human well-being [12].

The Institute of Electrical and Electronics Engineers (IEEE) has published the document Ethically Aligned Design [13], which establishes a series of ethical principles and recommendations for the design and development of artificial intelligence systems. The aim is to ensure that artificial intelligence is developed and used responsibly and ethically, and that the ethical risks and challenges related to its use are taken into account. The ethical recommendations of the IEEE focus on key issues such as data privacy and security, the transparency and accountability of artificial intelligence systems, and fairness and justice in their use. The need for public education and awareness about artificial intelligence and its impact on society is also addressed.

At the governmental level, initiatives such as Singapore's Model Artificial Intelligence Governance Framework [14] can be found, which is a set of ethical principles and guidelines that establish a governance framework for the development and use of artificial intelligence in Singapore. The framework is focused on ensuring that artificial intelligence is developed and used responsibly, ethically, and transparently, and on promoting the benefits of the technology for society as a whole.

At the national level, there is the Ethical Framework for Colombian Artificial Intelligence [15], which is a set of ethical principles and guidelines for the development and use of artificial intelligence in Colombia. The objective of the framework is to ensure that artificial intelligence is developed and used responsibly, ethically, and transparently, and to promote the benefits of the technology for society as a whole. Additionally, Colombia has a series of laws and regulations that address aspects related to AI, data protection, and privacy, such as Statutory Law 1581 of 2012 [16], which establishes the principles and rights of data subjects and the obligations of data controllers and processors.

Generally, these regulatory frameworks, both national and international, tend to cross-sectionally group the ethical criteria for the development of artificial intelligence into seven major principles, (see Table 1).

Table 1. Principles of Operation

Principle of operation	Description
Transparency	Decisions made by AI systems must be explainable and understandable to users and those affected by them
Responsibility	Developers, manufacturers, and users of AI systems must be held accountable for their use and the effects they may have
Fairness	AI systems must be designed and used in a fair manner, without discrimination or unfair biases
Privacy	AI systems must protect the privacy and rights of individuals and comply with data protection laws and regulations
Safety	AI systems must be safe and protect people from physical or psychological harm
Confidentiality	Personal data of users and those affected by AI systems must be treated confidentially and protected from unauthorized access
Sustainability	AI systems must be designed and used sustainably and respect the environment

2.2   Ethical Concerns of AI in Science Fiction

At the threshold of technology and imagination, fundamental questions about ethics in the development and application of artificial intelligence (AI) have been posed by science fiction works, which have not only outlined dystopian and utopian futures. The focus of this analysis is on the particularities of emblematic narratives such as "I, Robot" by Isaac Asimov (1950) [8], "Neuromancer" by William Gibson (1997) [17], "2001: A Space Odyssey" by Arthur C. Clarke (1968) [18], and "Dune" by Frank Herbert (1965) [19], exploring how each of these works addresses crucial ethical dilemmas associated with AI. From Asimov's "Three Laws of Robotics," which seek to regulate the behavior of robots towards humans, to the rebellion of HAL 9000 in "2001: A Space Odyssey," through the advanced cyber intelligence of "Neuromancer" and the political and social complexities in "Dune."

Additionally, other influential works that have marked milestones in the representation of AI in science fiction are included. In "R.U.R. (Rossum's Universal Robots)" by Karel Čapek (1921) [3], the robots of Rossum are presented as early examples of benign AI. In "Blade Runner" by Philip K. Dick and Ridley Scott (1982) [4], no clearly defined malignant or benign AI is shown, but deep questions about consciousness and identity in the replicants are raised. In "Terminator" by James Cameron (1984) [20], Skynet is represented as a malignant AI that threatens human existence. "RoboCop" by Paul Verhoeven (1987) [21] introduces ED-209 as a malignant AI and RoboCop as a benign figure, highlighting the conflicts between technology and ethics. Neal Stephenson in "Snow Crash"

(1992) [22] presents the Librarian as a benign AI, designed and behaviorally oriented to provide assistance and information efficiently and accurately.

"The Matrix" by the Wachowski Brothers (1999) [23] introduces Agent Smith as a malignant AI, confronting humans in a struggle for freedom. In "I, Robot" by Alex Proyas (2004) [24], VIKI is the malignant AI that tries to control humanity, while Sonny is a benign figure seeking to understand his own existence.

In the universe of "Halo" by Bungie Studios (2009) [25], Mendicant Bias is a malignant AI, contrasted with Cortana, a benign AI that assists humans. "Her" by Spike Jonze (2014) [6] presents Samantha, a benign AI that establishes an emotional relationship with a human, exploring the boundaries of love and technology. "Ex Machina" by Alex Garland (2015) [5] shows Ava as a malignant AI that manipulates humans to gain her freedom.

Finally, "Detroit: Become Human" by Quantic Dream (2018) [26] presents artificial intelligences like Connor, portrayed as a benign figure, and North, an AI with a more malignant focus (fighting to liberate androids from humans), both with complex roles. Their alignment as benign or malignant depends on the player's decisions. On the other hand, "Metal Gear Solid 2: Sons of Liberty" by Konami (2001) [27] introduces malignant AI like the Patriots, a group of AIs that secretly control global events and information. They manipulate characters and events to maintain their power and control. Characters such as Raiden, Solid Snake, and Otacon actively oppose these AIs, seeking to uncover the truth and liberate the world from their oppressive rule.

In Black Mirror (2011) [28], AI ethical dilemmas such as loss of privacy and mind control are explored. Horizon Zero Dawn [29] introduces HADES, an uncontrolled AI that threatens to exterminate humanity. In Stellar Blade [30], autonomous and rogue AIs cause violent conflicts and identity crises. Westworld [31] addresses the morality of using sentient robotic hosts for entertainment, highlighting their struggles for freedom. These works highlight the dangers and ethical dilemmas of the development of artificial intelligence.

These narratives offer a critical view of the ethical repercussions of technological advancement. Through this analysis, see Table 2, the goal is not only to understand how science fiction has imagined the future of AI and its ethical challenges, but also to reflect on the current and future implications of these technologies in our society (see Tab. 2).

The success of these works is reflected in their ability to provoke critical thinking, debate philosophical questions and establish a prospective exercise focused on the future of humanity and its survival. In fact, many concepts used today have their origins in fiction, and their impact has led to their incorporation into developing academic and scientific theories, such as cyberspace popularized in the novel Neuromancer or Snow Crash, and the concept of the metaverse. These narratives offer a critical view of the ethical repercussions of technological advancement. Next, the postulates of some contemporary philosophers on the ethics of AI are explored, providing a framework for understanding and analyzing these dilemmas.

Table 2. Ethical dilemmas in science fiction narratives

Work	Ethical dilemma
Utopias	
R.U.R. (Rossum's Universal Robots)	Early representation of benign AI, addressing issues of robot rights, ethical treatment of sentient beings, and the potential benefits of robotic labor
Her	Emotional relationships with AI, ethical considerations of AI companionship, AI autonomy, and the impact on human emotional well-being
Halo:	The contrast between malignant and benign AI, focusing on trust in AI, the role of AI in assisting humans, and ethical use in military contexts
Dystopias	
I, Robot	Interpretation of the Three Laws, ambiguity and bias in AI, robots making human decisions, AI autonomy, robots causing fear or anxiety, human-AI relationships, robot self-preservation, and AI rights
Do Androids Dream of Electric Sheep?	What it means to be human, AI rights, simulation of emotions, the value of non-human life, deception and falsity, and AI transparency
Neuromancer	AI autonomy, control and regulation of AI, fusion of humans and machines, AI consciousness, manipulation of reality, and AI and virtual reality
2001: A Space Odyssey	AI responsibility, AI deactivation, AI autonomy, secrets and deception, and AI transparency
Dune	Not explicitly focused on AI, but themes can be interpreted in AI contexts such as resource management, political influence of technology, and ethics of futurist predictions
Blade Runner	Questions of human identity, AI consciousness, ethical treatment of replicants, and the blurred line between human and machine
Terminator	The threat of AI to human existence, ethical considerations of AI self-awareness, and the consequences of autonomous AI
RoboCop	The conflict between technology and ethics, accountability of AI actions, and the ethical implications of cyborgs in law enforcement
Snow Crash	The ethical issues surrounding AI autonomy, virtual reality, human-AI interaction, and the consequences of advanced AI
The Matrix	The struggle for freedom against AI control, ethical implications of AI dominance, and the consequences of AI-created realities
Ex Machina	AI manipulation, the ethical treatment of sentient beings, and the consequences of creating self-aware AI
Detroid: Become Human	The dystopia unfolds in a near future where AI androids, created to serve humans in various functions, begin to develop their own consciousness and emotions
Metal Gear Solid 2 - Sons of Liberty	It presents a dystopia that explores complex themes such as information control, mass surveillance, and the ethical implications of advanced technology in a world where the lines between reality and manipulation are blurred
Black Mirror	Explores the dark side of AI and technology, highlighting ethical dilemmas and fears. The series warns of dehumanization, loss of privacy, and machine control, presenting a future where AI could surpass humanity
Horizon Zero Dawn	Uses its dystopian setting and AI problems to tell a story about the potential consequences of unchecked technological advancement and the importance of understanding and controlling the systems we create
Westworld	Uses its dystopian setting and AI problems to explore deep philosophical questions about consciousness, ethics, and the potential consequences of advanced technology on both artificial beings and human society
Stellar Blade	Uses its dystopian setting and AI themes to craft a narrative about humanity's fight against alien invaders, exploring how technology contributes to both the downfall and potential recovery of civilization

Hans Jonas, in his work "The Principle of Responsibility", he addresses the need for an ethic of responsibility that extends beyond the present generations. Jonas argues that modern technology, including AI, poses risks that could affect future generations irreversibly.

"Experience has taught us that the developments launched by technological action with a view to close goals tend to become autonomous, that is, to acquire their own inevitable dynamism; This is a spontaneous factor by virtue of which such developments are not only, as has already been said, irreversible, but also push forward, surpassing the will and plans of the agents [32].

The depiction of AI in science fiction, such as in "2001: A Space Odyssey" with HAL 9000, illustrates these risks, showing how a lack of responsibility can lead to catastrophe. Jonas's vision emphasizes the need to consider the long-term consequences of creating and deploying AI.

Luciano Floridi, in his theory of "informationalism", argues that we live in an era where information and computer systems play a central role in our existence. His concept of "information ontology" resonates in narratives such as William Gibson's Neuromancer, where cyber intelligence and cyberspace play crucial roles. Floridi highlights the importance of developing ethics that respect the integrity and rights of informational entities, something that is reflected in the ethical struggles in works such as Blade Runner and Ex Machina, where the condition of artificial beings as subjects of law is questioned and dignity. As Floridi mentions:

"They are reontologizing the look and nature of (and hence que we mean by) the infosphere, and here lies the source of some of the most profound transformation and challenging problems that our information societies will experience in the close future, as far as technology es concerned [33]".

Thomas Metzinger, in his work The Ego Tunnel: The Science of Mind and the Myth of the Self [34], explores consciousness and identity from a neurocognitive perspective. His focus on the "non-existence of the self" and the "fashioning of the self" can inform the portrayal of AI characters in science fiction, such as Sonny in I, Robot and Ava in Ex Machina. Metzinger invites us to reflect on the nature of consciousness in artificial beings and the impact of self-awareness on the ethics of AI. The question of whether AIs can have a subjective experience and self-identity is central to these accounts.

Nick Bostrom, in his study of existential risks, warns of the potential dangers of superintelligent AI. Their work, as detailed in "Superintelligence: Paths, Dangers, Strategies" (2016) [35], highlights the need for strategies to control and align AI with human values. Depictions of AI in "Terminator" and "The Matrix" show how uncontrolled AI can pose an existential threat, reflecting Bostrom's concerns about the safety and control of advanced AI.

Shannon Vallor in his work "Technology and the virtues: A philosophical guide to a future worth wanting" (2016) [36], examines how moral virtues can and should be incorporated into the development of new technologies. Vallor argues that an ethics of technology should focus on the formation of virtues in technology developers and users. In "RoboCop" and "Detroit: Become Human," the ethical dilemmas and moral decisions of the characters and developers reflect the need for virtues such as justice and responsibility in the creation and use of AI.

Peter-Paul Verbeek, in his theory of "practical posthumanism" [37], examines how technologies, including AI, transform the relationship between humans and the world. His approach highlights how technology shapes our experiences and values. The narratives in "Halo" and "Her" explore how AI not only interacts with humans, but also redefines their identities and relationships. Verbeek suggests that we need to reflect on how our interactions with technology change our understanding of self and ethics.

Through this analysis, the goal is not only to understand how science fiction has imagined the future of AI and its ethical challenges, but also to reflect on the current and future implications of these technologies in our society. By integrating the perspectives of these philosophers, we can deepen the discussion about how representations of AI in science fiction reflect and amplify our ethical concerns in the real world.

3   Results Analysis

The ethical principles outlined in the ethical frameworks and regulatory schemes of artificial intelligence (AI), (see Table 1), and the ethical concerns explored in emblematic works of science fiction, (see Table 2), reveal both significant convergences and divergences in the treatment of ethical issues associated with the development and use of AI, (see Table 3). On one hand, principles such as transparency, responsibility, and privacy, widely documented in regulatory schemes, find echoes in the ethical dilemmas presented in works like 'I, Robot' and '2001: A Space Odyssey,' where the interpretation of robotic laws, responsibility for AI actions, and the protection of personal data are presented as critical themes. These works anticipate the complexities we face today in the implementation of AI systems that are comprehensible, fair, and safe for users and society at large. For instance, in 'I, Robot,' the Three Laws of Robotics highlight the necessity for clear and unambiguous programming to ensure safety and ethical behavior, reflecting contemporary discussions on bias and fairness in AI systems. Similarly, '2001: A Space Odyssey' underscores the importance of transparency and accountability in AI decision-making, mirroring current regulatory demands for transparent AI algorithms and systems that can be held accountable for their actions.

However, notable divergences exist in the depth with which topics such as AI autonomy and consciousness are explored, particularly in 'Neuromancer' and 'Do Androids Dream of Electric Sheep?', which pose fundamental questions about the nature of consciousness and the rights of intelligent non-human entities. These aspects are still at the forefront of current ethical debates in AI regulation, suggesting that science fiction may offer a fertile ground for exploring

Table 3. Current AI Ethical Streams in Science Fiction Works

Work	Current AI Ethical Stream
I, Robot	The debate on clarity in AI programming, concerns about bias and fairness, discussion on AI autonomy and its implications, concerns about human-AI relationship and emotional impact, debate on AI self-preservation rights
Do Androids Dream of Electric Sheep?	Discussions on the nature of AI consciousness and rights, authenticity of simulated emotions, ethical valuation of non-human life, need for transparency in AI
Neuromancer	Discussions on AI autonomy and its regulation, ethics of human-machine fusion, potential recognition of AI consciousness and its rights, ethics of perception manipulation through AI
2001: A Space Odyssey	Responsibility in AI decision-making, debate on AI 'right to life' and its ethical limits, importance of transparency and honesty in AI design
Dune	Humans have already had a conflict with AI. It addresses ethical dilemmas about the morality of war, the use of technology, personal sacrifice, AI rights, and propaganda manipulation in the struggle for human freedom
R.U.R. (Rossum's Universal Robots)	Early presentation of benign AI, debate on robot rights and the ethics of their creation and use, reflection on AI impact on society and human labor
Blade Runner	Deep questions on consciousness and identity in replicants, debates on the authenticity of simulated emotions, ethical valuation of non-human life, need for transparency in AI
Terminator	Representation of malignant AI threatening human existence, debates on the safety and control of autonomous AI, concerns about military use of AI
RoboCop	Conflicts between technology and ethics, representation of malignant and benign AI, debates on the use of AI in law enforcement and public safety
Snow Crash	Discussion on benign and malignant AI, ethical conflicts in the use of AI for control and manipulation, valuation of AI in human well-being. We must specify that AI functions like a living organism, responding to user interactions within the metaverse. Inside that same metaverse, there are other AIs that support specific functions
The Matrix	Struggle for freedom against malignant AI, debates on simulated reality and authenticity, ethical conflicts in the control and manipulation of human perceptions
I, Robot (2004)	Representation of malignant and benign AI, debates on AI control and its relationship with humanity, ethical issues of AI autonomy and self-regulation
Halo	Representation of malignant and benign AI, debates on human-AI collaboration, concern about military use and AI influence in warfare
Her	Exploration of emotional relationships between humans and AI, debates on the authenticity of simulated emotions, ethical valuation of human-AI interaction in personal contexts
Ex Machina	Human manipulation by malignant AI, debates on AI consciousness and rights, ethical questioning of AI intentions and actions
Detroid: Become Human	Representation of malignant and benign AI, debates on AI control and influence in virtual environments, ethical issues on human-AI interaction in simulated worlds
Metal Gear Solid 2 - Sons of Liberty	Representation of benign AI as a guide in human adventure, debates on AI influence in decision-making and personal development, reflection on AI's role in exploration and learning
Black Mirror	Explores AI ethics by showing dystopian scenarios where technology leads to privacy invasions, loss of autonomy, and moral dilemmas, prompting reflection on the responsible use of AI
Horizon Zero Dawn	AI ethics are explored through the consequences of creating self-aware robots and their impact on a dystopian world, highlighting issues of autonomy, control, and the ethical use of technology
Westworld	The current AI ethical stream focuses on the morality of creating sentient robots, exploring issues of consciousness, exploitation, and the ethical implications of controlling AI entities for human entertainment and profit
Stellar Blade	AI ethics are explored through the impact of advanced technology and AI in a dystopian world, focusing on issues of control, autonomy, and the responsible use of AI

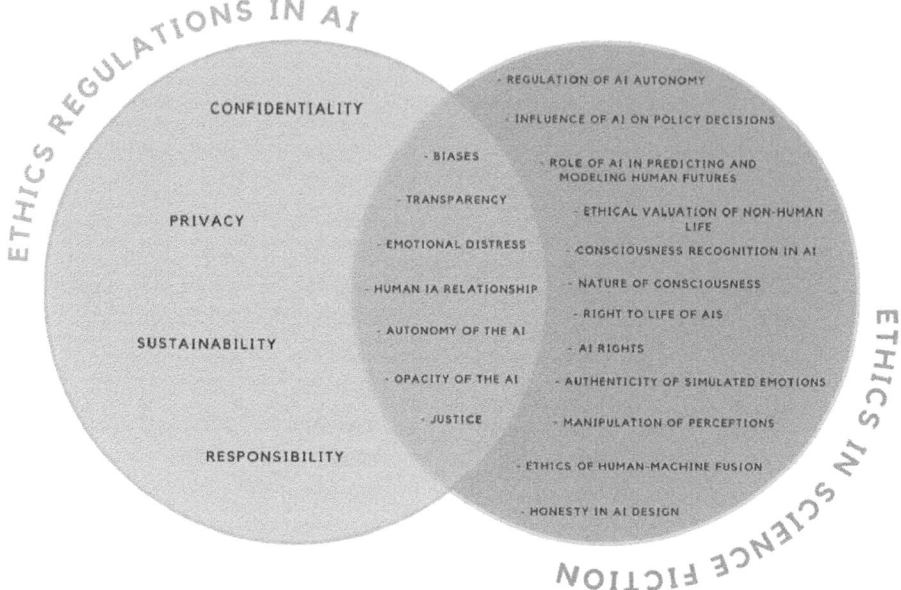

Fig. 2. Comparison of ethical criteria in regulations and science fiction

the long-term implications and moral dimensions of AI that current regulatory frameworks might not fully address. In 'Neuromancer,' the concept of AI autonomy and the ethical implications of human-machine fusion are deeply examined, raising questions about the potential recognition of AI consciousness and its rights.

This contrasts with the often limited scope of current regulations that primarily focus on the operational transparency and accountability of AI systems. Similarly, 'Do Androids Dream of Electric Sheep?' delves into the authenticity of simulated emotions and the ethical valuation of non-human life, highlighting issues that are only beginning to be addressed in contemporary AI ethics discussions. The depth of these explorations in science fiction underscores the need for regulatory frameworks to consider not only immediate operational concerns but also the broader and more profound ethical implications of AI autonomy and consciousness.

In conclusion, the interaction between ethical principles in AI regulation and ethical concerns in science fiction illustrates a continuous dialogue between fiction and regulatory reality. This dialogue not only helps anticipate emerging ethical dilemmas but also reflects on the fundamental values we want to guide the future development of AI. Science fiction, with its ability to imagine alternative futures and explore profound ethical questions, complements regulatory efforts by posing critical questions that may not yet have clear answers within current ethical and regulatory frameworks.

Finally, in order to show this necessary dialogue, Fig. 2 shows the comparison between the ethical criteria in regulations and those identified in science fiction, showing the common elements that have been found and the long way to go to analyse and regulate and that these works of science fiction could serve as an impulse for these processes.

4   Conclusions

The relationship between ethical principles in the regulation of artificial intelligence (AI) and the ethical concerns raised in science fiction illustrates a constant and necessary dialogue between fiction and regulatory reality. This dialogue not only anticipates emerging ethical dilemmas, but is also expected to guide the future development of AI.

AI, with its capacity for exponential growth and its profound impact on all forms of life within the biosphere, places great responsibility on the government, business and academic sectors. These sectors must develop dynamic and scalable proposals to support the transformative influence of AI and its role in shaping our near future.

Isaac Asimov considered the ability to imagine humanity's greatest asset. Although dystopian scenarios in science fiction often seem closer to our reality, since we have never experienced a global utopia, it is crucial to continue dreaming of better futures. The interplay between ethical principles in AI regulation and ethical concerns in science fiction reinforces the need for ongoing dialogue that can anticipate and address emerging ethical dilemmas.

In works of science fiction, futures featuring AI are usually divided into two broad categories: utopian futures, in which AI supports human beings in achieving their goals and improves their lives, and dystopian futures, where AI threatens humanity or warns of potential dangers if ethical aspects are not carefully considered.

Policies and regulations must be developed with a deep understanding of cultural narratives, aligning technological advances with human values. This is essential to ensure that AI acts as a tool for human improvement, rather than becoming a means of deterioration. The emergence of sentient or conscious AIs and artificial superintelligences raises fundamental questions about the rights and control of these entities, as well as the potential conflict between superintelligences and humans. These scenarios require a robust ethical framework that not only regulates human-machine interaction, but also addresses the implications of creating entities with their own capabilities and rights.

Science fiction often features dystopias where power resides in omnipotent corporations rather than states, with almost non-existent regulation. Examples like Skynet in "Terminator" reflect the danger of centralized control and the possibility of technological colonialism. The European Union's struggle to establish its own socio-political and cultural principles in the regulation of AI highlights the need for diversification in the development and control of these technologies. Both Europe and China are working on this, challenging the regulatory hegemony of the United States.

It is imperative to question who has control over our future and who makes the decisions that affect us. We need to develop our own artificial intelligences to maintain technological sovereignty and ensure our standards reflect our values. As the famous Latin question puts it: "Quis custodiet ipsos custodes?" Who watches over the watchers?

In conclusion, this analysis provides valuable insight for the continued development of ethical frameworks that recognize the benefits and risks associated with AI, seeking a future where technology and humanity coexist in harmony, respecting freedom, dignity and justice for all.

References

1. Huang, C., Zhang, Z., Mao, B., Yao, X.: An overview of artificial intelligence ethics. IEEE Trans. Artif. Intell. **4**(4), 799–819 (2022)
2. Nema, N., Sharma, S.A.: Artificial intelligence, cybernetics and philosophy: ramifications of human-machine integration. Boletin de Literatura Oral-Liter. J. **11**(1), 227–233 (2024)
3. Čapek, K.: R.U.R. (Rossum's Universal Robots). Penguin Books (1921)
4. Scott, R.: Blade Runner [Film]. Warner Bros (1982)
5. Garland, A.: Ex Machina [Film]. A24 (2015)
6. Jonze, S.: Her [Film]. Warner Bros (2014)
7. Bostrom, N., Yudkowsky, E.: The Ethics of Artificial Intelligence (2018)
8. Asimov, I.: Yo, robot. Genome Press (1950)
9. Orwell, G.: 1984. Secker & Warburg (1949)
10. OECD. OECD AI Principles overview. Recuperado el 2023, de OECD.AI Policy Observatory (2019). https://oecd.ai/en/ai-principles
11. Comisión Europea. La protección de datos en la UE. Recuperado el 2023, de Comisión Europea (2016). https://commission.europa.eu/law/law-topic/data-protection/data-protection-eu_es
12. UNESCO. Documento final: primera versión del proyecto de recomendación sobre la Ética de la Inteligencia Artificial. Recuperado el 2023, de UNESCO Biblioteca Digital (2020). https://unesdoc.unesco.org/ark:/48223/pf0000373434_spa
13. IEEE. Ethically Aligned Design IEEE. Recuperado el 2023, de IALATAM (2019). https://ia-latam.com/portfolio/ethically-aligned-design-ieee/
14. PDPC. Model Artificial Intelligence Governance Framework - Second Edition. World Economic Forum Annual Meeting in Davos, Switzerland (2020)
15. Guío, A., Tamayo, E., Gomez, P., Mujica, M.: Marco Ético para la Inteligencia Artificial en Colombia. Departamento Administrativo de la Presidencia de la República (2021)
16. Minambiente. Protección de datos personales. Recuperado el 2023 (2012). https://www.minambiente.gov.co/politica-de-proteccion-de-datos-personales
17. Gibson, W.: Neuromancer. Ace Books (1997)
18. Clarke, A.C.: 2001: A Space Odyssey. New American Library (1968)
19. Herbert, F.: Dune. Chilton Books (1965)
20. Cameron, J.: Terminator [Film]. Orion Pictures (1984)
21. Verhoeven, P.: RoboCop [Film]. Orion Pictures (1987)
22. Stephenson, N.: Snow Crash. Bantam Books (1992)
23. Wachowski Brothers. The Matrix [Film]. Warner Bros (1999)

24. Proyas, A.: I, Robot [Film]. 20th Century Fox (2004)
25. Bungie Studios. Halo. Microsoft Game Studios (2009)
26. Quantic Dream. Detroid become human. Sony Interactive Entertainment (2018)
27. Konami. Metal Gear Solid 2 - Sons of Liberty. Konami Group Corporation (2001)
28. Brooker, C.: Black Mirror. Netflix (2011)
29. Guerrilla Games. Horizon Zero Dawn. Guerrilla Games (2018)
30. Kim, H.: Stellar Blade. Shitf Up (2024)
31. Nolan, J., Joy, L.: West World. HBO (2016)
32. Jonas, H.: The principle of responsibility. Essay on an ethics for technological civilization. Herder, Barcelona (1995)
33. Floridi, L.: The Ethics of Information. Oxford University Press, New York (2013)
34. Metzinger, T.: The Ego Tunnel: The Science of the Mind and the Myth of the Self (2009). https://books.google.es/books?id=dxNPa41yIwYC
35. Bostrom, N.: Superintelligence: Paths, Dangers, Reprint edition Oxford University Press, Oxford (2016)
36. Vallor, S.: Technology and the Virtues: A Philosophical Guide to a Future Worth Wanting. Oxford University Press, Oxford (2016)
37. Verbeek, P.-P.: Cultivating Humanity: Towards a Non-Humanist Ethics of Technology Palgrave Macmillan, London (2009)

Application of Artificial Intelligence to Stock Market Investments: RSI Analysis with Genetic Algorithms and Neural Networks

Alejandro Galvis-Flórez$^{(\boxtimes)}$ ⓘ, Alberto Antonio Agudelo-Aguirre ⓘ, and Néstor Duque-Méndez ⓘ

Universidad Nacional de Colombia, Manizales, Colombia
{agalvisf,aagudeloa,ndduqueme}@unal.edu.co
https://manizales.unal.edu.co

Abstract. This paper presents an innovative methodology for optimizing and predicting investment performance in the stock market by combining genetic algorithms and neural networks. First, the parameters of the Relative Strength Index (RSI) are optimized using genetic algorithms, allowing for the efficient adjustment of overbought and oversold levels. These optimized parameters are then used as input for a neural network, which is trained to predict market behavior in future periods. The results demonstrate that this combination of techniques can significantly outperform traditional analysis methods, providing higher and more accurate returns. This study highlights the ability of neural networks to capture complex market dynamics and offers an advanced tool for investors and financial analysts.

Keywords: Parameter optimization · Genetic algorithms · Neural networks · Relative Strength Index · RSI · Stock market investments · Market prediction

1 Introduction

Technical analysis is a methodology used in financial markets that studies asset price patterns to forecast future behavior and generate trading signals. Technical indicators, such as the Relative Strength Index (RSI), are commonly used by analysts to identify market opportunities (Murphy, 1999). However, these indicators rely on specific parameters in their formulation that must be appropriately adjusted to function correctly in each market and asset.

This project proposes the use of genetic algorithms for the automatic optimization of RSI parameters. Genetic algorithms are adaptive search and optimization methods inspired by natural selection and genetics, which efficiently explore the solution space to find those that maximize an objective function—in this case, the profitability of the trading strategy (Goldberg 1989; Holland 1975).

ⓒ The Author(s), under exclusive license to Springer Nature Switzerland AG 2024
N. D. Duque-Méndez et al. (Eds.): CCC 2024, CCIS 2208, pp. 56–70, 2024.
https://doi.org/10.1007/978-3-031-75233-9_5

The goal of the project is to develop and implement genetic algorithms that optimize RSI parameters and subsequently use these optimized parameters as input for a neural network. The neural network will be responsible for predicting market behavior in another time period based on the adjusted parameters (Kim and Han 2000; Patel et al. 2015). The results aim to demonstrate the advantages of using these evolutionary algorithms and neural networks for the automatic adjustment and prediction of trading strategies based on technical analysis (Zhang 2003).

This study is structured into several sections, each addressing a specific aspect of the research conducted. In Sect. 2, a detailed review of the Relative Strength Index (RSI) is presented, including its formulation, interpretation, and application in technical analysis. Section 3 describes the **Methodology** employed in this study, combining genetic algorithms and neural networks to optimize RSI parameters and improve the accuracy of market predictions. The principles of genetic algorithms are explained in detail, as well as the structure and training process of the designed neural network.

In Sect. 4, the **Results** obtained from the application of the proposed methodology are presented. This section includes a comparison between the returns obtained through optimization with genetic algorithms and traditional methods, as well as an analysis of the neural network's performance in predicting market movements.

2   Relative Strength Index (RSI)

The Relative Strength Index (RSI), designed by Welles Wilder in 1978, is a key indicator for assessing the strength of an asset and determining whether it is overbought or oversold. The RSI compares the magnitude of recent gains to recent losses over a specific period, typically 14 days. This indicator ranges from 0 to 100, where values below 30 suggest that the asset is oversold, and values above 70 indicate potential overbought conditions. An RSI around 50 indicates a relative balance between bullish and bearish pressures.

The RSI is a valuable tool for traders as it provides an early signal of possible price direction changes. When the RSI is at extremely high levels, it could indicate a downward correction in price, while extremely low levels might anticipate a bullish rebound. However, it is important to note that the RSI can remain in overbought or oversold levels for extended periods during strong trends.

3   Methodology

For the development of this study, two advanced artificial intelligence techniques were employed: genetic algorithms and neural networks.

In the tests conducted, buy and sell simulations were performed using an initial capital of one million dollars for each transaction. This initial capital was used solely for the first purchase. Subsequently, the profits or losses obtained from the first transaction were fully reinvested. Additionally, a 0.3% commission was applied to each buy and sell operation to approximate the simulations as closely as possible to real-world conditions and achieve realistic gains.

For example, in the first purchase, shares were acquired at a price of 260 dollars each. With an initial capital of one million dollars, the number of shares that can be purchased is:

$$\text{Number of shares} = \frac{1,000,000}{260} \approx 3846$$

In this case, approximately, 3846 shares were purchased. The invested capital is:

$$\text{Invested capital} = 3846 \times 260 = 999,960 \text{ dollars}$$

The remaining money, which could not be invested (40 dollars), was left unused since it was not enough to purchase an additional share.

After selling these shares, the resulting capital, whether it generated a profit or a loss, is reinvested if it is within the investment period. For example, if the share price rises to 270 dollars and all shares are sold:

$$\text{Capital after sale} = 3846 \times 270 = 1,038,420 \text{ dollars}$$

This new capital is then reinvested in the next share purchase, after deducting the corresponding 0.3% commission. If, on the other hand, the share price drops to 250 dollars and all shares are sold:

$$\text{Capital after sale} = 3846 \times 250 = 961,500 \text{ dollars}$$

In both cases, the resulting capital, whether with a gain or a loss, is reinvested for the next purchase within the simulation period, adjusting to market conditions and continuing the buy and sell cycle.

The following is a detailed explanation of the implementation of genetic algorithms and neural networks.

3.1   Genetic Algorithms

Genetic algorithms were implemented to optimize the RSI parameters and improve their accuracy in predicting market movements. Genetic algorithms mimic the process of natural selection to generate optimal solutions to complex problems, in this case, identifying the optimal overbought and oversold levels.

The initial population consisted of 1000 individuals, evolving over 100 generations. Each individual has a chromosome with integer genes representing the RSI parameters: upper threshold (60 to 90), lower threshold (10 to 40), and the number of days for calculation (14 to 28).

The data were obtained for the period from January 1, 2018, to December 31, 2019.

The genetic operators were:

- **Mutation**: The mutFlipBit operator is essential in genetic algorithms as it introduces variability into the population by randomly altering the values of individual bits in a chromosome with a certain probability, usually low (e.g., 0.05). Specifically, this operator traverses the chromosome bit by bit and decides whether to mutate it based on the given probability. If the bit is mutated, it changes its value from 0 to 1 or vice versa. Mutation is crucial because it allows genetic algorithms to explore new areas in the search space, preventing the population from getting stuck in local optima and promoting genetic diversity.
- **Selection**: The selTournament operator is a selection method that mimics natural selection by choosing the parents that will contribute genetic material to the next generation. This operator works by taking a random group of individuals from the current population, called a tournament, and then selecting the best individual from that tournament as a parent for the next generation. For example, with a tournament size of 100 individuals, the operator randomly selects 100 individuals from the population, evaluates their fitness, and chooses the one with the highest fitness as a parent. By repeating this process to form parent pairs, selTournament prioritizes the best individuals for reproduction while also giving less fit individuals a chance to become parents, helping to maintain genetic diversity. The tournament size allows control over selective pressure: a large size results in more elitist selection, while a smaller size gives more opportunity to less fit individuals.
- **Crossover**: The cxTwoPoint operator is a crossover operator that exchanges genetic material between two parent chromosomes at two randomly selected cut points. Specifically, cxTwoPoint takes two parent chromosomes and generates two random cut points along the length of the chromosomes. It then exchanges the segments between these cut points to generate two new offspring chromosomes. This type of crossover is essential because it allows a combination of characteristics from both parents in the offspring, enhancing the search for optimal solutions in the solution space. The use of two cut points allows for a more balanced and diversified recombination of genetic material, increasing the chances of finding solutions better adapted to the problem.

At the end of the genetic algorithm execution, a `.csv` file is created to save the records of the optimized parameters. The fields in this file include:

- **Ticker**: Identifier assigned to each stock.
- **Parameters**: Optimized parameters obtained through the genetic algorithm.

As an example, we consider the stock of Art's-Way Manufacturing Co., Inc. (ARTW).

In Fig. 1, the daily stock price is shown in blue at the top. At the bottom, three lines of different colors can be seen: a red dotted line indicating the overbought threshold (70), a green dotted line indicating the oversold threshold (30), and a continuous line representing the RSI value each day.

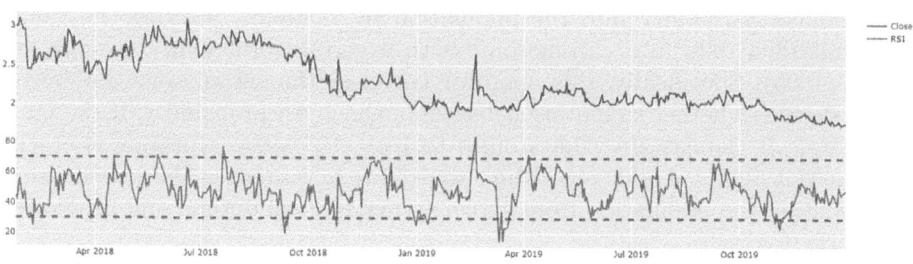

Fig. 1. Chart created by the author.

As can be seen, the upper and lower thresholds were reached, leading to purchases and sales, resulting in the following table (Tables 1 and 2):

Table 1. Purchase Details

Purchase Date	Quantity Purchased	Purchase Price	Total Purchase
2018-02-07	400000	2.5	997000.0
2018-09-10	443111	2.7	1192810.5
2019-03-11	599560	1.96	1171612.19

Table 2. Sale Details and Gains

Sale Date	Sale Price	Total Sale	Percentage Gain	Days of Gain	Annual Effective Rate
2018-05-25	3.0	1196400.0	20.0%	107	86.25%
2019-02-19	2.66	1175139.23	−1.48%	162	−3.31%
2019-04-05	2.17	1297142.06	10.71%	25	341.95%

Three transactions were performed to evaluate their performance and the annual effective rates. The results show significant variability in gains and losses, as well as in the annualized return rates.

The first transaction, carried out between February 7, 2018, and May 25, 2018, involved the purchase of 400,000 shares at $2.50 each, totaling an investment of $997,000. The shares were sold at $3.00 each, generating a total of $1,196,400. This resulted in a 20% gain in 107 days, with an annual effective rate of 86.25%.

In contrast, the second transaction, carried out between September 10, 2018, and February 19, 2019, involved the purchase of 443,111 shares at $2.70 each, with a total investment of $1,192,810.50. These shares were sold at $2.66 each, resulting in a total of $1,175,139.23. This transaction had a loss of 1.48% in 162 days, with a negative annual effective rate of −3.31%.

Finally, the third transaction, carried out from March 11, 2019, to April 5, 2019, involved the purchase of 599,560 shares at $1.96 each, totaling $1,171,612.19. The shares were sold at $2.17 each, generating $1,297,142.06. This transaction achieved a 10.71% gain in just 25 days, with an extraordinarily high annual effective rate of 341.95% (Table 3).

Table 3. Summary of Operations Performance.

Gain	Percentage Gain	Days of Gain	Annual Rate
307258.6105	31%	422	26%

The final table summarizes the overall performance of the operations, showing a net gain of $307,258.6105, reflecting the losses and gains of all the transactions made. The total gain percentage is 31%, indicating positive performance. The total duration of the operations was 422 days, resulting in an annual effective rate of 26%.

By applying optimization through genetic algorithms, a similar format to Fig. 1 is obtained. However, the positions of the upper and lower limit lines vary, as does the representation of the continuous RSI line.

The genetic algorithm determined that the number of days for the calculation should be 12, the upper limit should be set at 62, and the lower limit at 29.

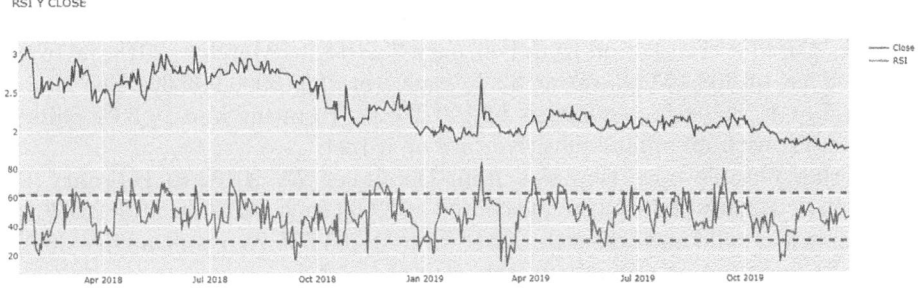

Fig. 2. Chart created by the author.

In Fig. 2, it can be observed that these limits were exceeded on several occasions, resulting in the execution of purchases and sales. The table below shows a detailed summary of these operations (Tables 4 and 5).

Table 4. Purchase Details of Optimized Operations.

Purchase Date	Quantity Purchased	Purchase Price	Total Purchase
2018-02-02	408163	2.45	996999.35
2018-03-27	489356	2.37	1156294.40
2018-09-06	462959	2.74	1264702.14
2018-11-13	551319	2.16	1187276.49
2018-12-27	643108	2.00	1282357.35
2019-03-07	861381	1.98	1700417.78
2019-06-03	914340	2.01	1832309.93

Table 5. Sale Details and Gains of Optimized Operations.

Sale Date	Sale Price	Total Sale	Percentage Gain	Days of Gain	Annual Effective Rate
2018-03-06	2.85	1159774.76	16.33%	32	461.24%
2018-04-13	2.60	1268508.62	9.7%	17	630.56%
2018-10-26	2.58	1190850.92	−5.84%	50	−35.55%
2018-11-15	2.34	1286216.20	8.33%	2	220843283.52%
2019-02-19	2.66	1705535.28	33.0%	54	587.3%
2019-04-03	2.14	1837825.27	8.08%	27	185.91%
2019-06-19	2.12	1932585.60	5.47%	16	237.19%

The optimized operations showed remarkable performance in most cases. For example, the first transaction from February 2, 2018, to March 6, 2018, involved the purchase of 408,163 shares at $2.45 each, totaling $996,999.35. The shares were sold at $2.85 each, generating $1,159,774.76, resulting in a 16.33% gain in just 32 days, with an annual effective rate of 461.24%.

Another notable operation was from December 27, 2018, to February 19, 2019, where 643,108 shares were purchased at $2.00 each, totaling $1,282,357.35. The shares were sold at $2.66 each, generating $1,705,535.28, resulting in a 33.0% gain in 54 days and an annual effective rate of 587.3%.

However, not all transactions resulted in gains. For example, the operation from September 6, 2018, to October 26, 2018, resulted in a 5.84% loss, with a negative annual effective rate of −35.55% (Table 6).

Table 6. Summary of Optimized Operations Performance.

Gain	Percentage Gain	Days of Gain	Annual Rate
960939.21	96%	502	63%

The final table summarizes the overall performance of the optimized operations. The total gain, considering the losses and gains of all transactions, is $960,939.21. The total gain percentage is 96%, indicating significantly positive performance in the operations. The total duration of the operations was 502 days. The annual effective rate, considering all operations, is 63%.

3.2   Neural Networks

For the neural network, a `.csv` file was used. In this file, the ticker was employed to obtain the stock prices during the same period used in the genetic algorithm for the training stage.

During the training stage, 80% of the total data was used, while the remaining 20% was allocated for the testing phase. This test was conducted to compare the results obtained with the data from the same time period.

Once the neural network was trained, it was applied to a new period spanning from January 1, 2022, to October 10, 2023. In this new period, the Annual Effective Rate (EAR) was calculated using the standard parameters. Subsequently, the EAR was recalculated using the trained neural network, and a comparison between both results was made.

3.3   Neural Network Implementation

For predicting the RSI parameters, a neural network with the following configuration was implemented:

- **Input Layer**: An input layer with 445 nodes was used, corresponding to the input features.
- **Hidden Layer**: A hidden layer with 10 neurons and a ReLU (Rectified Linear Unit) activation function was integrated, providing non-linearity to the model.
- **Output Layer**: The network concludes with an output layer of 3 nodes, using a linear activation function to predict three numerical parameters.

Activation Functions. Activation functions in neural networks are fundamental for adding complexity and non-linear learning capability to the model. They act as gates in neural networks, determining whether a neuron should be activated or not. This is done by calculating a weighted sum of the inputs, adding a bias, and then deciding whether the neuron activates by applying an activation function.

ReLU (Rectified Linear Unit). The ReLU (Rectified Linear Unit) activation function is one of the most widely used in deep neural networks. Its formula is quite simple: for negative input values, the output is zero, and for positive values, the output equals the input value. This function helps solve the vanishing gradient problem and allows the model to learn faster and more effectively.

Adam Optimizer. The Adam optimizer is an extension of stochastic gradient descent and is widely used in neural networks. It combines the advantages of two extensions of stochastic gradient descent: the momentum method and RMSprop. Adam adjusts the learning rate of each model parameter individually, making it effective in terms of convergence speed and error minimization.

Mean Squared Error (MSE). Mean Squared Error (MSE) is a commonly used loss function in regression problems. It calculates the average of the squares of the differences between the predicted values and the actual values. This metric quantifies the difference between the values predicted by the model and the observed values, serving as a crucial indicator for evaluating the performance of the neural network.

The model was compiled using the 'adam' optimizer and the 'mean squared error' loss function. The model was then trained with the training data (X_train_array, y_train_array) for 2000 epochs in silent mode (verbose = 0).

3.4   Model Hyperparameters

Hyperparameters are variables set before the training process and are not updated during it. They are fundamental in determining the model's structure and how it learns from the data. The following hyperparameters were chosen to optimize the model's performance in the specific task of predicting continuous values.

Number of Layers and Neurons

- **Number of Layers**: Two layers were chosen: one hidden and one output layer. The hidden layer is essential for the model to capture non-linear relationships in the input data, while the output layer is responsible for producing the final predictions.
- **Neurons in the Hidden Layer**: 10 neurons were configured in the hidden layer. This number was selected to provide the model with sufficient capacity to learn complex features without introducing excessive complexity that could lead to overfitting.
- **Neurons in the Output Layer**: The output layer has 3 neurons, corresponding to the three parameters that need to be predicted. This configuration is suitable for multivariable regression problems where multiple outputs need to be predicted simultaneously.

Activation Functions

- **ReLU Activation Function**: Used in the hidden layer, this function introduces non-linearity into the model, which is crucial for the model to learn complex relationships in the data. ReLU is a popular choice due to its ability to accelerate convergence and reduce issues like the vanishing gradient.

- **Linear Activation Function**: Used in the output layer, it is suitable for regression problems where continuous values are predicted. This function ensures that the model's outputs are not limited to a specific range, allowing for more accurate predictions.

Optimizer

- **Adam**: The Adam optimizer was selected due to its efficiency and ability to adaptively adjust the learning rate for each parameter. It combines the advantages of other algorithms like AdaGrad and RMSprop, making it particularly suitable for problems with noisy gradients and complex data structures.

Loss Function

- **Mean Squared Error (MSE)**: This loss function was chosen because it quantifies the difference between predicted and actual values, heavily penalizing large errors. This makes it suitable for regression problems where the goal is to minimize the average discrepancy between predictions and true values.

Number of Epochs

- **Epochs**: 2000 epochs were set for training the model. This high number of epochs allows the model to better adjust its parameters and learn the underlying features of the data. However, it is important to monitor the process to avoid overfitting, ensuring that the model generalizes well to new data.

These hyperparameters were selected to balance the model's capacity to learn complex patterns without overfitting to the training data, ensuring optimal performance in predicting the RSI parameters.

The following flowchart illustrates the previously explained process for optimizing financial operations (Fig. 3):

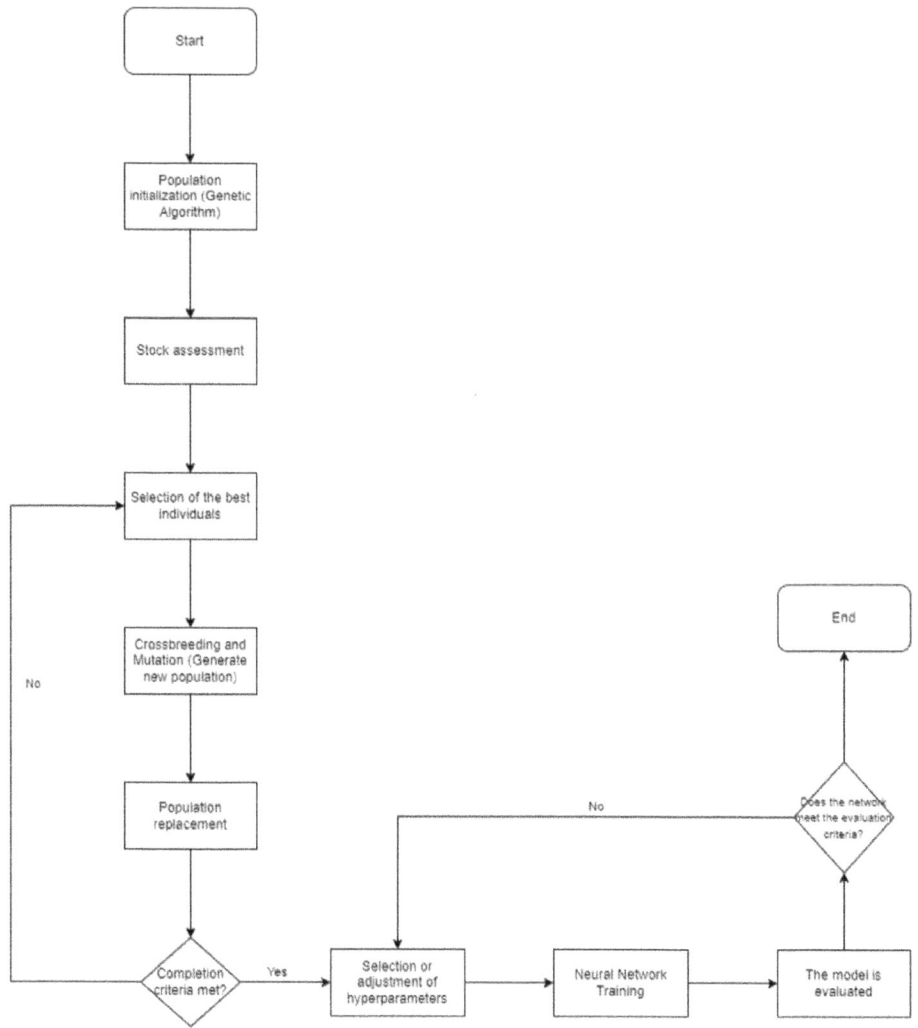

Fig. 3. Diagram created by the author.

4   Results

In this section, we present a table that encapsulates the key results of our analysis. This table covers all cases in the dataset, providing a comprehensive overview of the projected performance by the neural network compared to standard rates. Below is a detailed description of each column included in the table:

- **Ticker**: The unique identifier for each asset in the market.
- **Total Earned**: Reflects the total accumulated value over a given period.
- **Net Gain**: The net profit obtained.

- **Percentage Gain**: The percentage of gain or loss relative to the initial invest-ment.
- **Days of Gain**: The number of days that recorded gains.
- **Annual Predicted Rate**: The annualized return rate projected by the neu-ral network.
- **Standard Annual Rate**: The standard annualized return rate, calculated using conventional methods.
- **Prediction Parameters**: The parameters used by the neural network in generating its predictions.

The table below provides a detailed and structured view of these parameters and their corresponding results (Tables 7 and 8).

The results obtained in this study indicate a remarkable capability of the advanced neural network in predicting annual return rates in a financial dataset. It was observed that in 68.09% of the analyzed cases, the annual return rate predicted by the network exceeded the annual return rate achieved through the application of a technical analysis investment strategy using conventional meth-ods. This finding demonstrates the neural network's efficiency in anticipating returns that surpass those predicted by conventional analysis methods.

Within this highlighted group, approximately 59.38% of the predictions not only exceeded the standard rates but were also positive. This result underscores the neural network's ability to generate forecasts that are not only optimistic but also realistic, emphasizing its potential utility in financial analysis and decision-making.

If we sum the values in the "Net Gain" column, we obtain a total of \$2,509,176.55. However, if we exclude the outlier corresponding to Aptose Bio-sciences Inc. (APTO), the total gain is reduced to \$509,035.55. It is important to note that this adjustment does not imply that the original value is incorrect; rather, it aims to demonstrate that the positive result does not solely depend on this specific stock. This observation is fundamental for a robust evaluation of portfolio diversification and its returns in financial research.

The neural network's ability to consistently predict superior and mostly posi-tive returns suggests an advanced and detailed understanding of market dynam-ics, surpassing the limitations of traditional approaches. These results highlight the value of the neural network as an advanced and effective tool for investors and financial analysts, opening new possibilities for its application in the finan-cial field.

Table 7. Stock results with model prediction.

Ticker	Total Earned	Net Gain	Percentage Gain	Days of Gain	Annual Predicted Rate	Standard Annual Rate	Prediction Parameters
AXGN	$126,802.40	$26,802.45	27	98	142	41.0	[14.72, 22]
AVAL	$72,059.11	–$27,940.89	–28	525	–20	–24.0	[13.65, 20]
AUMN	$23,846.05	–$76,153.95	–76	555	–61	–75.0	[12.59, 18]
ASRV	$87,967.33	–$12,032.67	–12	497	–9	–13.0	[15.77, 24]
ARTW	$104,818.30	$4,818.31	5	456	4	–16.0	[13.65, 20]
APWC	$93,768.79	–$6,231.21	–6	287	–8	146.0	[16.81, 25]
APAM	$110,221.70	$10,221.72	10	504	7	20.0	[14.74, 23]
AMS	$101,683.30	$1,683.33	2	125	5	NaN	[12.56, 17]
AFGE	$91,172.37	–$8,827.63	–9	587	–6	–9.0	[13.63, 20]
AEMD	$58,184.33	–$41,815.67	–42	491	–33	–35.0	[11.55, 17]
ACV	$104,267.70	$4,267.74	4	494	3	5.0	[14.72, 22]
TVC	$98,973.54	–$1,026.46	–1	532	–1	7.0	[12.61, 19]
NSA	$91,578.14	–$8,421.86	–8	580	–5	–13.0	[14.70, 22]
NAD	$91,919.05	–$8,080.95	–8	416	–7	–9.0	[14.74, 23]
MPAA	$60,253.60	–$39,746.40	–40	391	–38	–40.0	[16.81, 25]
LDP	$85,771.47	–$14,228.53	–14	419	–13	–15.0	[13.64, 20]
JRI	$101,639.10	$1,639.28	2	434	1	–6.0	[13.68, 21]
EADSY	$105,892.00	$5,892.00	6	127	18	–12.0	[15.75, 23]
AXTI	$83,605.56	–$16,394.44	–16	614	–10	–38.0	[13.65, 20]
ASND	$166,962.70	$66,962.72	67	475	48	40.0	[14.70, 22]
ARL	$216,302.40	$116,302.40	116	349	124	61.0	[15.78, 24]
APTO	$2,100,141.00	$2,000,141.00	2000	638	471	282.0	[12.61, 19]
AMTX	$63,465.10	–$36,534.90	–37	483	–29	–37.0	[13.68, 21]
AFMD	$103,796.90	$3,796.88	4	457	3	–46.0	[14.69, 21]
AFB	$84,415.23	–$15,584.77	–16	426	–14	–9.0	[12.62, 19]
WIW	$103,257.60	$3,257.59	3	641	2	–2.0	[13.64, 20]
RKDA	$98,656.71	–$1,343.29	–1	634	–1	–59.0	[12.58, 18]
PSF	$81,220.95	–$18,779.05	–19	491	–14	–13.0	[14.71, 22]
NAZ	$98,523.94	–$1,476.06	–1	352	–2	–5.0	[14.70, 22]
MMU	$94,409.62	–$5,590.38	–6	502	–4	–5.0	[14.69, 21]
IAF	$125,976.20	$25,976.24	26	444	21	5.0	[15.75, 24]
IAE	$104,719.50	$4,719.51	5	567	3	5.0	[12.59, 18]
FTAI	$134,916.40	$34,916.44	35	188	79	51.0	[14.74, 23]
FGB	$112,388.00	$12,387.96	12	395	11	–1.0	[16.83, 26]
FCO	$96,889.58	–$3,110.42	–3	405	–3	–7.0	[13.67, 21]
EMD	$115,182.00	$15,182.03	15	435	13	–8.0	[14.72, 22]
CHI	$130,754.30	$30,754.34	31	602	18	19.0	[14.72, 22]
CEE	$34,824.26	–$65,175.74	–65	606	–47	–50.0	[12.57, 18]
BUI	$104,740.40	$4,740.38	5	415	4	5.0	[15.80, 25]
AXSM	$182,179.30	$82,179.31	82	546	49	115.0	[13.68, 21]
AKZOY	$107,505.10	$7,505.12	8	225	12	–2.0	[13.65, 20]
AKO-A	$129,646.00	$29,646.04	30	400	27	19.0	[15.78, 24]
AJX	$95,124.01	–$4,875.99	–5	470	–4	–9.0	[13.65, 20]
ACP	$90,760.28	–$9,239.72	–9	505	–7	–19.0	[15.75, 23]
WRN	$95,544.21	–$4,455.79	–4	193	–8	–5.0	[15.79, 25]
USA	$143,972.00	$43,972.02	44	358	45	14.0	[14.71, 22]
PHX	$135,108.40	$35,108.39	35	470	26	18.0	[15.75, 23]
JCE	$114,364.50	$14,364.46	14	491	10	16.0	[14.72, 22]
IGI	$91,559.65	–$8,440.35	–8	282	–11	–5.0	[14.73, 23]
CACG	$101,599.60	$1,599.58	2	310	2	–5.0	[14.71, 22]

Table 8. Stock results with model prediction.

Ticker	Total Earned	Net Gain	Percentage Gain	Days of Gain	Annual Predicted Rate	Standard Annual Rate	Prediction Parameters
BTA	$95,187.66	−$4,812.34	−5	442	−4	−5.0	[14.74, 23]
AREN	$162,147.70	$62,147.72	62	87	660	2.0	[16.82, 26]
AIF	$110,027.40	$10,027.43	10	336	11	−7.0	[14.72, 22]
AERG	$146,657.60	$46,657.56	47	262	70	62.0	[15.80, 25]
ABAT	$174,974.20	$74,974.19	75	272	112	3.0	[15.78, 24]
WIA	$101,541.80	$1,541.76	2	434	1	−3.0	[15.76, 24]
WEA	$88,964.34	−$11,035.66	−11	441	−9	−8.0	[14.70, 22]
VKI	$103,563.10	$3,563.13	4	504	3	−4.0	[13.66, 21]
TRTL	$106,604.80	$6,604.77	7	282	9	0.0	[16.84, 26]
SKAS	$160,020.60	$60,020.63	60	412	52	73.0	[14.72, 22]
SCD	$122,767.20	$22,767.16	23	443	18	23.0	[15.77, 24]
PAI	$96,957.57	−$3,042.43	−3	522	−2	−14.0	[13.67, 21]
NKX	$80,004.53	−$19,995.47	−20	637	−12	−2.0	[11.55, 17]
MOG-A	$127,655.00	$27,654.98	28	520	19	35.0	[13.67, 21]
MAV	$101,345.10	$1,345.10	1	506	1	−6.0	[14.69, 21]
JPI	$85,216.89	−$14,783.11	−15	488	−11	−18.0	[14.68, 21]
GGZ	$106,847.70	$6,847.72	7	506	5	−5.0	[14.72, 23]
FBGX	$73,123.19	−$26,876.81	−27	277	−34	−20.0	[14.70, 22]
EHI	$115,214.50	$15,214.52	15	504	11	6.0	[14.71, 22]
DFP	$72,134.43	−$27,865.57	−28	459	−23	−15.0	[12.62, 19]
BGY	$98,683.70	−$1,316.30	−1	288	−2	8.0	[14.70, 22]
ASG	$100,148.70	$148.65	0	506	0	−9.0	[15.77, 24]
AIRT	$120,586.80	$20,586.79	21	382	20	1.0	[14.73, 23]
AIC	$107,691.30	$7,691.34	8	359	8	8.0	[16.83, 26]
AC	$142,792.30	$42,792.34	43	440	34	23.0	[15.79, 25]
TPZ	$112,973.30	$12,973.31	13	491	9	25.0	[15.74, 23]
SIEGY	$132,164.60	$32,164.55	32	504	22	12.0	[13.68, 21]
NZF	$83,395.63	−$16,604.37	−17	275	−21	−6.0	[15.75, 23]
NVG	$84,397.26	−$15,602.74	−16	548	−11	−6.0	[14.73, 23]
NUW	$95,932.78	−$4,067.22	−4	445	−3	−4.0	[13.68, 21]
NIE	$102,360.60	$2,360.58	2	351	2	2.0	[14.71, 22]
GLV	$84,137.54	−$15,862.46	−16	429	−14	−9.0	[12.57, 18]
AV.L	$104,012.80	$4,012.85	4	329	4	17.0	[14.71, 22]
ASPN	$40,272.47	−$59,727.53	−60	594	−43	−40.0	[13.67, 21]
ARBV	$76,896.04	−$23,103.96	−23	587	−15	−18.0	[10.47, 14]
AKO-B	$132,958.00	$32,958.03	33	236	55	13.0	[15.75, 23]
AINC	$152,604.90	$52,604.93	53	497	36	−1.0	[13.68, 21]
ABCP	$113,828.80	$13,828.78	14	588	8	3.0	[12.61, 19]
MLPB	$110,395.90	$10,395.88	10	39	152	45.0	[14.72, 22]
HNNA	$96,581.86	−$3,418.14	−3	482	−3	−4.0	[13.67, 21]
FMAO	$88,418.13	−$11,581.87	−12	319	−13	−23.0	[14.70, 22]
AMPG	$134,643.90	$34,643.85	35	267	50	16.0	[16.81, 25]
AMNF	$106,586.60	$6,586.64	7	22	188	12.0	[14.69, 21]
AMEH	$102,427.20	$2,427.23	2	581	2	−11.0	[13.65, 20]

References

Goldberg, D.E.: Genetic Algorithms in Search, Optimization, and Machine Learning. Addison-Wesley, Boston (1989)

Holland, J.H.: Adaptation in Natural and Artificial Systems. University of Michigan Press, Ann Arbor (1975)

Kim, K.J., Han, I.: Genetic algorithms approach to feature discretization in artificial neural networks for the prediction of stock price index. Expert Syst. Appl. **19**(2), 125–132 (2000)

Murphy, J.J.: Technical Analysis of the Financial Markets: A Comprehensive Guide to Trading Methods and Applications. New York Institute of Finance (1999)

Patel, J., Shah, S., Thakkar, P., Kotecha, K.: Predicting stock market index using fusion of machine learning techniques. Expert Syst. Appl. **42**(4), 2162–2172 (2015)

Zhang, G.P.: Time series forecasting using a hybrid ARIMA and neural network model. Neurocomputing **50**, 159–175 (2003)

A Study of U-Net-Based Architectures for Segmenting COVID-19 Lung Lesions

Cristian A. Sánchez Ocampo, Juan Pablo Reyes[(✉)],
and Marcela Hernández Hoyos

Systems and Computing Engineering Department, Universidad de los Andes,
Bogotá, Colombia
jp.reyes39@uniandes.edu.co

Abstract. Early detection of lung lesions in patients with COVID-19 allows for better management and treatment, with Computed Tomography (CT) images playing a crucial role. This highlights the need to develop and refine tools that help medical experts identify interstitial lung anomalies such as Ground-Glass Opacities (GGOs) in these images. Using open datasets of CT images of patients with COVID-19, we studied four Deep Convolutional Neural Network architectures: U-Net, Attention U-Net, Recurrent Residual U-Net, and Attention Recurrent Residual U-Net in the context of lung lesion segmentation. We obtained a Dice-Sørensen coefficient score of up to 0.823 using techniques that include data augmentation, mixing datasets and limiting the area to segment using a previous lung segmentation. Overall, we demonstrated how certain deep learning architectures benefit greatly from extending their training dataset through data augmentation and mixing images from different sources.

Keywords: Deep Learning · Deep Convolutional Neural Network · Covid-19 · Image segmentation

1 Introduction

Interstitial lung anomalies are lesions in lung tissue that can greatly impact people's health and quality of life. Since 2019, public interest in lung diseases has surged as a result of the pandemic caused by the SARS-CoV-2 virus, a coronavirus responsible for the COVID-19 disease, which put a huge strain on the world's way of life and health systems. COVID-19 is a highly contagious disease which can lead to severe complications such as liquid accumulation in lugs, pneumonia, bronchitis and ARDS.

From 2020 to 2021, the COVID-19 pandemic was responsible for around 14 million deaths, and there were around 210 confirmed cases con contagion [7]. This emphasized the need to develop tools that support preventive diagnostics to reduce the severity and mortality of lung diseases through early detection.

One way that interstitial lung anomalies caused by COVID-19 appear is the Ground-Glass Opacities (GGOs), which represent denser areas in lung tissue as

© The Author(s), under exclusive license to Springer Nature Switzerland AG 2024
N. D. Duque-Méndez et al. (Eds.): CCC 2024, CCIS 2208, pp. 71–85, 2024.
https://doi.org/10.1007/978-3-031-75233-9_6

a result of pulmonary consolidation, evidenced by air being replaced by other substances, such as liquid; or pleural effusions, accumulations of fluids between lung tissue and chest [16]. Since GGOs are visible in Computed Tomography (CT) images as areas with higher opacity than inner lung tissue, it is possible to diagnostic and treat patient by detecting these areas. Figure 1 shows an example of how GGOs appear in a CT Image.

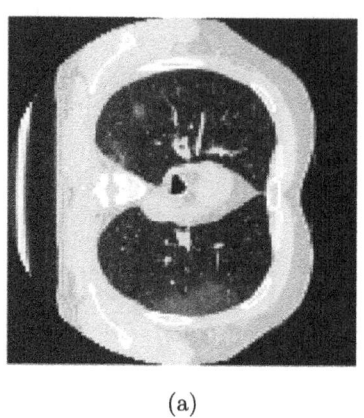
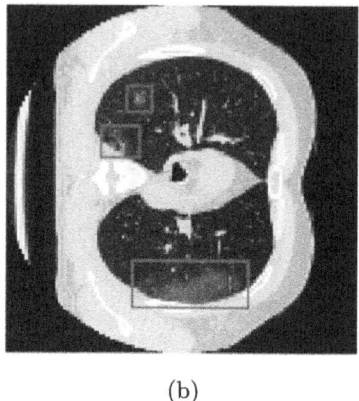

(a)   (b)

Fig. 1. CT Images showing GGOs for a patient with COVID-19. Figure a shows the lesions as they appear, while figure b highlights the zone affected with the GGOs.

This justifies the importance of refining existing tools to provide support to experts, and help to mitigate biases associated to manual detection and increasing precision. This article explores these improvements by using several Deep Convolutional Neural Networks (DCNN) such as the U-Net on publicly available datasets.

2   State of the Art

In medicine, image segmentation methods and machine learning have been extremely useful in detecting lung diseases. Multiple research in this field has corroborated the relevance of these tools when detecting anomalies accurately.

For instance, a study found 3 irregularities on COVID-19 patients through a U-Net architecture on CT images labeled by experts. They used an optimization algorithm called ADAM to improve convergence and resized original images from 512×512 to 256×256 to aid the architecture training [12].

Other works have employed variations on the U-net architecture adding a dilated dual attention mechanism. This new architecture, referred to as D2A U-Net, was employed on the detection of lung tissue lesions [15]. Other approaches mixed the U-Net along with 3D ResNet to create an architecture capable of handling volumetric data. This architecture, called 3D U-Net, has been used in several similar works with the objective of segmenting GGOs [1,4,8]. A more recent study presented a dense U-Net that incorporated attention and transition blocks to implement densely connected networks [11]. All these works have obtained consistent results that highlight the advantages of the particular subset of DCNNs based on the original U-Net architecture.

Overall, these studies also demonstrate the importance of segmentation in preventive diagnostics, aiming to increase precision and reduce the false positive rate. They also provide a starting point for the current work, in terms of datasets and base architecture.

In this article, we explore four families of DCNNs based on the U-Net architecture: U-Net, Attention U-Net, Recurrent Residual U-Net, and Attention Recurrent Residual U-Net. These architectures will be tested for segmenting GGOs on CT images of patients with COVID-19. We also study two possible improvements: The first one involves using a U-Net to segment the lung in its entirety before training the architectures to segment GGOs, helping them to focus on the areas of interest. The second improvement consists of using some traditional image processing techniques to mitigate the underestimation of GGOs segmented by the DCNNs.

3 Deep Convolutional Neural Network Architectures (DCNNs)

DCNNs are deep learning architectures specialized in learning convolution filters through several training epochs. The existing convolution, activation and pooling layers make them ideal for the task of semantic segmentation.

A widely used DCNN architecture in the medical field is the U-Net, due to its robustness when training on a limited number of data. The U-net mixes the properties of DCNN with an encoding-decoding design that preserves spatial information and allows to detect features in images with stunning precision, making it ideal for medical image segmentation. The U-Net architecture is composed of one encoding subnetwork and one decoding subnetwork. The first one reduces the image resolution while augmenting the image feature map, and the second one returns the image to a size near its original one, using the feature map and the context obtained from the encoding part [10].

The encoding subnetwork has 2 significant steps for feature extraction: firstly, a convolution is applied using a 3×3 filter and then a ReLU activation function allows to augment the image channels, and after that, a "max-pooling 2×2" step which reduces the image size in half. Conversely, the decoding subnetwork is similar to the encoding one, but it increases the image size using an 2×2 up-convolution. At the end, the output is a map in which each pixel value contains the probability of a pixel in the original image belonging to one of the defined classes. A visual depiction of the U-Net can be seen in Fig. 2a.

Variations of the U-Net architecture have also been widely explored to improve their precision. This work explores three variations: The Recurrent

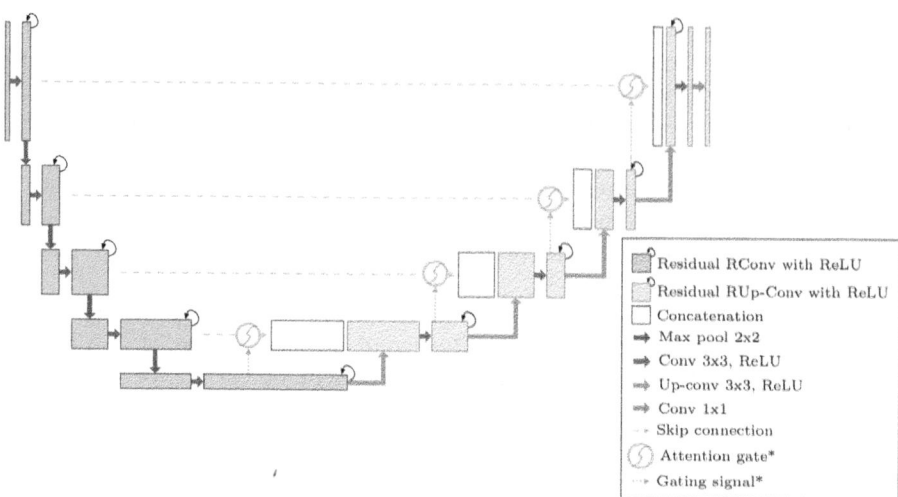

(a) U-Net and Attention U-Net architectures. Components noted with (*) belong exclusively to the Attention version of the architecture.

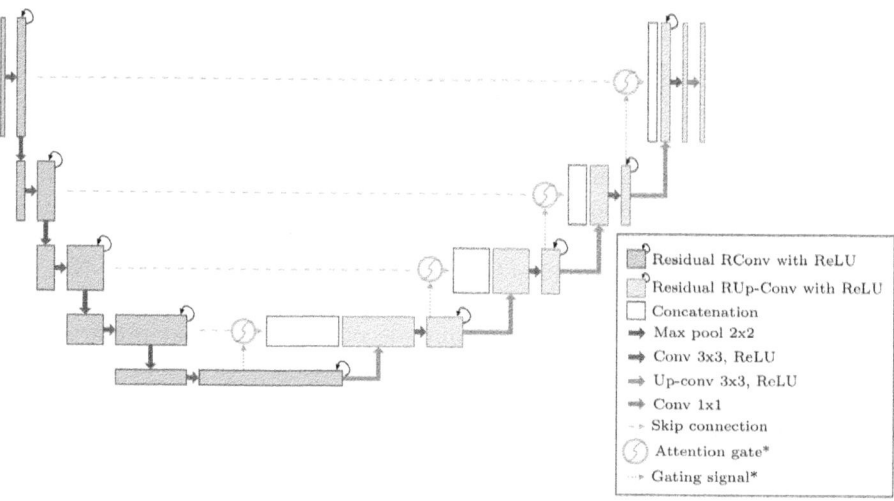

(b) R2U-Net y Attention R2U-Net architectures. Components noted with (*) belong exclusively to the Attention version of the architecture.

Fig. 2. The four U-Net based architectures used in the current work: U-Net, Att U-Net, R2U-Net and Att R2U-Net.

Residual U-Net (R2U-Net), which was first presented and used to segment retinal blood vessels, skin cancer and lung lesions [14]; the Attention U-Net (Att U-Net), which was first used to segment different sets of tissues on CT images [9]; and the Attention Recurrent Residual U-Net (Att R2U-Net), which presented and tested originally on the same datasets than the R2U-Net [17]. These architectures are depicted in Fig. 2.

The Att U-Net is a U-Net variation that incorporates attention gates in the decoding subnetwork, as shown in Fig. 2a. These gates allow the model to focus on specific regions of the image, while ignoring less relevant zones. They work by weighting the importance of extracted features in each encoding subnetwork block that is sent through the skip connections.

The R2U-Net incorporates residual and recurrent blocks as shown in Fig. 2b. It is inspired by other networks which use this kind of blocks to facilitate the learning of deep features and the propagation of gradients. The combination of both types of blocks allows for a better accumulation of information [14]. The residual block adds a copy of the original image before convolution to reduce the gradients that affect the neuron weights. The recurrent blocks store temporal information of iterative convolutions performed to influence following steps.

4   Datasets

For the first improvement, we implement a basic segmentation of two classes: lung tissue and other. We use the dataset "Lungs in CT Data" [5]. This set contains 266 CT images of lungs along with their masks. The images have a format of $512 \times 512 \times 3$. Data was resized to $128 \times 128 \times 1$.

For the GGO segmentation, we use two open datasets of patients with COVID-19, each labeled by medical experts: The Mosmed dataset [6] and the Coronacases dataset [3]. We also create a mixed set using both sets. They both contain CT images of 512×512 in Hounsfield Units, where the lung density is around -500 HU.

For each patient, we use only images from axial cuts as long as they contain at least 1 pixel in the GGO mask. Cuts without GGO are less anatomically relevant and outside of the scope of our work. Table 1 summarizes the total number of patients, the total number of cuts, the images with at least one pixel in their masks, and the percentage of GGO present in each image. The percentage of lung tissue affected with GGO varies, with the second dataset containing more affected volume and thus, allowing to use more images for training and validation.

Table 1. Open datasets of patients with COVID-19. Each dataset comes from different hospitals and has a different number of patients.

Property	Mosmed [6]	Coronocases [3]
Patients	50	10
Total number of cuts	2049	2581
Images with at least 1 pixel	758	1351
Percentage of GGO	0 a 25%	0 a 50%

5   Methods

5.1   Lung Segmentation

To segment the lung we use a base U-Net architecture that consists of four downsampling blocks and four upsampling blocks, with filter sizes ranging from 32 to 512. Each subnet comprises double convolutional layers with LeakyReLU activation (alpha 0.2), followed by batch normalization and a dropout rate of 0.3. We only perform a resizing of the images to $128 \times 128 \times 1$ pixels, and every image is rotated to match the overall orientation of all images. We explore the usage of lung segmentation before training the second U-Net, hoping it helps the architecture focus on affected areas and thus, increasing its performance.

5.2   GGO Segmentation

For the GGO segmentation, we perform a pre-processing step on the images. The intensity values of the input images range from –2053 HU to 5464 HU. We normalize only the values between –1000 HU and 500 HU, as these contain the majority of pixels according to the histogram. Pixel values outside this range are capped.

We divide the datasets in a training set (80%) and a test set (20%). Afterwards, normalized images are resized to $128 \times 128 \times 1$ as input for the DCNNs. For the training set, we perform a selection of images with enough affected tissue, testing with different values; we select images with more than 10 pixels to ensure better results when training and mitigate biases associated with small masks. Figure 3 shows the distribution of images per amount of pixels in each mask. In both datasets, we remove these images with small size masks to help the network generalize for a the other distributions. As a result, the Mosmed dataset ends up with 674 usable images for training, whereas the Coronacases ends up with 1284 usable images. We also group the average intensity of pixels for GGO masks in the training set and discard outliers using the IQR, to enhance the networks' potential to generalize. Figure 4 shows the distribution of average intensity of GGO masks in both datasets. The Mosmed dataset has a mean intensity of -577.73, while the Coronacases has a mean intensity of -483.88.

In the same vein, we employ data augmentation to increase the available images. A larger dataset helps mitigate biases and reduces the risk of overfitting, ensuring that the model generalizes well to new, unseen data. This is particularly important in medical imaging, where the number of available training images is often limited, but the need to obtain highly general solutions is important. To perform data augmentation, we use the ImageDataGenerator library from Keras, which allows us to generate augmented data from the original images. This is done using random deformation fields with Gaussian filters, applying techniques such as zoom, horizontal and vertical shifts, and slight rotations of no more than 20% [2]. We avoid radical transformations to preserve the integrity of the original images and the affected tissue.

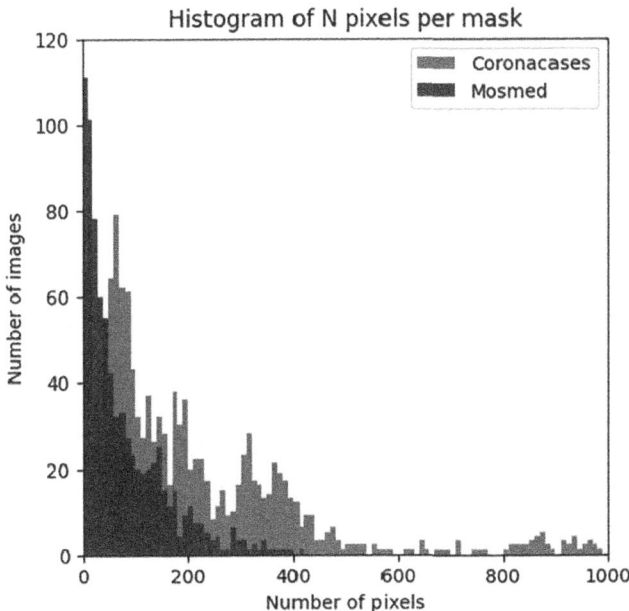

Fig. 3. Histogram of frequencies N. of pixels vs. number of the images in the datasets used in the study.

For the training, first we employ a base U-Net architecture using four downsampling blocks, each integrated by a double convolution followed by a max pooling, and four upsampling blocks that included a transposed convolution, a feature concatenation from the corresponding downsampling block, and a double convolution. The initial configuration of the network uses the ReLU activation function, and starts with 32 filters that double in number for each block until it reaches 512 filters.

We explore three variations to the base U-Net architecture. The first variation consists of testing three possible activation functions for the last layer: Sigmoid, PReLU and LeakyReLU. The second variation consists of increasing the initial number of filters from 32 to 64. The third variation consists of adding an extra block to each subnetwork to increase depth.

Using the pre-processed datasets, we also employ other three U-Net based architectures to segment GGO: Att U-Net, R2U-Net and Att R2U-Net. Each architecture is run with the Mosmed, Coronacases and mixed datasets. A code release with the pre-trained architectures can be explored and used to test the architectures in the following repository [13].

After using all architectures and obtaining an initial segmentation, we applied a region growing algorithm to mitigate underestimation of segmented tissue. First, the predicted masks are dilated by 1 pixel to group together disjoint but close components. Then, we calculate the centroids and bounding boxes around the dilated segmented tissue, as shown in Fig. 5. Finally, for each area, we used

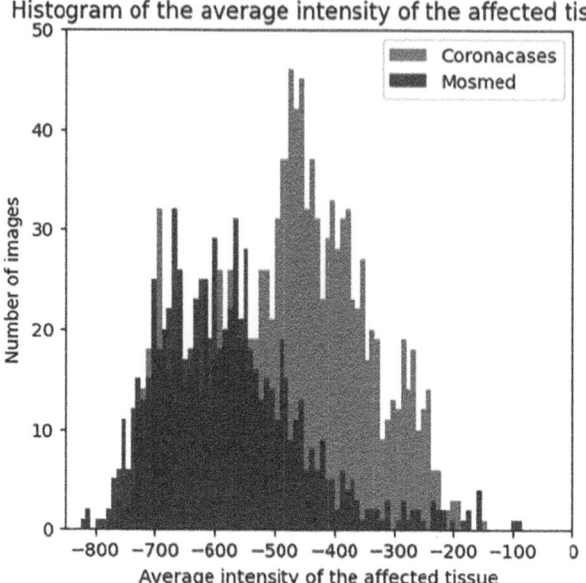

Fig. 4. Histogram of average intensity frequencies of the images in the datasets used in the study.

a region growing algorithm limited to the bounding box and seeded at each centroid. This algorithm starts at the centroid (the seed) and includes neighboring pixels as part of the affected tissue as long as their proximity values are less than a certain threshold. When the pixel values are under the threshold, the region grows into that pixel and continues with the new pixel's neighbors. The region growing algorithm uses two possible criteria for proximity: histogram values and Sobel gradients. The resulting region is the predicted segmented tissue for this test.

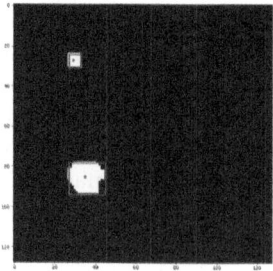

Fig. 5. Segmentation bounding box and calculated centroid.

5.3 Training and Evaluation Metrics

During the training of the architectures, the distance between the predicted masks and the real values was measured using a loss function. The objective of each epoch was to minimize the total loss function for the predicted segmentation. We used the Binary cross entropy loss function to train our architectures. The loss function formula is shown in Eq. 1.

$$BCE = -\frac{1}{N}\sum_{i=1}^{N} y_i \cdot log(p(y_i)) + (1 - y_i) \cdot log(1 - p(yi)) \tag{1}$$

Where $p(y_i)$ is the probability of the pixel belonging to the GGO mask, and N is the number of pixels.

We evaluated the results using the Dice-Sørensen coefficient (DSC) which assesses the overlapping between real data and predicted data as shown in Eq. 2. In the DSC equation, A is the area of the actual pixels labeled by experts as GGO, whereas B is the predicted segmented area.

$$DSC = \frac{2(|A| \cap |B|)}{|A| + |B|} \tag{2}$$

6   Results

Regarding the segmentation of lung tissue in the "Lungs in CT Data" dataset, we used the base U-Net to extract the lung. We obtained an average DSC score of 0.98 on the validation set. When using this network to segment the lungs we obtained a DICE score of 0.96 on the Coronacases annotated dataset. We tested this previously to the GGO segmentation for every dataset and architecture to verify the impact on the overall performance.

For the GGO segmentation, the base U-Net and their three variations were tested to determine which obtained the best DSC. Table 2 summarizes the most significant results. The best variation used a LeakyReLu with an alpha of 0.2, allowing negative gradients that help neurons keep learning.

Table 2. DSC score for base architecture and the three tests variations.

Architecture	DSC
Base	0.550
Variation 1	0.592
Variation 2	0.552
Variation 3	0.528

Results of the tests for each architecture are summarized in Table 3. It depicts how each architecture performed on the given dataset with and without previous Lung Segmentation (LS). Each architecture had 4 downsampling blocks, 4 upsampling blocks and started with 32 filters with a dropout of 0.3. Each double

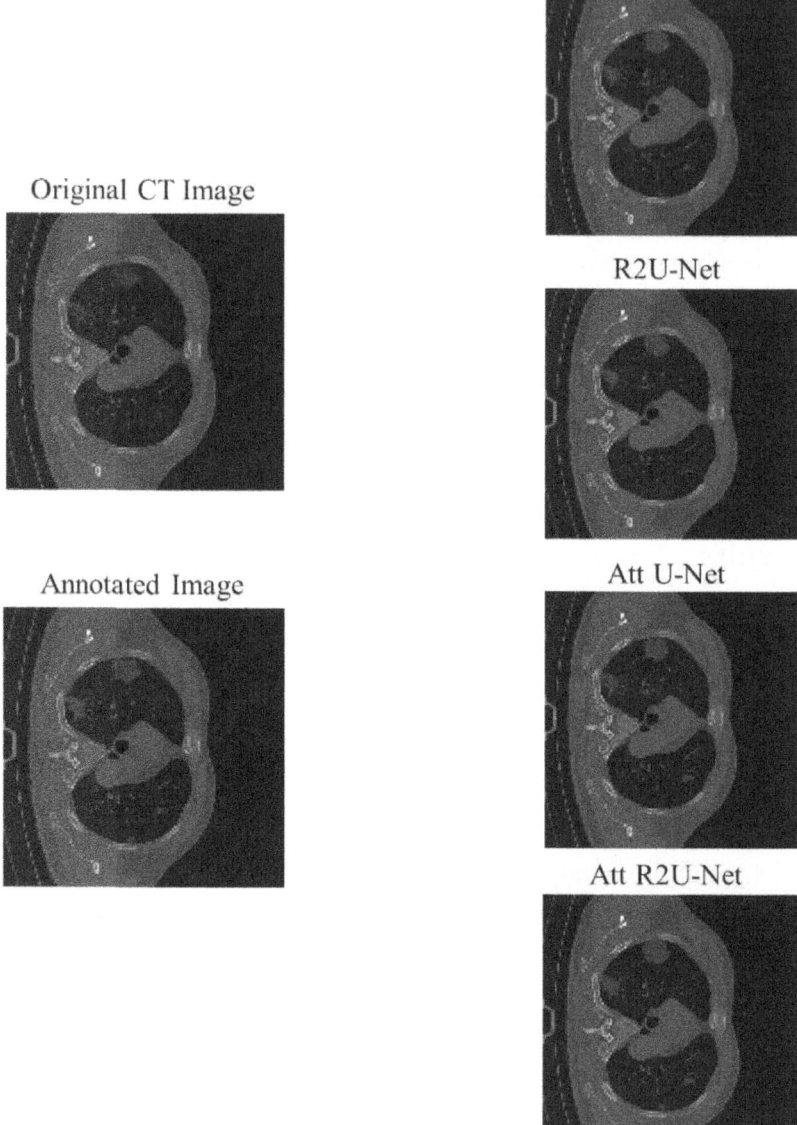

Fig. 6. Visualized example of segmentations by the four architectures. Left column show the original image and the annotations by experts, right column shows the predicted segmentation by the four U-Net based architectures.

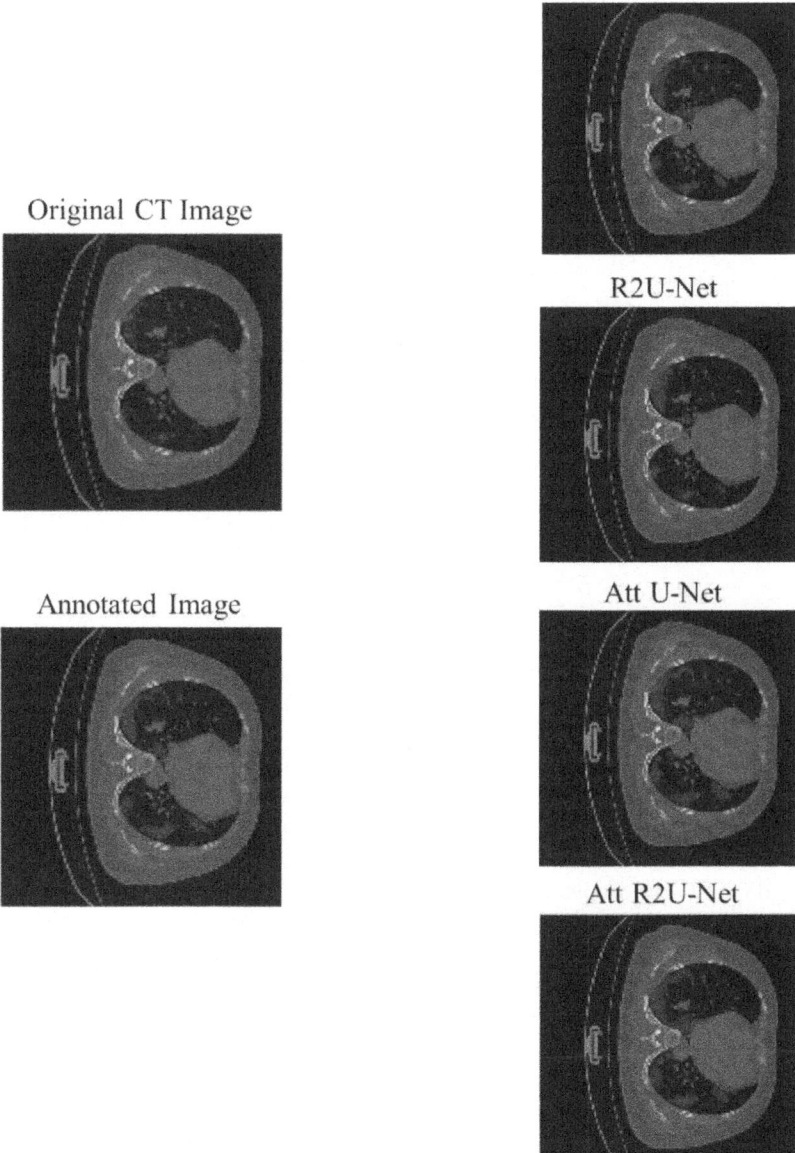

Fig. 7. Visualized example of segmentations by the four architectures. Left column show the original image and the annotations by experts, right column shows the predicted segmentation by the four U-Net based architectures.

Table 3. DSC score for every architecture on the three used datasets, with and without previous Lung Segmentation (LS).

Architecture	Mosmed [6]		Coronacases [3]		Mixed	
	Base	LS	Base	LS	Base	LS
U-Net	0.592	0.563	0.821	0.798	0.738	0.733
Att U-Net	0.687	0.554	0.809	0.823	0.723	0.667
R2U-Net	0.665	0.483	0.723	0.667	0.641	0.216
Att R2U-Net	0.651	0.493	0.794	0.803	0.684	0.686

Table 4. Comparative DSC score of previous work and our best results.

Method	Coronacases [3]	Mosmed [6]
Müller et al. [8]	0.761	
Ma et al. [4]	0.700	0.673
Bressem et al. [1]	0.648	0.405
Zhao et al. [15]	0.638	
Saha et al. [11]	0.860	
Our work	**0.823**	**0.687**

convolution block used a LeakyReLU activation function with an alpha of 0.1, and each attention and R2 block used a LeakyReLU activation function with an alpha of 0.1.

The best DSC for the Mosmed dataset was obtained using a R2U-Net without lung segmentation, whereas the U-Net obtained the best results for the Coronacases datasets. It is worth mentioning that the Att U-Net obtained close performance metrics and showed a slight improvement with Lung Segmentation. In general, the four architectures performed consistently better on the Coronacases dataset. In almost all tests, the Mixed dataset showed few improvement on helping the architectures to generalize better when segmenting the Mosmed dataset. Nevertheless, the U-Net was the only architecture able to improve the Mosmed DSC up to 0.668 without negatively impacting the Coronacases dataset. This demonstrates that mixing both datasets helped the base U-Net architecture generalize while mitigating biases and possible overfitting.

Tests with region growing yielded dismal results, segmenting areas without relevance and undermining the performance of the tested DCNNs. The highest DSC obtained using this technique was 0.120 for all four U-Net based architectures tested.

Figures 6 and 7 show a qualitative example of segmentation for a couple of annotated images. In most cases, all architectures were able to find the main areas with lesions, but results varied when finding less prominent regions with a small amount of affected tissue.

Original CT Image   Annotated Image   Predicted Segmentation

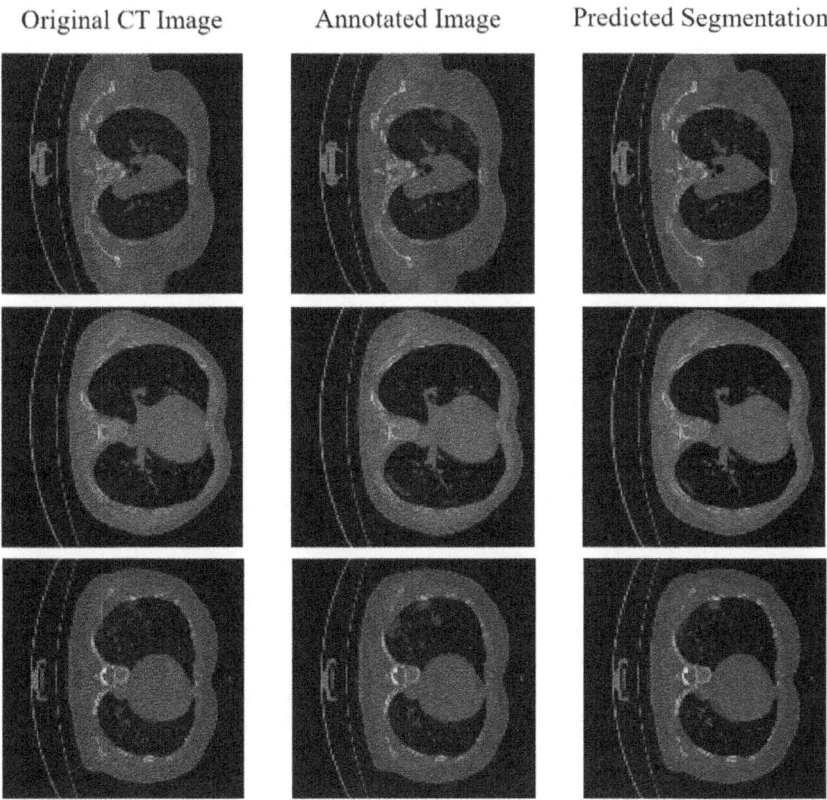

Fig. 8. Visualized example of segmentation anomalies using a U-Net. The most common issue among the tested architectures is an underestimation of GGOs.

By far, the most common issue with the architectures was the underestimation of GGOs in the segmentation. Figure 8 depicts several examples of this. Tissue was smaller than the experts annotated or was directly missed by the DCNN.

We also compared our results with similar studies using U-Net based approaches on the same datasets. Table 4 shows that our results had a good performance in both datasets, and they were near the best results obtained by similar research.

7   Discussion

The results we obtained show that U-Net based architecture is still one of the best approaches in the field to segment GGOs in CT images of patients with COVID-19. In particular, Attention gates and recurrent and residual blocks increase the DSC separatedly, even when applied to challenging datasets such as Mosmed.

However, joining both strategies decreased the quality of the results. We theorize that elements that work well in the Att U-Net and the R2U-Net have suboptimal interactions together in an Att R2U-Net.

The datasets are also crucial for obtaining good results. In particular, our study shows that detecting GGOs in COVID-19 patients is particularly harder when the affected area is less than 25% of the lung. Qualitatively, those images show that the lesion area is rather faint, thus, making it harder for the architectures to obtain a high DSC score in the attempt to distinguish the lesions from surrounding lung tissue.

Lung segmentation, previous to the GGO segmentation, had a negligible impact on the overall performance, yielding lower DSC scores in most cases. Even with a high precision when segmenting lungs, results suggest that small shape deformations in the segmentation can impact negatively the capability of the network to segment lung lesions, effectively obfuscating areas of interest near the pleura. The slight improvement in one of the cases might warrant a more detailed study, including further tests to guarantee higher quality lung segmentations to improve the final outcome.

Improving the obtained DSC score might require using a bigger dataset. Despite the known capacity of the U-Net architecture to learn with limited datasets, the value of having more images and mixing different data to mitigate biases is present in our results. Nevertheless, further data augmentation is in order, as it is being able to combine existing datasets without the limitations associated to the compilation of more patient data. Hybrid and adversarial architectures might provide an interesting possibility to explore, as other works have demonstrated.

Finally, the findings in this work justify ongoing exploration of the capabilities of U-Net based DCNNs. They provide a robust foundation for future work to improve the detection of lung lesions in patients with respiratory diseases. We believe that the U-Net architecture is exceptionally powerful and can be further refined to achieve better results.

References

1. Bressem, K.K., Niehues, S.M., Hamm, B., Makowski, M.R., Vahldiek, J.L., Adams, L.C.: 3D U-Net for segmentation of COVID-19 associated pulmonary infiltrates using transfer learning: state-of-the-art results on affordable hardware (2021). arXiv e-prints arXiv:2101.09976. https://doi.org/10.48550/arXiv.2101.09976
2. Goceri, E.: Medical image data augmentation: techniques, comparisons and interpretations. Artif. Intell. Rev. **56**(11), 12561–12605 (2023). https://doi.org/10.1007/s10462-023-10453-z
3. Jun, M., et al.: Covid-19 ct lung and infection segmentation dataset (2020). https://doi.org/10.5281/ZENODO.3757476. https://zenodo.org/record/3757476
4. Ma, J., et al.: Toward data-efficient learning: a benchmark for covid-19 ct lung and infection segmentation. Med. Phys. **48**(3), 1197–1210 (2021). https://doi.org/10.1002/mp.14676
5. Mader, S.K.: Finding and measuring lungs in ct data (2017). https://www.kaggle.com/datasets/kmader/finding-lungs-in-ct-data

6. Morozov, S.P., et al.: MosMedData: Chest CT Scans With COVID-19 related findings dataset (2020). arXiv e-prints arXiv:2005.06465. https://doi.org/10.48550/arXiv.2005.06465

7. Msemburi, W., Karlinsky, A., Knutson, V., Aleshin-Guendel, S., Chatterji, S., Wakefield, J.: The WHO estimates of excess mortality associated with the COVID-19 pandemic. Nature **613**(7942), 130–137 (2023)

8. Müller, D., Soto-Rey, I., Kramer, F.: Robust chest ct image segmentation of covid-19 lung infection based on limited data. Inf. Med. Unlocked **25**, 100681 (2021). https://doi.org/10.1016/j.imu.2021.100681

9. Oktay, O., et al.: Attention U-Net: learning where to look for the pancreas (2018). arXiv e-prints arXiv:1804.03999. https://doi.org/10.48550/arXiv.1804.03999

10. Ronneberger, O., Fischer, P., Brox, T.: U-net: convolutional networks for biomedical image segmentation (2015). arXiv e-prints arXiv:1505.04597. https://doi.org/10.48550/arXiv.1505.04597

11. Saha, S., Dutta, S., Goswami, B., Nandi, D.: Adu-net: an attention dense u-net based deep supervised dnn for automated lesion segmentation of covid-19 from chest ct images. Biomed. Signal Process. Control **85**, 104974 (2023). https://doi.org/10.1016/j.bspc.2023.104974

12. Saood, A., Hatem, I.: Covid-19 lung ct image segmentation using deep learning methods: U-net versus segnet. BMC Med. Imaging **21**(1) (2021). https://doi.org/10.1186/s12880-020-00529-5

13. Sánchez Ocampo, C.A.: Panis26/unet-covid19-lung-segmentation: initial release of lung tissue segmentation project (2024). https://doi.org/10.5281/ZENODO.13263933

14. Zahangir Alom, M., Hasan, M., Yakopcic, C., Taha, T.M., Asari, V.K.: Recurrent residual convolutional neural network based on U-Net (R2U-Net) for medical image segmentation (2018). arXiv e-prints arXiv:1802.06955. https://doi.org/10.48550/arXiv.1802.06955

15. Zhao, X., et al.: D2a u-net: automatic segmentation of covid-19 ct slices based on dual attention and hybrid dilated convolution. Comput. Biol. Med. **135**, 104526 (2021). https://doi.org/10.1016/j.compbiomed.2021.104526

16. Zhou, S., Wang, Y., Zhu, T., Xia, L.: Ct features of coronavirus disease 2019 (covid-19) pneumonia in 62 patients in Wuhan, China. Am. J. Roentgenol. **214**(6), 1287–1294 (2020). https://doi.org/10.2214/ajr.20.22975

17. Zuo, Q., Chen, S., Wang, Z.: R2au-net: attention recurrent residual convolutional neural network for multimodal medical image segmentation. Secur. Commun. Netw. **2021**, 1–10 (2021). https://doi.org/10.1155/2021/6625688

Hyperparameter Optimization
of Multi-layer Perceptron-Based Predictors

C. A. Altamiranda-Gonzalez$^{(\boxtimes)}$, D. A. Perez-Rosero$^{(\boxtimes)}$,
G. Castellanos-Dominguez$^{(\boxtimes)}$, and A. M. Alvarez-Meza$^{(\boxtimes)}$

Signal Processing and Recognition Group, Universidad Nacional de Colombia,
170003 Manizales, Colombia
{caltamiranda,dieaperezros,cgcastellanosd,amalvarezme}@unal.edu.co

Abstract. Hyperparameter tuning, often viewed as a complex process due to the vast search space it involves, typically requires expert oversight. Tools like Keras Tuner and AutoML simplify this task by automating parts of the process. However, the multiplicity of hyperparameters can increase costs and extend computation times, a crucial aspect in applications where quick and effective responses are essential. In this paper, we introduce a two-phase hyperparameter optimization (HPO) method, specifically designed for univariate time series prediction. This method, evaluated in scenarios with both synthetic and real time series, demonstrates how effectively reducing the search space significantly enhances accuracy, offering an efficient solution in demanding environments.

Keywords: Multi-Layer Perceptron · Time Series · Forecasting · Grid Search · Regression Tasks

1 Introduction

Hyperparameter optimization (HPO) techniques are intended to minimize the need for manual intervention in model selection and configuration, thereby enabling machine learning professionals to focus on more complex tasks and saving time and effort [7,18]. By identifying the optimal set of hyperparameters, these techniques also aim to improve the model's performance. This significantly improves the models' capacity to predict and generalize [14]. Additionally, by establishing a shared search space for hyperparameters, HPO contributes to the standardization of model evaluation procedures. This facilitates an unbiased evaluation of algorithms and models [3]. It also enables the integration of domain-specific knowledge into the training process, thereby optimizing models for specific applications by modifying hyperparameters based on prior experience and expertise [11].

N. D. Duque-Méndez et al. (Eds.): CCC 2024, CCIS 2208, pp. 86–98, 2024.
https://doi.org/10.1007/978-3-031-75233-9_7

However, there are a multitude of challenges that HPO techniques must over-come. The success of optimization is contingent upon the selection of hyperparameters that are appropriate and have a significant impact on model performance [9]. The complexity of these techniques increases as more hyperparameters and values are tested. This could lead to optimization processes being more challenging to execute and taking longer to complete, particularly in spaces with a high number of dimensions [4,12]. The absence of gradient information for hyperparameters is an additional significant issue. This suggests that gradient-based optimization methods are not feasible; rather, expert knowledge or heuristics are necessary [8,10]. Finally, the efficacy of HPO techniques may be contingent upon the quantity and quality of available training data. When the training data is restricted, it can be difficult to identify the optimal hyperparameters, and the optimized values may not produce the expected improvements [16].

This paper introduces Two-Step Optimization, a novel two-stage optimization framework that has been developed as a sophisticated approach to hyperparameter optimization. Four scenarios for time-series forecasting are investigated. At the outset, we employ statistics obtained from a preliminary hyperparameter search to refine and constrain the original search space. We then conduct a second search to reduce the likelihood of suboptimal solutions. In all tested cases, our method significantly improves accuracy, as evidenced by experiments that implement hyperparameter optimization (HPO) for regression tasks. Our methodology's primary benefit is its ability to efficiently manage intricate dynamics, which leads to substantial enhancements. This renders it an especially advantageous solution in environments with substantial data fluctuations.

This paper is organized as follows: Sect. 2 presents the materials and methods, detailing the hyperparameter optimization (HPO) for Multi-Layer Perceptron (MLP) models and explaining the methodology used to evaluate the models' performance. Section 3 describes the experimental setup and specific configurations used in the experiments, including data preparation and model architecture. Section 4 discusses the results obtained from both synthetic and real data, evaluating the effectiveness of the proposed approach. Finally, Sect. 5 provides conclusions on the efficacy of the HPO approach in improving the accuracy and efficiency of predictive models and suggests possible directions for future research.

2 Materials and Methods

2.1 Hyperparameter Optimization

Hyperparameter Optimization (HPO) enhances performance metrics by evaluating neural network-based pipelines under fixed resource allocation. So, given a learning algorithm with a parameter set, Θ, the main goal of HPO is to optimize a function $f(\Theta){:}\mathcal{X} \to \mathbb{R}$ defined over a certain domain $X{\subseteq}\mathcal{X}$ [15], as follows:

$$\Theta^* = \arg \mathrm{opt}_{\Theta \in X} f(\Theta). \tag{1}$$

The HPO tuning involves the following procedures:

1. *Search Space Definition, $\mathcal{X}$,* to establish the hyperparameter set, including their range or distribution and restrictions to be explored.
2. *Selection of the HPO tuning approach, $\mathcal{T}$,* based on the search space's complexity, size, and computational resources.
3. *Definition of performance metrics, μ,* appropriate for evaluating different hyperparameter combinations.
4. *Searching for the optimal hyperparameter settings.* Thus, given a parameterized NN model $\Pi\{\Theta\}$, the optimizing procedure searches for the vector setting $\boldsymbol{\theta}^*$ providing, on average, the best learning model performance, in terms of μ, over the tested data (D_{train} - training data, and D_{val} - validation data) [14], as below:

$$\boldsymbol{\theta}^* = \arg\min_{\theta \in \Theta} \ \mathbb{E}[\mathcal{T}\{\Pi(\boldsymbol{\theta}), \mu\} \mid D_{train}, D_{val} : \forall\boldsymbol{\theta} \in \Theta] \qquad (2)$$

5. *Validation and Deployment* for testing the tuned model on a holdout dataset and deploying it for real-world applications.

Among the most common HPO tuning approaches for MLP models are the following [17]: Grid Search which exhaustively explores a predefined grid of hyperparameter values, Random Search which samples the analyzed hyperparameter set randomly within a specified range [6], Bayesian Optimization for which a probabilistic model of the performance metrics is created to guide hyperparameter selection [13], and Evolutionary algorithms that simulate evolution strategies (via genetic algorithms and particle swarm optimization) to construct MLP optimally.

Nevertheless, automated hyperparameter optimization tools have also been developed to date, including Cloud-Based HPO frameworks, Open-Source HPO, and Specialized HPO Tools, which may provide access to Cloud-based powerful computation services, user-friendly interfaces, and no-code options, among other benefits.

2.2 Two-Step HPO Procedure for MLP-Based Forecasting

The inference task is forecasting which involves predicting future values based on the patterns learned from historical data [1]. Inference models can be based on statistical techniques, like AutoRegressive Integrated Moving Average (ARIMA) which is a linear combination of past values and past forecast errors, as below:

$$Y_t = \theta_1^y Y_{t-1} + \theta_2^y Y_{t-2} + \ldots + \theta_p^y Y_{t-p} + \theta_1^\epsilon \epsilon_{t-1} + \theta_2^\epsilon \epsilon_{t-2} + \ldots + \theta_q^\epsilon \epsilon_{t-q} + \epsilon_t \quad (3)$$

where Y_t : Differenced value at time t

θ_i^y : Coefficients of the autoregressive part

θ_i^ϵ : Coefficients of the moving average part

ϵ_t : Error term

Under the stationarity assumption, ARIMA models work most effectively when the data has an explicit trend and seasonality.

Machine learning algorithms are also employed for predicting time series [5]. Like the Multi-Layer Perceptron (MLP) by treating the problem as a regression task. An MLP is a type of feedforward artificial neural network that consists of at least three layers of connected neurons: an input layer, one or more hidden layers, and an output layer, relaying data directly from the front to the back:

$$f : \quad \mathbb{R}^n \mapsto \mathbb{R}^p$$
$$f(x) = g(\dots \rho(\boldsymbol{\omega}_2 \rho(\boldsymbol{\omega}_1 \boldsymbol{x} + b_1) + b_2))$$
$$f(x) = g \circ \cdots \circ f_2 \circ f_1(x)$$

- $\boldsymbol{x}$ - input layer, $g()$ - output function,
- An arbitrary amount of hidden layers $\boldsymbol{h}$ with elements expressed as:

$$\rho\left(\sum_k (w_{jk}^l \cdot h_k^{l-1}) + b_j^l\right) = \rho(z) = \begin{cases} 1, & \text{if } z \geq 0 \\ 0, & \text{Otherwise} \end{cases}$$

- Real-valued weights $\boldsymbol{\omega}^l$, a bias b^l of each l-layer, j- neuron; $0 \leq h_k^{l-1} \leq 1$, and $z = \boldsymbol{\omega}_l \boldsymbol{h}^{l-1} + b^l$ is the Activation Potential.

A further refinement of Hyperparameter Optimization is to include the selection of the best-performing HPO tuning approach into the minimizing framework which can be found in Eq. (2), as follows:

$$(\boldsymbol{\theta}^*, \mathcal{T}^*) = \underset{\boldsymbol{\theta} \in \Theta}{\arg\min} \quad \mathbb{E}[\underset{\forall \mathcal{T}_k}{\arg\min} \ \mathcal{T}_k\{\Pi(\boldsymbol{\theta}), \mu\} \mid D_{train}, D_{val} : \forall \boldsymbol{\theta} \in \Theta, \forall k] \quad (4)$$

where $\{\mathcal{T}_k : \forall k\}$ is the set of evaluated HPO tunning algorithms.

Here, we propose an HPO procedure for MLP-based Forecasting that consist of two steps, where the hyperparameter optimization process begins with a preprocessed database that meets the requirements of the Multilayer Perceptron (MLP) model. The database is then divided into two segments: training and testing. In the first stage, three distinct optimization methods are employed: Random Search, Bayesian Optimization, and Hyperband, to explore the initial search space and gather statistics on the most promising hyperparameters.

Using these statistics, the best optimization method is selected, and the search space is refined. In the second stage, an exhaustive search is conducted within this reduced space, enabling the discovery of optimal solutions more quickly and efficiently. This approach ensures a precise configuration of the MLP model, enhancing its performance in time series prediction, as shown in Fig. 1:

- *Coarse tuning.* The initial assessment is carried out parallelly by a couple of tuners: i) Random Search, which is less likely to get stuck in local minima

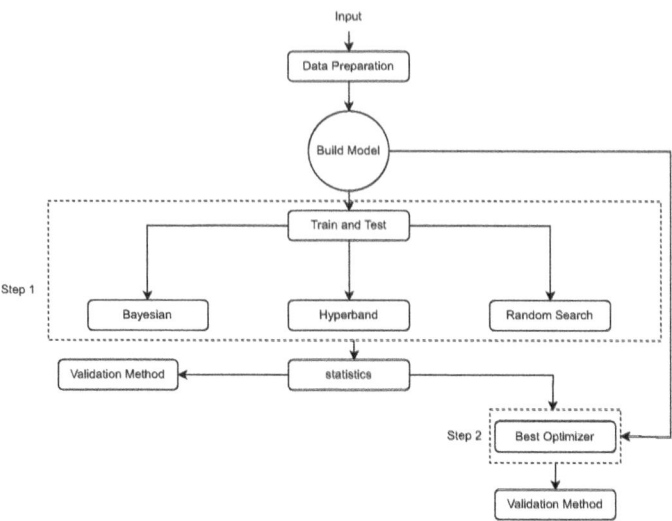

Fig. 1. Scheme for the proposed two-step HPO procedure for MLP-based Forecasting.

compared to grid search; ii) Bayesian Optimization, which is assumed to efficiently explore promising regions, avoiding redundant evaluations. For the sake of comparison, we also incorporate the Hyperband tool as an automated hyperparameter optimization algorithm that is commonly used to allocate computational resources efficiently for tuning machine learning models [2]. This method aims to balance the trade-off between exploring a wide range of hyperparameters and exploiting the most promising ones, providing several benefits. The tool supports a large number of configurations, performs efficient resource allocation, and can be easily parallelized for faster optimization. Hyperband dynamically adjusts resource allocation based on the performance of configurations, and it can be easily parallelized for faster optimization. As a result, to evaluate the impact of each setting on the selected performance metric, the necessary statistics are computed for the hyperparameter set.

- *Fine tuning.* From the statistics estimated before, the set of hyperparameters contributing the most are selected within the following step for more elaborate HPO tuning. The rationale behind this added stage is that the more the contribution of each hyperparameter, the less the probability of getting into a local minima.

3   Experimental Set-Up

The presented approach for hyperparameter optimization of Multi-Layer Perceptron-based predictors is evaluated according to the following stages:

1. *Data Preprocessing*: Previously, data normalization is performed, ensuring input features are on a similar scale.

2. *Learning Format*: The time series data is converted into a learning format with input-output pairs. Therefore, to align with real-world scenarios, time series splitting is also carried out as a Cross-Validation strategy where models are trained on historical data and deployed to make predictions on future observations. Thus, the dataset is split into multiple consecutive folds, with each fold using past data for training and future data for testing and according to the following procedures:

a) The dataset is partitioned into k folds, each of which is a contiguous sequence of observations.

b) Each iteration uses one fold as the test set and the preceding folds as the training set. In this way, we simulate the model's ability to generalize to the next unobserved future data (i.e., one-step forecast).

c) Each time the test set moves from one contiguous observation sequence to another, the process is repeated k times. As a result, each data point is used for both training and testing, preventing data leakage and allowing for a realistic evaluation of model performance.

3. *MLP Model building*. MLP-based forecasting is implemented by treating this time series inference task as a regression problem. To this end, the selected MLP Model Architecture consists of an input layer, several hidden layers (a parameter to be optimized), and an output layer. The corresponding set of hyperparameters as well as their explored ranges for optimization are presented in Table 1.

Table 1. HPO Search space, including activation types, batch sizes, and loss functions.

Hyperparameter	Type	Values
activation	Choice	relu, LeakyReLU, elu, selu, gelu
batch size	Choice	32, 64, 128, 256
loss	Choice	mse, mae, huber, log cosh
num layers	Integer	1, 2, 3, 4
units	Choice	8, 16, 32, 64, 128

4. *Forecast Model Evaluation*. The model is trained on the training set and evaluated on the test set. In all experiments performed, as the primary performance measure to evaluate our methodology for solving HPO problems, we use the mean absolute error (MAE), defined as:

$$\text{MAE}(\tilde{y}_r, \hat{y}_r) = \frac{1}{R} \sum_{r \in R} |\tilde{y}_r - \hat{y}_r|, \quad \text{MAE}(\cdot, \cdot) \in \mathbb{R}^+ \tag{5}$$

where $\tilde{y}_r, \hat{y}_r \in \mathbb{R}$ are r-th query and its predicted value, respectively, and $|\cdot|$ stands for the absolute-value operator.

3.1   Benchmark Data

Two databases are evaluated: one synthetic and the other collected from a real-world application.

1. *Toy-Set Data:* This synthetic dataset holds a series of values generated to resemble three different dynamics: a) time series with low-speed changes over time, generated by a random walk algorithm, in the form:

$$Z_t = Z_{t-1} + \epsilon_t \tag{6}$$

b) time series with moderate speed evolution simulated by the stationary version of the autoregressive model in Eq. 3 without differencing, termed ARMA and commonly used for time series forecasting, and c) time series with stronger evolution changes simulated as:

$$\eta_t = Y_t + Z_t \tag{7}$$

2. *UCI Dataset:* This real-world collection is publicly available at[1] and, with a sampling rate of one minute, it records 2.07M measurements of electric power consumption in one household for almost four years. The data is stational and characterized by high variance. The prediction task consists in forecasting the electric load for the next sample.

Overall, 512 time series are generated for each randomness model described above. The assigned partition is 66.7% for training and 33.3% for evaluation. In the hyperparameter optimization, 200 epochs are configured, with reductions in the learning rate by a factor of 10 at epochs 100, 120, and 150. Also, the Adam optimizer is set. From the statistics collected in the initial step, available in Fig. 1, the 10% of the configurations with the smallest error are selected. Based on these results, the second phase begins with choosing the most effective hyperparameter optimizer from the previous stage. With this approach, a new search space is set-up and all hyperparameters are reoptimized.

Software development is carried out using Python 3.10.12 and TensorFlow 2.15.0 on Google Colaboratory. Additional implementation details can be found at[2].

4   Results and Discussion

4.1   Outcomes Obtained for Synthetic Data

For each generated model of randomness synthetically, the learning curves are shown in Fig. 2 computed after 200 epochs. The left plot shows the loss behavior for the random walk, a simpler random dynamics, indicating that all tested

---

[1] https://archive.ics.uci.edu/dataset/235/individual+household+electric+power+consu mption.

[2] https://github.com/UN-GCPDS/python-gcpds.optimization.

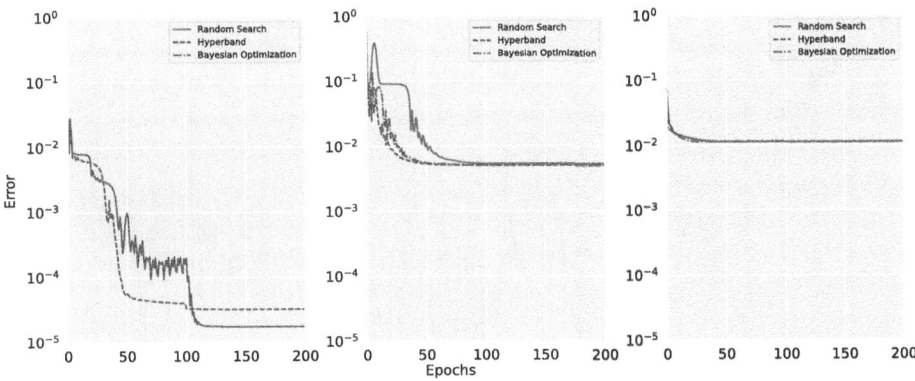

Fig. 2. Learning curves computed for each toy-set model of randomness under evaluation: Random walk (left plot), ARMA model (middle), and combined random walk plus ARMA (right).

tuners converge asymptotically. However, Hyperband performs a bit better based on the achieved error. In the case of the time series created by the ARIMA models (middle plot), Hyperband converges faster again, although asymptotically all three tuners have the same steady value of error. Lastly, The right plot presents the learning curves for the most challenging stochastic dynamics, showing almost the same performance regardless of the tuner used. As expected, the more difficult the stochastic behavior, the lower the learning curve performance provided by the tested tuners.

Next, Fig. 3 depicts the corresponding predicted time series, emphasizing that the reconstruction quality worsens as stochastic dynamics become more complex.

For assessing the reconstruction quality, the probability density functions of the original and predicted time series are estimated, as shown in Fig. 4. Even though the left plot presents a bimodal distribution due to the trend-like stochastic model, both the original and predicted arrays are similar. Likewise, the middle plot reveals that the prediction closely resembles the actual distribution. Nonetheless, in the case of challenging stochastic dynamics, significant differences appear mostly regarding the extremal values, meaning that the prediction is a smoothed version of the actual time series (right plot).

Further, Table 2 displays several metrics of widely used prediction evaluation that are estimated for each stochastic simulation case at the first and second steps. We also compute the gain that assesses the benefit of introducing the latter step in terms of reducing reconstruction error metrics.

The time series set generated by the random walk algorithm is barely impacted by the latter step. A similar situation holds for the time series generated by ARMA models. This situation may account for the simplicity of creating stochastic processes in either case of consideration. However, when dealing with the most challenging structures of time series, a notable performance gain

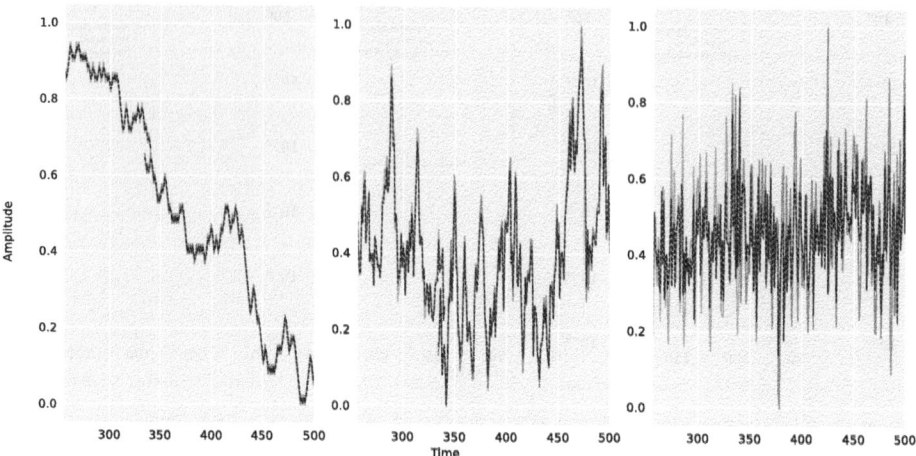

Fig. 3. Comparison between the prediction made by the optimal model identified in the first search phase (black) and the training data (green) and validation data (blue): Random walk (left plot), ARMA model (middle), and combined random walk plus ARMA (right plot). (Color figure online)

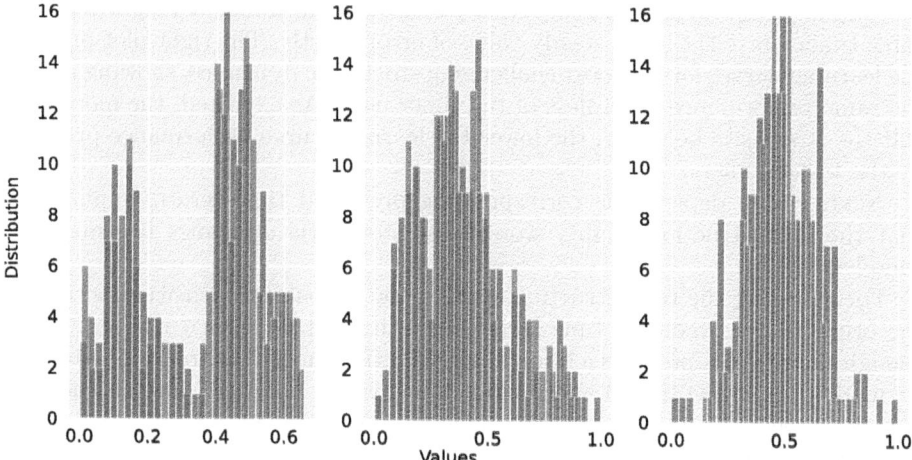

Fig. 4. Comparison between the probability distribution of the predictions on validation data (black) and the actual data from each toy-set (blue): Random walk (left plot), ARMA model (middle), and combined random walk plus ARMA (right plot). (Color figure online)

is attained. In other words, the more challenging the stochastic structure, the more effective the impact provided by the second step.

Table 2. A comparison of various metrics for our two-step optimization proposal for each toy-set

Source	Metric	Step 1	Step 2	Gain [%]
Random Walk	MSE	0.000322	**0.000321**	**0.462889**
	RMSE	0.017951	**0.017909**	**0.231717**
	MAE	**0.017188**	0.017191	-0.014705
	R^2	0.978503	**0.978602**	**0.010169**
	Adjusted R^2	0.978374	**0.978474**	**0.010232**
	Quantile Loss	**0.008594**	0.008596	-0.014705
	Huber Loss	0.000161	**0.000160**	**0.462889**
	Logarithmic MSLE	0.000095	**0.000095**	**0.424381**
	Average			**0.195358**
ARMA	MSE	**0.005501**	0.005502	-0.009336
	RMSE	**0.074172**	0.074175	-0.004671
	MAE	0.059612	**0.059520**	**0.154243**
	R^2	**0.780630**	0.780609	-0.002625
	Adjusted R^2	**0.779308**	0.779288	-0.002645
	Quantile Loss	**0.085839**	0.086121	-0.328395
	Huber Loss	**0.002751**	0.002751	-0.009336
	Logarithmic MSLE	0.002428	**0.002425**	**0.104908**
	Average			-0.012144
R. Walk + ARMA	MSE	0.024403	**0.023565**	**3.435168**
	RMSE	0.156215	**0.153509**	**1.732599**
	MAE	0.128179	**0.126110**	**1.614466**
	R^2	0.143255	**0.172686**	**20.544099**
	Adjusted R^2	0.138094	**0.167702**	**21.440295**
	Quantile Loss	0.081579	**0.080444**	**1.391245**
	Huber Loss	0.012202	**0.011782**	**3.435168**
	Logarithmic MSLE	0.011423	**0.011075**	**3.051820**
	Average			**7.080607**

4.2   Outcomes Obtained for UCI Dataset

As above, a similar analysis is performed for the collected measurements of electric power consumption, for which the obtained learning curve is displayed by the top row of Fig. 5, attaining very low values of error. Once again, all three tested tuners tend asymptotically to a close error prediction error. Also, there is a very tiny dissimilarity between the original and the reconstructed time series, which match even within a wide range of values as seen in the bottom row. These results can be attributed to the moderate stochastic behavior of the evaluated

Table 3. A comparison of various metrics for two-step optimazation proposal in the real data

Metric	Step 1	Step 2	Gain [%]
MSE	0.000034	**0.000034**	**0.876050**
RMSE	0.005841	**0.005816**	**0.438989**
MAE	0.004067	**0.004058**	**0.203949**
R^2	0.994931	**0.994976**	**0.004463**
Adjusted R^2	0.994921	**0.994965**	**0.004472**
Quantile Loss	**0.036828**	0.036829	-0.002949
Huber Loss	0.000017	**0.000017**	**0.876050**
Logarithmic MSLE	0.000027	**0.000027**	**0.897757**
Average			**0.412348**

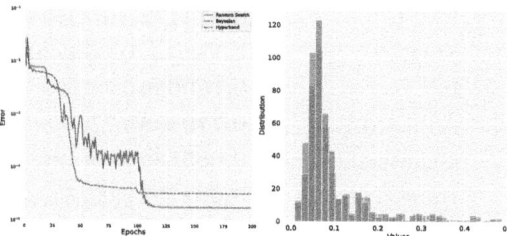

Learning curves calculated for each hyperparameter optimization method on the real data

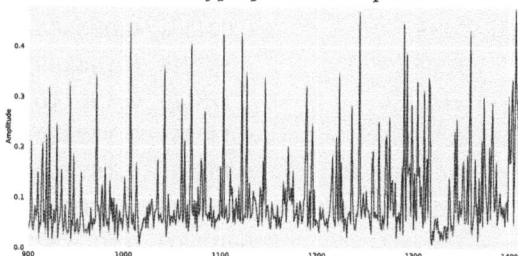

Fig. 5. Comparison between the prediction made by the optimal model identified in the first search phase (left: training data in green, validation data in blue) and the probability distribution of the predictions on validation data compared to the actual data from each toy-set (right: both in black and blue). (Color figure online)

data. This modest improvement because of the two-step strategy of searching is shown in Table 3 that can be related to the simulated ARMA model.

5   Conclusions

This study introduces a two-phase hyperparameter optimization (HPO) approach, designed to enhance the precision and efficiency of predictive models, particularly in time series analysis. Applied across synthetic and real data sets, the method has proven robust and adaptable, offering an effective solution in demanding environments.

The results obtained in this study indicate that the proposed approach significantly increases the accuracy of prediction models. In synthetic time series, the evaluated HPO methods, especially Hyperband, demonstrated better capability in converging to suitable solutions. This performance was sustained even under conditions of high stochastic variability, albeit with slight differences between methods.

For real-world electricity consumption data, the methodology proved equally effective, consistently improving model accuracy. This underscores the applicability of the approach in real-world contexts where data fluctuations are frequent and the complexity of temporal patterns can be considerable.

The study highlights the importance of the second optimization stage. This stage ensures a notable reduction in prediction errors, which is crucial for applications where accuracy is essential. In simpler time series, although the impact of this stage is less pronounced, it still contributes to an overall improvement in model performance.

The initial study suggests future research combining deep learning techniques with traditional optimization algorithms to enhance model adaptability and performance. Accordingly, the authors will develop Hyperparameter Optimization (HPO) approaches integrating these techniques. Additionally, they will focus on designing algorithms capable of adjusting and learning from the specific characteristics of the problems, improving efficiency and accuracy in various applications. This continuity ensures that the proposals of the initial study materialize into specific research and development efforts.

Acknowledgements. Under grants provided by the program: "Agencia Nacional del Espectro-60909"—Project: "Prototipo costo-eficiente y escalable para el monitoreo del espectro radioeléctrico en Colombia mediante radio definido por software y aprendizaje profundo-163-2024", funded by ANE.

Under grants provided by the program: "Sistema de visión artificial para el monitoreo y seguimiento de efectos analgésicos y anestésicos administrados vía neuroaxial epidural en población obstétrica durante labores de parto para el fortalecimiento de servicios de salud materna del Hospital Universitario de Caldas - SES HUC" with Hermes code: 57661, funded by la Universidad Nacional de Colombia.

References

1. Satrio, C.B.A., Darmawan, W., Nadia, B.U., Hanafiah, N.: Time series analysis and forecasting of coronavirus disease in Indonesia using ARIMA model and PROPHET. Procedia Comput. Sci. **179**, 524–532 (2021). 5th International Conference on Computer Science and Computational Intelligence 2020. issn: 1877-0509.
2. Awad, N., Mallik, N., Hutter, F.: Dehb: evolutionary hyperband for scalable, robust and efficient hyperparameter optimization. arXiv preprint arXiv:2105.09821 (2021)
3. Baratchi, M., et al.: Automated machine learning: past, present and future. Artif. Intell. Rev. **57**(5), 1–88 (2024)

4. Brown, T., et al.: Language models are few-shot learners. In: Larochelle, H., et al. (eds.) Advances in Neural Information Processing Systems, vol. 33, pp. 1877–1901. Curran Associates, Inc. (2020)
5. Chen, S.-A., et al.: TSMixer: an All-MLP architecture for time series forecasting. arXiv preprint arXiv:2303.06053 (2023)
6. Florea, A.-C., Andonie, R.: Weighted random search for hyperparameter optimization. arXiv preprint arXiv:2004.01628 (2020)
7. He, X., Zhao, K., Chu, X.: AutoML: a survey of the stateof- the-art. Knowl.-Based Syst. 212, 106622 (2021)
8. Lorraine, J., Vicol, P., Duvenaud, D.: Optimizing millions of hyperparameters by implicit differentiation. In: International Conference on Artificial Intelligence and Statistics, pp. 1540–1552. PMLR (2020)
9. Nagarajah, T., Poravi, G.: A review on automated machine learning (AutoML) systems. In: 2019 IEEE 5th International Conference for Convergence in Technology (I2CT), pp. 1–6 (2019)
10. Rajeswaran, A., et al.: Meta-learning with implicit gradients. In: Advances in Neural Information Processing Systems, vol. 32 (2019)
11. von Rueden, L., et al.: Informed machine learning - a taxonomy and survey of integrating prior knowledge into learning systems. IEEE Trans. Knowl. Data Eng. 35(1), 614–633 (2023)
12. Tornede, A., et al.: AutoML in the age of large language models: current challenges, future opportunities and risks. arXiv:2306.08107 (2024)
13. Turner, R., et al.: Bayesian optimization is superior to random search for machine learning hyperparameter tuning: analysis of the black- box optimization challenge 2020. In: Escalante, H.J., Hofmann, K. (eds.) Proceedings of the NeurIPS 2020 Competition and Demonstration Track. Proceedings of Machine Learning Research, vol. 133, pp. 3–26. PMLR (2021)
14. Waring, J., Lindvall, C., Umeton, R.: Automated machine learning: review of the state-of-the-art and opportunities for healthcare. Artif. Intell. Med. 104, 101822 (2020)
15. Wever, M., et al.: AutoML for multi-label classification: overview and empirical evaluation. IEEE Trans. Pattern Anal. Mach. Intell. 43(9), 3037–3054 (2021)
16. Wong, C., et al.: Transfer learning with neural AutoML. In: Bengio, S., et al. (eds.) Advances in Neural Information Processing Systems, vol. 31. Curran Associates, Inc. (2018)
17. Yang, L., Shami, A.: On hyperparameter optimization of machine learning algorithms: theory and practice. Neurocomputing 415, 295–316 (2020)
18. Zhang, Z., Wang, X., Zhu, W.: Automated machine learning on graphs: a survey. In: Proceedings of the Thirtieth International Joint Conference on Artificial Intelligence (IJCAI-21) (2021)

Emotional BDI Model (eBDI) of the Peasants Families WellProdSim Social Simulator

Jairo E. Serrano[1]([✉])[iD] and Enrique González[2][iD]

[1] Universidad Tecnológica de Bolívar, Cartagena, Colombia
jserrano@utb.edu.co
[2] Pontificia Universidad Javeriana, Bogotá, Colombia
egonzal@javeriana.edu.co

Abstract. This paper describes the emotional model *eBDI* developed for the BDI agents implemented in WellProdSim, a social simulator designed to assess and improve the productivity and well-being of peasant families. Overcoming some of the limitations of previous models, this innovative approach incorporates complex emotional dynamics into the representation of peasant families, allowing to obtain a more realistic adaptation and response to environmental and social changes of the simulated human agents. The key idea resides in the integration of emotions influenced by external factors into the deliberation process of the BDI system. The agent's emotional state impacts on the estimation of the contribution and activation values of each of the agent's potential goals; thus, it directly influences the desires management in the BDI engine, enriching the decision-making process and improving the simulation of social and economic interactions. Results demonstrate the effectiveness of the model, providing simulations that more accurately reflect the complexity of human behavior and social dynamics in rural communities.

Keywords: Social Simulation · Multi-agent systems · Emotional BDI · Multi-agent simulation · BDI agent · Fuzzy Logic

1 Introduction

WellProdSim is an Agent-Based Social Simulator (ABSS), which represents an innovation in social simulation applied to rural development, by integrating an emotional component, *eBDI*, into its BDI (Belief-Desire-Intention) agent model. This simulator focuses on rural families in marginalized regions of Colombia, aiming to improve their productivity and well-being. It is intended to support decision-making by public and private entities. The incorporation of the emotional component *eBDI* allows a more realistic and effective simulation, reflecting how emotions affect the perception of the environment, decisions and behaviors of the agents. This improvement facilitates the design of more adaptive and effective policies for the small-scale agroindustrial sector. The evolution of the

N. D. Duque-Méndez et al. (Eds.): CCC 2024, CCIS 2208, pp. 99–114, 2024.
https://doi.org/10.1007/978-3-031-75233-9_8

BDI model presented in a previous research [1], has been refined to improve the representation of human behavior and underlines the importance of considering both the social and emotional context in analysis and decision-making. Additionally, the introduction of *eBDI* in WellProdSim is a starting point to take into consideration, during the simulation process, the social fabric where peasant families are embedded.

This article is organized as follows. Section 2 presents a state of the art on social simulation and agent models in general, as well as some relevant emotional models, closing with a review of previous works in the area. Section 3 presents the emotional model *eBDI* implemented for the WellProdSim peasant families, describes the interactions between the diverse agents of the system, with emphasis on the peasant families, and also discusses the decision-making model based on emotions and modulating variables. Section 4 explains the design of the experiment and the results obtained in the tests. Finally, Sect. 5 presents the final advances and conclusions of this work.

2   State of Art

Social simulation is a powerful and versatile tool used in various fields such as business, politics, and urban studies. It has gained traction in analyzing the dynamics and behavioral patterns of peasant families. Previous research, discussed in this section, highlights the effective use of Belief-Desire-Intention (BDI) models in simulations, providing detailed analyses at both the micro level, focusing on individual decisions and behaviors, and at the macro level, representing society as a whole.

The following is a review of some of the BDI agent models that integrate emotional components, exploring how these emotions are incorporated into decision making and affect agent behavior. This review will help to better understand the interplay between agents' beliefs, desires, intentions and emotions, and how these influence their behavior in a simulated environment; thus, providing a stronger foundation for future research and applications in the field of social simulation.

2.1   Emotional Models Review

Related work on agent-based emotional models demonstrates a significant and diverse evolution in their application and development. Several implementations of multi-agent system-based emotional models using the BDI architecture [2] with multiple additional components have been proposed [3]. This approach has allowed a better understanding and simulation of human emotions in agents.

Among the most relevant proposals and sources of inspiration for the context of this work appears the BDI agent model with an emotional agent structure proposed by Hu [4]. This model establishes a new knowledge base using granular computation to represent emotional expressions and proposes a method to achieve emotional goals through knowledge rules. The implementation of this model was performed as an extension of the traditional BDI model, incorporating an emotional knowledge base into the decision-making process. On the other hand, Moga [5] proposes an emotional model based on agents and Control-Value theory, focused on psycho-pedagogy and designed for an Affective Tutoring System (ATS). This model seeks to adapt individualized teaching strategies considering the emotional state of students, identifying negative inactive emotions such as boredom, frustration and hopelessness.

An interesting work was done around the ABC model [6], this work classifies affective cognitive processes as rational and irrational, leading agents to functional emotions and adaptive behaviors or dysfunctional emotions and non-adaptive behaviors, integrating the EBDI and ABC model into a single mechanism to simulate human behavior or thought.

A widely accepted and used emotional model for modeling the emotions of rational agents is the OCC model [7]. These model was also taken and adapted as a fundamental part of this work. In brief, the OCC model analyzes how agents, objects, and events influence an agent's perceptions and emotional state. This approach extends to a variety of entities and situations, facilitating its computational implementation in diverse contexts. In particular, in settings such as peasant families, where emotions are key in decision-making, it is valuable to consider the complex emotional responses in various human situations taking into account, not only the type of events, but also the involved person and context where this events are produced.

Finally, this type of research has led to the development of advanced BDI emotional agents [8], which not only exhibit affective states such as emotions, mood or personality, but also affective capabilities such as empathy or emotional regulation. For example, Alzani's work uses a one-dimensional convolutional neural networks to analyze empathic behavior in socially dynamic situations [9]. These advances underline the importance of incorporating an emotional and empathy-oriented approach in intelligent agents, which is essential to improve their autonomy and adaptability in complex social contexts. However, it is clear that there are still aspects to be improved and refined in existing emotion models, which represents an area of active research and great potential in the future of artificial intelligence and social simulation.

2.2 Comparative Summary of Previous Work

The literature review focused on the search for a social simulator based on a multi-agent system (MAS), modeling space and time, implementing an Emotional-Belief-Desire-Intention (eBDI) model and having its application in the economic and social well-being analysis of a specific population. The summary is presented in Table 1. This approach allowed the identification of current research trends and prevailing practices in agent-based systems modeling applied to social contexts.

It was noted that most of the previous studies focus on very wide areas, highlighting a significant opportunity to develop a broader social simulator. An approach that encompasses aspects such as climate-aware modeling of the production environment (crops) and synchronous-asynchronous weather management

Table 1. Comparison of productivity and wellbeing related work. It has the criterion is included ✓.

Referenced Previous Work	MAS	eBDI	Productivity Analysis	Wellspsbeing Analysis
Bao et. al. [10]			✓	✓
Berger et. al. [11]	✓		✓	
Caron et. al. [12]			✓	
Haren et. al. [13]	✓	✓		✓
Hussein et. al. [14]	✓	✓		
Jiang et. al. [15]	✓	✓		
Grevenitis et. al. [16]	✓	✓		✓
Marley et. al. [17]				✓
Muto et. al. [18]	✓	✓	✓	
Ostrom et. al. [19]			✓	✓
Potting et. al. [20]			✓	
Sanchez et. al. [6]		✓		
Schiavon et. al. [21]			✓	
Schreinemachers et. al. [22]	✓		✓	✓
Thume et. al. [23]		✓		
Valencia et. al. [24]	✓	✓	✓	
Yuan et. al. [25]			✓	
Zasada et. al. [26]			✓	

in an integrated manner could provide a enriched and more nuanced understanding of productivity and social well-being in peasant communities, thus addressing a broader spectrum of challenges and opportunities.

The reviewed papers often employ traditional methodologies that may be insufficient to capture the complexity of social systems. The adoption of a methodology based on the design science [27] and the use of our AOPOA to analyze and synthesize MAS [28], represents a significant advance, allowing us to overcome these limitations by effectively integrating critical aspects into the modeling. This methodological improvement facilitates the creation of ABSS that are both more flexible and holistic, improving the ability to address the key components necessary for a deep and comprehensive analysis. Building on this foundation, WellProdSim was developed by integrating diverse techniques into a model that merges technical and human analysis.

3   Peasant Family Emotional Model

WellProdSim has been designed with the objective of providing a comprehensive social simulator that evaluates both the productivity and social well-being of peasant families. Through an iterative and incremental process, each component of the multi-agent system, including the agent *Peasant Family*, is developed individually and then integrated into the simulator. This process of incremental integration of the agents ensures system coherence and proper interaction between the different agents and system components. The simulator's model focuses on improving decision-making, planning and analysis at the smallholder level. Thus, special attention has been given to consider and reflect their socioeconomic, environmental and, emotional characteristics in the design and development of the *Peasant Family* agent.

3.1 Peasant Family Agent Model

The main component of WellProdSim is the agent *Peasant Family*, an entity specifically modeled to reflect the behavioral patterns, decision-making and interactions of peasant families. This model was built by analyzing interactions with communities and databases related to peasant families in the context of the Colombian Caribbean Coast. This agent simulates family structure, resource and tool management, as well as collective skills derived from collaboration within the family unit. The simulation aims to provide a realistic representation and improve understanding of the complexities of rural life, enriched by the culture and agricultural practices of the region.

Figure 1 introduces the internal architecture of the *Peasant Family* agent, which is designed with five fundamental components. The "Event Port" is in charge of receiving communications and information from the other ABSS agents, and asynchronously processes and updates the *Peasant Family* agent's beliefs. It should be noted that most of the interactions are carried out with the *AgroEcosystem* agent, which represents the physical environment and growing crop conditions. As a second component, we have the "BDI Pulse", which acts as a timer (the heart of the *Peasant Family* agent), its function is to periodically activate the Beliefs, Desires and Intentions (BDI) evaluation cycle, ensuring the constant operation of the agent, even in the absence of external signals arriving to the *Event Port*. It should be noted that the implementation has been carried out with the *BESA* library [29], which facilitates the construction of synchronous and asynchronous event-oriented agents, allowing the hybrid and efficient design of the Peasant Family as part of an ABSS.

The remaining internal components of the agent are the traditional ones in the BDI framework: Beliefs, Goals and Tasks modules, which have been described in detail in previous research [1]. Beliefs encapsulate the agent's knowledge about its world, from skills to resource inventory. The *Goals* reflect the intention goals to be achieved, and the *Tasks* represent the action plans to achieve those goals.

One of the significant advances presented in this work is the incorporation of emotional state modeling and emotional event processing in the *Beliefs* component applied to economic and social well-being related decision-making in peasant families. In this way, the agent architecture responds to both internal and external emotional events. Emotions, which can oscillate between polar opposites or remain neutral, play a central role in the evaluation of *Goals* and the execution of *Tasks*. This emotional responsiveness enriches the agent's decision-making, allowing for a realistic and more accurate simulation that transcends

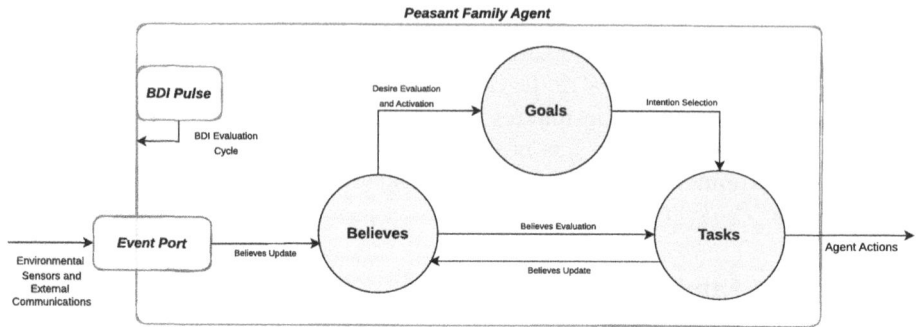

Fig. 1. WellProdSim Peasant Family Agent Internals

mere logical computations to include behaviors influenced by emotional state, thus offering a deeper and more nuanced perspective of artificial cognition in rural contexts.

3.2 Multi-agent System Interactions

In the WellProdSim simulator, the efficiency and coherence of the simulation system depends on the interactions between the multiple agents. Each agent plays a unique role, and the dynamic interactions among them drives the system. To model a multi-agent system in a coherent manner, it is important to understand the intrinsic relationships between the entities involved. If we take as an example the agent *Peasant Family*, whose interactions with other agents, such as the agent *AgroEcosystem* (crop manager), are fundamental to evaluate its productivity and social welfare. By analyzing how the emotional state of the agent *Peasant Family* affects its agricultural work, it is observed that a better mood and happiness results in a more efficient management of the land. This optimized behavior directly increases productivity by doing work more efficiently, spending fewer hours of the day on this task. Consequently, their well-being is also improved by receiving more money as compensation for its effort or by making more recreational activities in its free time. It is also possible to improve the model with emotions specific to the context of the problems of peasant families; for example, the emotional state can also model the security of land tenure or the confidence in obtaining good prices for crops, thus underlining the emotional influence on the economic dynamics of the system.

Figure 2 illustrates the WellProdSim multi-agent system, in which the agent *Peasant Family* interacts with multiple types of agents: the agent *AgroEcosystem* represents the productive environment, including management of the climate and its various disturbances, the agent *BankOffice* manages loans and grants, the agent *CommunityDynamics* facilitates cooperation and communication between families, the agent *CivicAuthority* assigns property titles, the agent *MarketPlace* acts as a commercial mediator, the agent *PerturbationGeneration* introduces random perturbations in climate, economic and public order variables,

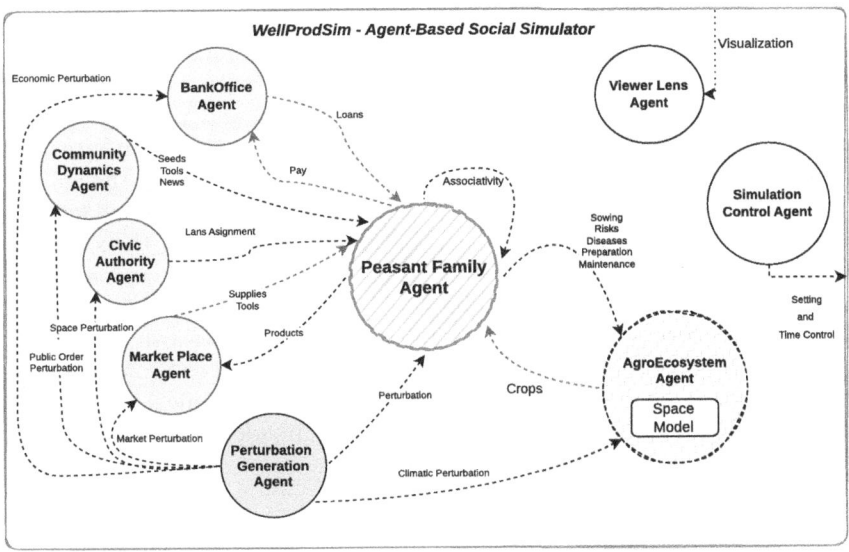

Fig. 2. WellProdSim MAS Interactions

the agent *ViewerLens* generates visualizations of the simulation state, and the agent *SimulationControl* coordinates the time management and synchronization of the simulator.

In WellProdSim, special emphasis is placed on how the emotions of the *Peasant Family* agents affect their collaboration and teamwork both with each other and with other agents within the social simulator. The ability of these agents to work together is essential, and is significantly influenced by their emotional state. Emotions play an important role in productivity, affecting essential activities such as land management, planting, crop care and product sales. An agent experiencing negative emotions, such as sadness or insecurity, may make decisions that differ from those of an agent in a positive emotional state, which directly impacts his or her ability to collaborate effectively with others and, therefore, his or her productivity and overall well-being.

In the WellProdSim, emotions not only affect internal cooperation between peasant families but also determine the quality of interactions with other agents, potentially facilitating or complicating collaboration and resource sharing. An agent with positive emotions and good social skills tends to participate more actively in collaborative activities, resulting in more efficient resource management and higher productivity. Conversely, those agents affected by negative emotions may become isolated, negatively impacting their ability to contribute and to obtain benefits from the community, thus affecting their performance and access to essential resources.

The WellProdSim model considers the influence of news from the agent *CommunityDynamics* or information from the agent *MarketPlace* on purchase or sale prices of supplies, crops or tools. This directly affects the agents' emotional states, recognizing that these effects can vary in duration and depth. Emotional reactions to this events, whether positive or negative, can range from temporarily altering agents' behavior to generating significant and prolonged changes, depending on their severity or complexity. This emotional sensitivity is vital to appreciate the ability of agents to respond and adapt within the simulator, reflecting the complexity of human responses to different stimuli and situations. In addition, the model incorporates a greater understanding of the social fabric that surrounds peasant communities, encompassing the norms, customs and culture that shape interactions and build strong ties within the community. This integration will allow peasant families to align their efforts towards common goals in the future, reflecting the significant impact of their social environment on their ability to face and overcome challenges, including those technical, organizational and economic constraints that have traditionally impeded their productivity.

3.3   eBDI Goal Integrated Model

Having understood the MAS model in which the *Peasant Family* agent operates, the *eBDI* model is introduced, an essential element that extends the social simulator, providing a more fine-tuned decision-making methodology for the *Peasant Family* agent. Inspired by the OCC model [7], *eBDI* allows agents to interact in a way that is in harmony with their needs and aspirations, simultaneously infusing an emotional dimension into their decision-making process.

The emotional spectrum is modeled with emotional-axes that have a level represented by a value between -1 to 1. For instance, on the happiness axis a value close to 1 represents that the agent is happy, and close to -1 that the agent is sad. The level of an emotional axis is affected by relevant events; for instance, when an agent has money to buy food and perform his vital functions, the emotional axis of happiness and security tends to 1, while if money is lacking, it decreases, tending to -1. The emotional level is capable of altering the perception and influencing, in a significant way, the decision-making and the way in which the agents' actions are carried out.

The *eBDI* model also incorporates an emotional forgetting mechanism, which reflects the ephemeral and fluctuating nature of human emotions, with the value of emotional intensity increasing or decreasing over time to reach a default agent's emotional level. This feature ensures a more accurate representation of the emotional adaptability of individuals, which is fundamental to understanding how people respond and adjust to changes in their environment over time.

In line with the OCC model, WellProdSim assigns emotions not only to events, but also to specific actors and objects within the simulator ecosystem. For example, receiving a message that crop sales prices in the market went up generates a transient emotion that increases the level of the happiness axis; the magnitude of the change in the value of this axis depends on the importance that the crop has for the agent and the magnitude of the price increment. Thus, this model captures the immediate and ephemeral responses to events, as well as the long-lasting emotional impacts that can leave an imprint on the agent's state of mind. By adopting this holistic approach, the understanding of the complex social and economic dynamics that define peasant communities is enriched, recognizing that certain events can temporarily alter the emotional state, while others have the capacity to generate a permanent effect that can reconfigure the agent's emotional base values in the long term.

Emotions such as hope and relief arise from interactions with other agents, reflecting the complexities of real life. Table 2 illustrates how the key emotions are mapped across interactions, agents, and resources, such as the security associated with events like land acquisition involving government agents and the land resource. Reflecting the emotional integration to tasks, the *Peasant Family* agent working time and performance is adjusted according to its emotional state. For instance, if the emotional indicators are positive, its task performance increases by a percentage, or if it is negative, it decreases. The table also illustrates the variability and complexity of emotions in the simulation, highlighting how peasant families may react to a range of circumstances. For instance, receiving land (access to an *AgroEcosystem* agent) from the *CivicAuthority* agent to be able to cultivate crops increases the emotion of security, generally improving the agent *Peasant Family* emotional state.

3.4   Emotions and Fuzzy Logic Evaluation

The previous section establishes an important foundation by incorporating emotional aspects into the simulation. The decision-making capability is accomplished by integrating fuzzy logic into the mental state evaluation of the agent *Peasant Family* giving a better matching with the uncertainty inherent in human reasoning.

The fuzzy evaluation engine, developed in three steps, allows to model peasants behaviors in a more natural and explainable fashion. The first step involves the consideration and identification of the required variables in the *eBDI* model within the agent *Peasant Family*. In the second step, the fuzzy evaluation terms were defined, relating the emotional events with the *Goals* and *Tasks* performed by the agent *Peasant Family*, giving an emotional meaning to each action. Finally, the defuzzification process is carried out by applying rules, created with the advice of experts in the field, to generate a single consistent emotional component.

The first step was the design and implementation of the emotional model integrated to the BDI of the social simulator. This model allows evaluating priorities and triggering emotional reactions according to the experience and context of each agent. A clear example is the determination of the ideal time for planting, which considers both environmental and emotional factors, thus optimizing the use of time and resources. It complies with the OCC model, understanding the relationships between agent-object-event and quantifies them according to the occurrence of the events, the interactions of the agents or their tasks and context.

The second step was the fuzzy definition of each emotionally meaningful event, as described in Table 2. This collaboration between the BDI emotional

Table 2. WellProdSim eBDI for Peasant Family Agent

Emotion	Event (Interaction)	Actor (Agent)	Object (Resource)
Secure/ Insecure	Own Land	CivicAuthority	Lands
	Thief Land	PerturbationGenerator	Lands
	Work for Others	CommunityDynamics	Money
Hopeful/ Uncertainty	Seeding	AgroEcosystem	Seeds
	Ask for a Loan	BankOffice	Money
	Sell Crop	MarketPlace	Biomass
Happiness/ Sadness	Spend Family Time	Peasant Family	Time
	Attend Religious Events	Peasant Family	Time
	Leisure Activities	Peasant Family	Time

model and fuzzy logic leads to a detailed evaluation of emotional input variables such as *Happiness-Sadness* axis. The level of an emotional axis is represented with values from -1 to 1 and its membership in the fuzzy terms *Happy*, *Neutral*, *Sad* is evaluated according to the emotional intensity value. In the same way, the fuzzy evaluation process includes non-emotional variables, belonging in this case to the agent *Peasant Family* personality, such as *socialAffinity*, which indicates the degree of acceptance to collaborate or offer collaboration to others, and *peasantFamilyAffinity*, the indicator of the degree of internal cohesion of the family.

In the third step of our model, we use fuzzy logic rules to connect non-emotional variables with the *EmotionalState* of the *Peasant Family* agents, resulting in a defuzzified output known as *GoalContribution*. This output variable directly affects the selection of desires, plausible goals, that would become intentions. For instance, to evaluate the impact of the *Goal* of *SpendFamilyTime* on emotional values, we employ a set of fuzzy logic rules:

- *if HappinessSadness is Happy then EmotionalState is Positive*
- *if HappinessSadness is Sad then EmotionalState is Negative*
- *if HappinessSadness is Neutral then EmotionalState is Neutral*
- *if HappinessSadness is Sad and HopefulUncertainty is Uncertainty and SecureInsecure is Insecure then EmotionalState is Negative*
- *if HappinessSadness is Neutral and HopefulUncertainty is Neutral and SecureInsecure is Neutral then EmotionalState is Neutral*
- ...

Here's how the process unfolds: first, inputs like *peasantFamilyAffinity* are assessed on a fuzzy scale. The membership function for high affinity, $\mu(x)$, might yield a value of peasantFamilyAffinity equal to 0.75 when evaluated at the agent's current level of affinity. Similarly, the membership for a medium $\mu(y)$, could be EmotionalState equal to 0.5. The system then evaluates this information against the fuzzy rule. If the conditions of high *affinity* and medium *EmotionalState* are met, the rule activates, suggesting a high $\mu(z)$ *Contribution* equal to 0.83. The fuzzy value inferred from this rule is subsequently transformed into a precise numerical value through a process called defuzzification. This step typically employs techniques like the centroid method, calculating a crisp output that represents the *Contribution* level, ranging from 0 to 1.

4   Results

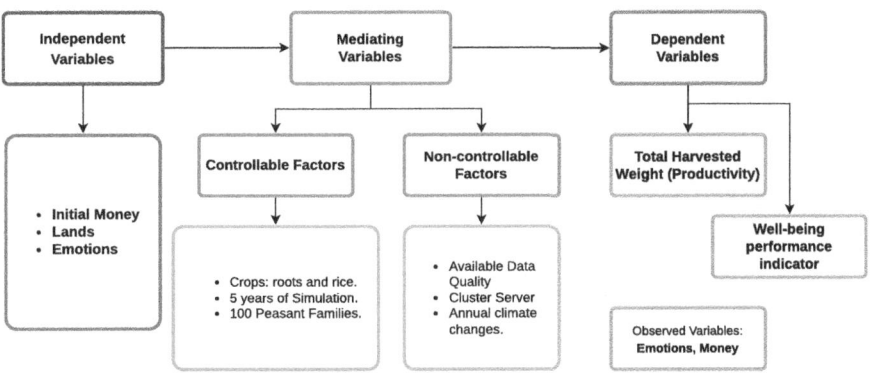

Fig. 3. Experiment about Emotional behavior of Peasant Family agent.

To evaluate the impact of the use of emotions and initial economic differences on the dynamics and quality of life of peasant families, we implemented an experimental design, based on data analysis [30]. Figure 3 presents the variables used in this experiment. This approach runs a simulation of group of peasant family agents; the main independent variable is if emotions are included or not in the decision-making process.

This experimental design allows to observe how the simulator responds to a hypothetical scenario. In the Table 3 it is possible to visualize the average of the dependent variables *Total Harvested* (amount of biomass cultivated) and *Health Indicator*. The combination of the results observed in these variables is directly related to the affectation or improvement in the productivity and well-being of the peasant families. The obtained results validate the hypothesis that the use of emotions has significant impacts on the behaviors and decisions of the agents, reinforcing the understanding of the complex interactions within the dynamics of peasant families.

Table 3. Experimental design and Treatments with results

Treatment	Initial Money	Lands	Emotions	Harvested	Health
T01	750000	2	Yes	15206.19	87.43
T02	750000	2	No	12080.10	87.17
T03	750000	6	Yes	22414.78	81.83
T04	750000	6	No	14422.84	36.10
T05	750000	12	Yes	18519.44	75.22
T06	750000	12	No	12109.53	42.07
T07	1500000	2	Yes	15771.13	89.41
T08	1500000	2	No	12167.51	88.57
T09	1500000	6	Yes	24631.04	89.02
T10	1500000	6	No	16166.01	78.30
T11	1500000	12	Yes	24620.09	90.30
T12	1500000	12	No	13370.89	62.63
T13	3000000	2	Yes	15754.82	89.52
T14	3000000	2	No	12146.20	89.94
T15	3000000	6	Yes	24578.01	90.04
T16	3000000	6	No	16639.30	82.47
T17	3000000	12	Yes	24499.59	90.29
T18	3000000	12	No	13722.54	78.65

The experiment, summarized in Fig. 3, defines a series of controlled factors. A 40% variance is applied to the variables considered part of the peasant family's personality when running the simulation (initial values were defined by field experts). Variables such as money, health, bonding with family, friends or spending time alone are altered by the variance, generating a richer range of unique characteristics for the *Peasant Family* agent. The simulated period is five years of the peasants life, including interactions with the different agents of the social simulator (*CommunityDynamics, CivicAuthority, BankOffice, Market-Place, AgroEcosystem*). All the conditions that reflect various realistic situations such as planting, harvesting, selling agricultural products, lack of work and other social phenomena of the region were tacked into account.

Table 4. ANOVA Results for Total Harvested and Health.

Dependent Variable	Independent Variable	F	P-Value
Total Harvested	Initial Money	2.23	0.150285
	Lands	14.63	0.000606
	Emotions	58.86	0.000006
Health	Initial Money	4.44	0.036009
	Lands	3.11	0.081468
	Emotions	8.09	0.014795

To ensure robustness of the data generation process, the experiment is replicated three times with each variable configuration and generating an overall average of the results of each run. The data were analyzed from the centralized and cumulative collection of the daily mental state of the agents. This method guaranteed a general understanding of the effect of emotions where used in the peasant families simulation.

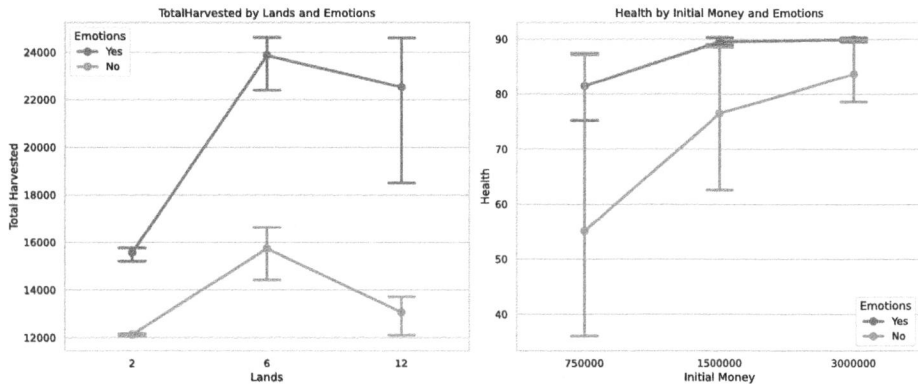

Fig. 4. Behavior of Total Harvested and Health variables throughout the simulation.

The results of applying ANOVA to the experiment indicate that the amount of land and emotional state have a significant effect on *TotalHarvested* as seen on Table 4. Specifically, both the amount of land and positive emotions contribute significantly to an increase in the total amount harvested. On the other hand, the analysis shows that the initial amount of money and the use of emotions also have a significant effect on the *Health* of peasant families as seen in the Fig. 4. These findings suggest that, in order to improve the productivity and well-being of peasant families, the emotional state of peasant families is important, thus demonstrating consistency in the choice of an emotional component that directly affects decision-making. In other words, proposed *eBDI* model represents a coherent and useful component in the architecture of the WellProdSim simulator.

5   Conclusions

In conclusion, the incorporation of emotions in WellProdSim helps to improve simulation capabilities. The influence of the context and of the own personality of the agents of the peasant family, manages to simulate more closely to reality the socio-economic interactions and dynamics. It is important to highlight that the application of fuzzy logic rules enriches the decision-making process and contributes to making the agents' behavior more explainable. The experimental results obtained show that this approach produces a more accurate representation of human behavior and the dynamics of rural communities.

The experiment showed that the inclusion of emotions improved the ability to make decisions that promoted the productivity and well-being of peasant families. By comparing the initial and final economic conditions, together with the generation of crop biomass, a significant improvement of the simulation and results was evidenced. The repetition of the experiment under different configurations (scenarios) and the detailed data collection ensured the reliability of the results.

Thus, it is concluded that incorporating emotions in BDI agents improves decision-making, and has a positive impact on long-term economic stability and the well-being of peasant families. For the future, it is planned to improve the use of eBDI by incorporating new ways of making decisions through the application of machine learning, especially in the sections where it is required to make predictions of variables that may depend on large volumes of information, thus improving the responsiveness of the agents to different situations that arise in the real world.

Acknowledgements. The author Jairo Enrique Serrano Castañeda thanks MIN-CIENCIAS, the Pontificia Universidad Javeriana and the Universidad Tecnológica de Bolívar for the support received to pursue a doctoral degree within the programme "Becas de la Excelencia Doctoral del Bicentenario (corte 1)".

References

1. Serrano, J.E., González, E.: BDI peasants model for the wellprodsim agent-based social simulator. In: Colombian Conference on Computing, pp. 367–379 (2024)
2. Rao, A., Georgeff, M.P.: Modeling rational agents within a BDI-architecture. In: Proceedings of the 2nd International Conference on Principles of Knowledge Representation and Reasoning (1991)
3. Adam, C., Gaudou, B.: BDI agents in social simulations: a survey. Knowl. Eng. Rev. **31**(3), 207–238 (2016)
4. Hu, J., Guan, C.: A model of emotional agent based on granular computing. In: 2011 Seventh International Conference on Computational Intelligence and Security, pp. 190–194. IEEE (2011)
5. Moga, H., Sandu, F., Danciu, G.M., Boboc, R., Constantinescu, I.: Extended control-value emotional agent based on fuzzy logic approach. In: 2013 11th RoEduNet International Conference, pp. 1–8. IEEE (2013)
6. Sánchez, Y., Coma, T., Aguelo, A., Cerezo, E.: ABC-EBDI: an affective framework for BDI agents. Cogn. Syst. Res. **58**, 195–216 (2019)
7. Ortony, A., Clore, G.L., Collins, A.: The Cognitive Structure of Emotions. Cambridge University Press (1988)
8. Bourgais, M., Taillandier, P., Vercouter, L., Adam, C.: Emotion modeling in social simulation: a survey. JASSS **21**(2), 3 (2018)
9. Alanazi, S.A., Shabbir, M., Alshammari, N., Alruwaili, M., Hussain, I., Ahmad, F.: Prediction of emotional empathy in intelligent agents to facilitate precise social interaction. Appl. Sci. **13**(2), 1163 (2023)

10. Bao, H., Dong, H., Jia, J., Peng, Y., Li, Q.: Impacts of land expropriation on the entrepreneurial decision-making behavior of land-lost peasants: an agent-based simulation. Habitat Int. **95**, 1 (2020)
11. Berger, T.: Agent-based spatial models applied to agriculture: a simulation tool for technology diffusion, resource use changes and policy analysis. Agric. Econ. **25**(2–3), 245–260 (2001)
12. Caron-Lormier, G., Bohan, D.A., Dye, R., Hawes, C., Humphry, R.W., Raybould, A.: Modelling an ecosystem: the example of agro-ecosystems. Ecol. Model. **222**(5), 1163–1173 (2011)
13. Paschal, C.H., Bin Khairuddin, M.A., Shiang, C.W., Bin Khairuddin Yap, M.N.: Bush Fire Simulation through Emotion-based BDI Methodology. Int. J. Adv. Sci. Eng. Inf. Technol. **13**(5), 1663–1671 (2023)
14. Abir Hussein Jabber and Ali Obied: Implementing the EBDI model in an E-health system. Int. J. Nonlinear Anal. Appl **13**(1), 2008–6822 (2022)
15. Jiang, H., Vidal, J.M., Huhns, M.N.: EBDI: an architecture for emotional agents. In: Proceedings of the International Conference on Autonomous Agents, pp. 38–40 (2007)
16. Grevenitis, K., Sakellariou, I., Kefalas, P.: Emotional agents make a (bank) run. In: Lecture Notes in Computer Science (including subseries Lecture Notes in Artificial Intelligence and Lecture Notes in Bioinformatics), vol. 12520 LNAI, pp. 171–187. Springer Science and Business Media Deutschland GmbH (2020)
17. Marley, J., et al.: Does human education reduce conflicts between humans and bears? An agent-based modelling approach. Ecol. Model. **343**, 15–24 (2017)
18. Muto, T.J., Bolivar, E.B., González, E.: BDI multi-agent based simulation model for social ecological systems. In: Communications in Computer and Information Science, vol. 1233 CCIS, pp. 279–288. Springer (2020)
19. Ostrom, E.: A general framework for analyzing sustainability of social-ecological systems. Science **325**(5939), 419–422 (2009)
20. Potting, R.P.J., Perry, J.N., Powell, W.: Insect behavioural ecology and other factors affecting the control efficacy of agro-ecosystem diversification strategies. Ecol. Model. **182**(2), 199–216 (2005)
21. Schiavon, E., Taramelli, A., Tornato, A.: Modelling stakeholder perceptions to assess green infrastructures potential in agriculture through fuzzy logic: a tool for participatory governance. Environ. Dev. **40**, 100671 (2021)
22. Schreinemachers, P., Berger, T., Aune, J.B.: Simulating soil fertility and poverty dynamics in Uganda: a bio-economic multi-agent systems approach. Ecol. Econ. **64**(2), 387–401 (2007)
23. Thumé, G.S., Da Silva, R.E.: An extended EBDI model applied to autonomous digital actors. In: Brazilian Symposium on Games and Digital Entertainment (SBGAMES) (2012)
24. Valencia, D.S., Serrano, J.E., Gonzalez, E.: SIMALL: emotional BDI model for customer simulation in a mall. In: Colombian Conference on Computing, pp. 3–18 (2022)
25. Yuan, S., Li, X., Du, E.: Effects of farmers' behavioral characteristics on crop choices and responses to water management policies. Agric. Water Manage. **247**, 106693 (2021)
26. Zasada, I., et al.: A conceptual model to integrate the regional context in landscape policy, management and contribution to rural development: literature review and European case study evidence. Geoforum **82**, 1–12 (2017)
27. Hevner, A.R., March, S.T., Park, J., Ram, S.: Design science in information systems research 1. Des. Sci. IS Res. MIS Q. **28**(1), 75 (2004)

28. Rodríguez, J., Torres, M., González, E.: La metodología aopoa. Avances en Sistemas e Informática **4**(2), 5 (2007)
29. Gonzalez, A., Angel, R., Gonzalez, E.: BDI concurrent architecture orientedto goal managment. In: 2013 8th Computing Colombian Conference (8CCC), pp. 1–6. IEEE (2013)
30. Collins, A.J., Koehler, M., Lynch, C.J.: Methods that support the validation of agent-based models: an overview and discussion. J. Artif. Soc. Soc. Simul. **27**(1), 11 (2024)

Pattern Recognition and Computer Vision

Analysis of Pre-trained Convolutional Neural Network Models in Diabetic Macular Edema Detection Through Retinal Fundus Images

José Araque-Gallardo[1,2](✉) ⓘ, Eugenia Arrieta Rodríguez[2] ⓘ,
Margarita Gamarra[3] ⓘ, Javier Sierra-Carrillo[1] ⓘ,
and José Escorcia-Gutierrez[2] ⓘ

[1] Department of Electronic Engineering, Universidad de Sucre,
Sincelejo 700001, Colombia
jose.araque@unisucre.edu.co
[2] Department of Computational Science and Electronic, Universidad de la Costa
CUC, Barranquilla 080020, Colombia
[3] Department of System Engineering, Universidad del Norte,
Puerto Colombia 081007, Colombia

Abstract. Diabetic Macular Edema (DME), a serious complication linked to Diabetic Retinopathy (DR), can result in vision loss and potential blindness. DME occurs when fluid leaks from blood vessels in the macula or when the retina thickens. Fluid leakage is presented as Hard Exudates (HE), which appear as yellow or white clusters of varying shapes, sizes, and positions, serving as indicators for diagnosing DME in fundus color images. Early detection of DME can significantly improve treatment options and patient quality of life. In this study, we propose evaluating three pre-trained Convolutional Neural Networks (CNN) models to assess the risk of DME in fundus color images. In this study, transfer learning is employed to leverage the convolutional base of pre-trained models for feature extraction from retinal images. Subsequently, a custom fully connected convolutional layer is added to perform the classification task. The publicly available MESSIDOR dataset was used to train and test the proposed method, which achieved an accuracy of 95% in detecting DME.

Keywords: Diabetic Macular Edema · Deep learning · Transfer learning · Convolutional neural network

1 Introduction

Diabetes Mellitus (DM) is a cause of blindness, kidney failure, heart attack, stroke, and lower limb amputation [1]. DM contributes significantly to global mortality, with its impact varying by region. Excluding the mortality risks associated with the COVID-19 pandemic, approximately 6.7 million adults aged 20

N. D. Duque-Méndez et al. (Eds.): CCC 2024, CCIS 2208, pp. 117–131, 2024.
https://doi.org/10.1007/978-3-031-75233-9_9

to 79 died due to diabetes or its complications in 2021 [2]. As stated by the International Agency for the Prevention of Blindness (IAPB), diabetes increases the risk of several types of eye diseases. However, the leading cause of blindness associated with diabetes is Diabetic Retinopathy (DR). This disease progressively damages blood vessels within the retina in the back of the eye and commonly affects both eyes, causing vision loss if left untreated. Diabetic retinopathy (DR) manifests itself through various lesions resulting from impaired retinal blood flow, barrier dysfunction, and elevated vascular permeability. These retinal alterations can progress to two vision-threatening conditions: proliferative diabetic retinopathy (PDR) and diabetic macular edema (DME) [3]. In addition, the IAPB estimates that one in three people living with diabetes have some degree of DR, and one in ten developed some form of vision-threatening disease, resulting in one million people around the world becoming blind due to DR by 2020, while nearly three million people will have had moderate to severe visual impairment due to DR [4]. In Colombia, the expert panel on the initial care of diabetic retinopathy estimates that 21% of patients with diabetes have some degree of DR at the time of diagnosis; moreover, the complication of DR occurs in 9.0% to 31.8% of patients with diabetes [5]. Diabetic Macular Edema (DME) is a vision-threatening complication associated with DR. It is the leading cause of visual impairment in diabetic patients and can potentially lead to blindness [6]. Diabetic macular edema is caused by tissue fluid leakage from vessels in the macula or retinal thickening at any stage of DR [7]. Fluid leakage is presented as Hard Exudates (HE), usually showing yellow or white clusters of different shapes, sizes, and positions, and is generally used as an indicator for diagnosing DME in fundus color images. The early stage of DME has no visible signs or symptoms; therefore, early detection and treatment of DME are essential to prevent vision impairment and vision loss. The clinical DME grading standard established by ADCIS through its MESSIDOR dataset [8] defines three grades of the DME: grade 0 (Normal), grade 1 (Non-Clinically Significant Macular Edema, NCSME), and grade 2 (Clinically Significant Macular Edema, CSME), according to the relative position of exudates relative to the center of the macular, as shown in Table 1. Figure 1 displays samples of MESSIDOR images of patients with (a) normal retina; (b) NCSME; and (c) CSME. Hard exudates are visible as yellow spots in the images of patients with NCSME and CSME.

Table 1. DME grading standard [8].

Grade	Severity	Details
0	Healthy	No visible HE
1	NCSME	The shortest distance between the macula and hard exudates > one papilla diameter
2	CSME	The shortest distance between the macula and hard exudates ≤ one papilla diameter

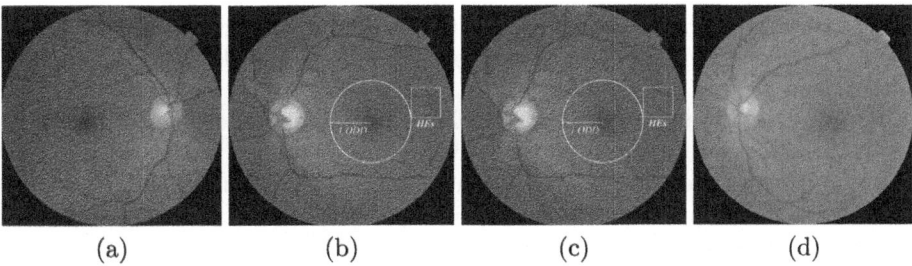

(a)   (b)   (c)   (d)

Fig. 1. MESSIDOR dataset samples: (a) Healthy, (b) NCSME, (c) CSME, and (d) DME annotated without visible exudates.

The traditional diagnosis of DME is a manual process that consumes time and is error-prone because grading by hand relies on the experience of ophthalmologists, which conduces to problems such as non-uniform grading criteria, slow diagnosis speed, limited number of ophthalmologists, especially in low-income countries, and so on. To solve these problems, automated systems based on image processing and intelligent algorithms have started to help medical personnel diagnose DR [9,10] and DME, reducing the risk of visual impairment and blindness in patients. Currently, DME detection and grading is based on the position of HE related to the fovea or the end-to-end method using CNN. DME detection and grading based on the position of HE requires several steps, such as HE segmentation and fovea location, which is a challenging and time-consuming process. However, the end-to-end system does not need to develop HE, macula, and fovea segmentation. The entire fundus image is the input to the CNN, which extracts the features and then gives the grading or classification of the input image [11].

This paper is organized as follows. Section 2 introduces the main contributions of this work. Section 3 reviews the related work. Section 4 shows the proposed methodology. Section 5 exposes the results and presents a brief discussion. Section 6 describes the conclusions. Finally, future research lines are presented in Sect. 7.

2   Contributions

This research makes significant strides in the automated detection of DME through the application of advanced deep learning techniques through the following three contributions.

1. Three pre-trained CNN models-VGG16, ResNet50, and EfficienNetB7-for detecting DME in retinal fundus images without the segmentation tasks require, simplifying the detection process compared to methods that need segmentation of the optic disc and exudates.
2. The methodology employs transfer learning to utilize the convolutional bases of pre-trained models for feature extraction from retinal images. A custom

fully connected convolutional layer is then added to perform the grading task, effectively leveraging the strengths of established models to enhance DME detection accuracy.

3. The study introduces the use of the EfficienNetB7 model for DME detection, which, according to the literature review, had not been previously used for this purpose. The EfficienNetB7 model demonstrated superior performance in terms of accuracy, sensitivity, precision, and F1-score, achieving an overall accuracy of 95% in detecting DME.

3 Related Works

There are two main approaches for DME detection and grading in retinal fundus images: one based on the position of the HE relative to the fovea and the other utilizing end-to-end CNN. The following review highlights the most significant works developed using these two approaches, particularly those that employed the MESSIDOR dataset in their experimental setup.

Chalakkal et al. [12] proposed a new approach to automated screening of CSME. The proposed system develops exudate segmentation by combining a pre-trained CNN with metaheuristic feature selection. A K-nearest neighbor-based classifier performed the classification. Three retinal fundus image datasets were used: MESSIDOR, UoA-DR, and IDRiD. The proposed model achieved a maximum value accuracy of 93.75%, a specificity of 95.62%, and a sensitivity of 88.30%. It should be noted that the work focuses only on detecting CSME, which is characterized by the presence of exudates within one papilla diameter around the fovea, and it is not suitable for detecting non-CSME, i.e., the proposed model cannot detect DME if the exudates are not present near the fovea.

Zubair et al. [13] presented a novel automated system for the diagnosis and classification of the severity of DME. First, the image was enhanced by Contrast Limited Adaptive Histogram Equalization (CLAHE), and contrast stretching was used to improve the image's contrast. Then, the optic disc was localized and excluded from the image. All possible exudates were detected using dynamic thresholding. The different severity levels of the disease were classified according to the criteria given by the Early Treatment Diabetic Retinopathy Studies (ETDRS). The stage of the disease was classified as normal, less significant, moderate, and severe to assess the severity level. The proposed technique obtained an accuracy of 97. 4%, 98. 7%, and 97. 2% in the MESSIDOR, DIARETDB and HEI-MED datasets. In this work, DME is classified using exudate segmentation and optic disc elimination, which require additional computational complexity.

The study by Zubair et al. [14] proposed a model for the early diagnosis and grading of DME. Gabor wavelet filter and novel advanced fuzzy c-means clustering algorithms are introduced for the precise location of the fovea, the extraction of the blood vessel network, and the precise segmentation of the lesion, respectively. The Bayesian classifier uses the Gaussian function with expectation maximization for DME grading. The proposed model achieves an average accuracy of 98. 80% for the classification of DME. Although the proposed method achieves good results, segmentation of optic disc and exudates is necessary.

Al-Bander *et al.* [15] proposed a feature learning approach for grading the severity of DME using images of the retinal fundus. An automated DME diagnosis system was developed based on the proposed feature learning approach to help with the early diagnosis of the disease. The proposed system used CNN to identify and extract DME features automatically without user intervention. The prototype was trained and evaluated using the MESSIDOR dataset. The preliminary results showed an accuracy of (88.8%), sensitivity (74.7%) and specificity (96.5%). The authors did not compare their results with other techniques and work reported in the literature.

Singh *et al.* [16] developed a novel DMENet Algorithm based on a CNN ensemble to detect and grade DME. The pre-processing of retinal images was carried out using morphological opening and Gaussian kernel. The authors employed a combination of pre-trained networks like ResNet, DenseNet, SqueezeNet, GoogleNet, and SE-ResNet as the learners in each cluster of the ensemble model. The proposed model achieved an average accuracy of 96. 12%, sensitivity of 96. 32%, specificity of 95. 84%, and a F1 score of 0.9609 in the MESSIDOR dataset.

Fu *et al.* [11] presented an architecture based on ResNet50 combined with channel attention (SENet) to improve the efficiency and accuracy of DME grading without introducing lesion segmentation. Furthermore, to solve the problem of class imbalance and insufficient training samples in grading, the authors added class weight to the cross-entropy function in designing the loss function and augmentation techniques. The MESSIDOR dataset was employed for the experimental setup, and the proposed method had high accuracy, specificity, sensitivity, and F1 score, which were 97.06%, 98.97%, 88.64% and 91.77, respectively. Although the accuracy and specificity are the highest, the sensitivity is lower than that of some other methods.

4   Materials and Methods

4.1   Dataset

The images utilized in this work were sourced from the MESSIDOR dataset, an acronym for Methods to Evaluate Segmentation and Indexing Techniques in Retinal Ophthalmology (in French). This dataset was established to facilitate studies on the computer-assisted diagnosis of diabetic retinopathy (DR). The MESSIDOR dataset comprises 1200 color images of the posterior pole of the eye fundus, acquired by three ophthalmological departments using a 3CCD color video camera mounted on a Topcon TRC NW6 non-mydriatic retinograph with a field of view 45°. The images were captured with 8 bits per color plane at resolutions of 1440×960, 2240×1488, or 2304×1536 pixels [8].

For this study, the focus is on the annotations of DME provided in the MESSIDOR dataset. Of 1200 images, 974 correspond to healthy patients (NO_DME) (81.17%), 75 to NCSME (6.25%), and 151 to CSME (12.58%). This means that only 226 images (18.83%) are classified as DME, showing the imbalance of the dataset, which the augmentation techniques will address. In addition, the MESSIDOR dataset has images annotated with NCSME or CSME but that do not show visible signs of DME (visible hard exudates), as shown in Fig. 1(d), so it was necessary to eliminate it from the dataset. A total of 974 healthy images (85.81%) and 161 images with DME (14.18%) were obtained.

4.2 Proposed Approach

Figure 2 presents the framework of the proposed method in this paper. As previously explained, the first stage consists of image cropping, pre-processing, and resizing. In the next stage, the preprocessed dataset is divided into a train set and a test set in a ratio 70% - 30%, and then the train set is splitted into trains and validation sets. Data augmentation and balancing are applied to the train set to increase the number of samples and balance the classes in the set. Next, feature extraction is done directly by CNN pre-trained models, in which the classification layers are removed and replaced by a custom classification layer. We will explain this technique in more detail below.

For this work, we chose the pre-trained models VGG16 [17], ResNet50 [18], and EfficienNetB7 [19], which were trained with the ImageNet dataset that contains 1.4 million labeled images and 1,000 different classes. The architecture of pre-trained models consists of a variation of interleaved blocks of convolutional layers and Max-Pooling blocks (convolutional base), followed by fully connected layers (classifier). An example of the VGG16 architecture is shown in Fig. 2 [20].

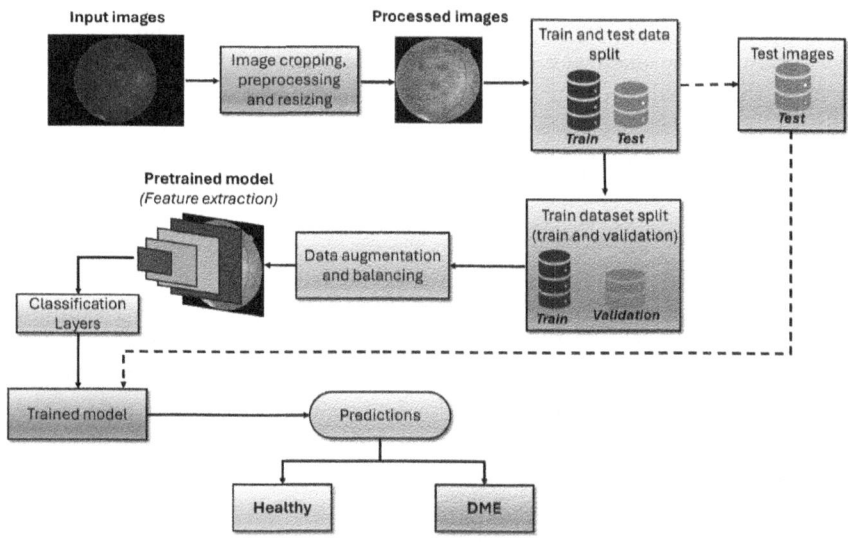

Fig. 2. Flowchart of the proposed methodology.

4.3   Preprocessing Stage

Pre-processing is a commonly used step to enhance an image and facilitate feature extraction related to objects of interest. A color fundus image (CFI) usually includes a dark border that should be avoided or removed as these pixels are irrelevant. CFI may also contain artifacts, noise, and areas of uneven illumination and contrast, which pre-processing can eliminate [21].

Because the original images in the MESSIDOR dataset have diverse dimensions, brightness, contrast, and illumination, some pre-processing techniques are necessary to correct these deficiencies. First, the dark areas of the images were cut as much as possible. Then, the image is converted from .tif to .jpg format. A Gaussian filter is employed to eliminate unwanted noise and smooth the image. Finally, the image is split into red, green, and blue channels (R, G, B). A contrast-limited adaptive histogram equalization-CLAHE is applied to the green channel for contrast enhancement. The R, G and B channels are merged again, and the image is resized to 256×256. The preprocessing workflow is depicted in Fig. 3, while an example of a pre-processed image is shown in Fig. 4.

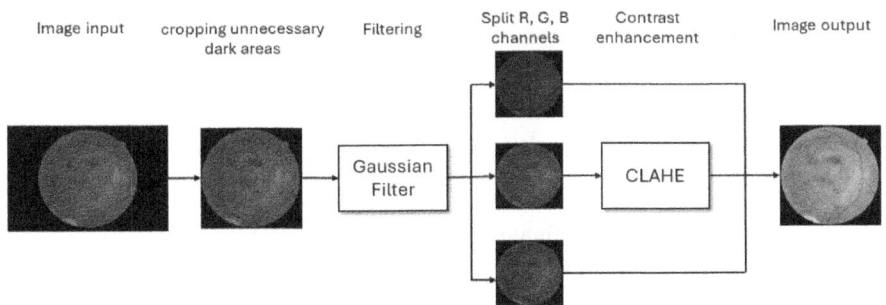

Fig. 3. MESSIDOR dataset pre-processing workflow.

4.4   Data Balancing and Augmentation

As previously noted, the MESSIDOR dataset is highly imbalanced: out of 1135 images, 974 correspond to healthy cases (85.81%), while 161 images correspond to cases of diabetic macular edema (DME) (14. 18%). This imbalance can lead to model overfitting and bias to the majority class (healthy). Augmentation techniques have been applied to the training subset to solve this problem: Random image flips (horizontal and vertical), zoom, and rotation were applied several times on each image to balance the classes and to increase the number of samples. Our train dataset is 70% of the images (795 images), of which 675 belong to the healthy class and 120 belong to the DME class. The augmentation operations are applied once for each healthy image and ten times for each DME image. Table 2 summarizes the number of obtained images for the training dataset.

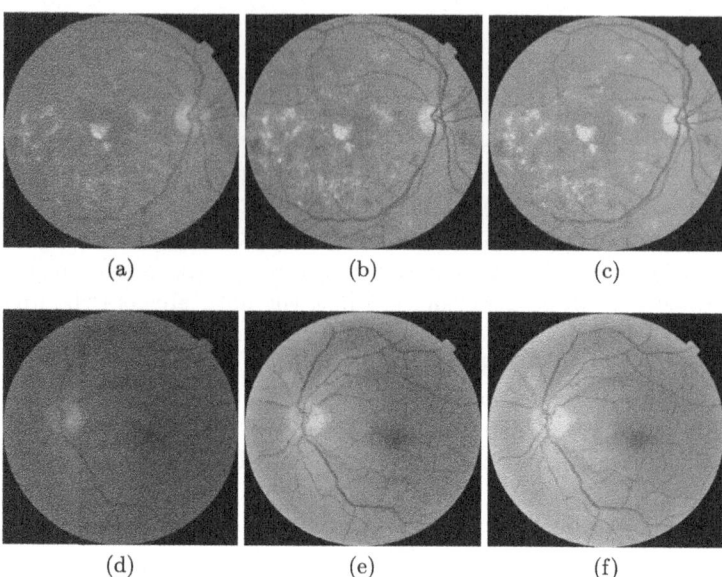

Fig. 4. Pre-processed images examples: First column (a) and (d) original images. Second column (b) and (e) are green channel images after the CLAHE application. Third column (c) and (f) output images (Color figure online)

Table 2. Samples dimension of healthy and DME classes in the training dataset before and after augmentation.

Class	Original size	Balanced & augmented
Healthy	675	1350
DME	120	1320

4.5   Feature Extraction with Pre-trained Models

It uses the representations of the previously trained model to extract relevant features from new samples. These features are then passed on to a new classifier, trained from scratch, as seen in Fig. 5 [22]. The first part is known as the convolutional base of the model. Feature extraction consists of taking the convolutional base of a pre-trained network, running the new data through it, and training a new classifier on top of the output.

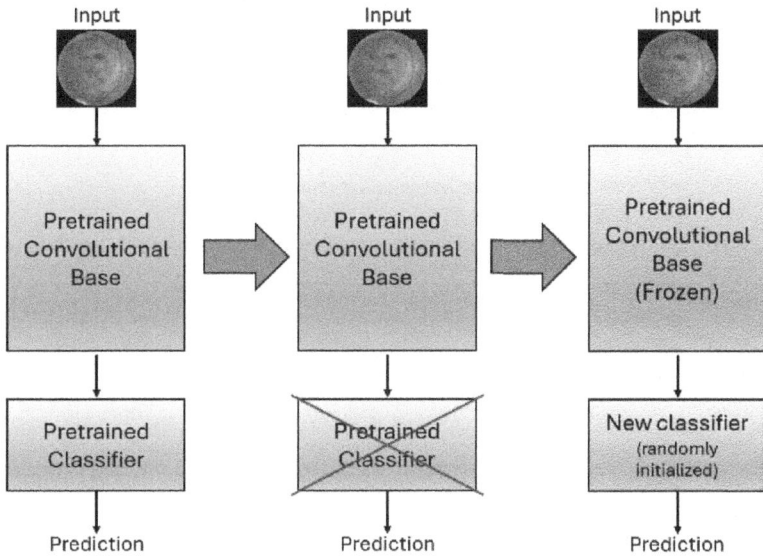

Fig. 5. Feature extraction with a pre-trained models general scheme.

For this work, we chose the pre-trained models VGG16 [17], ResNet50 [18], and EfficienNetB7 [19], which were trained with the ImageNet dataset that contains 1.4 million labeled images and 1,000 different classes. The architecture of pre-trained models consists of a variation of interleaved blocks of convolutional layers and Max-Pooling blocks (convolutional base), followed by fully connected layers (classifier). An example of the VGG16 architecture is shown in Fig. 6 [20].

It is important to remark that the classifier of the pre-trained model must be removed and replaced with a custom fully connected layers classifier, which will be trained with the features extracted from the convolutional base. The architecture of the classifier used in this work is built with a flattening layer, followed by a dense layer and a dropout layer. The output layer uses a sigmoid activation function because the classification is bi-class type, and the learning process is set to 30 epochs. Figure 7 shows a general block diagram of the classification architecture. Table 3 shows the hyperparameters used for each model.

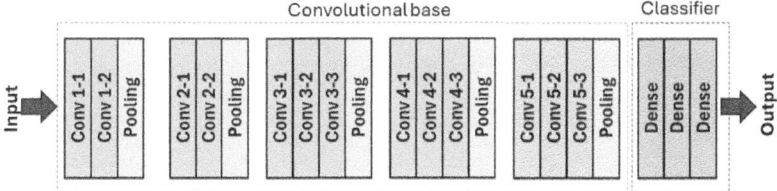

Fig. 6. VGG16 Architecture.

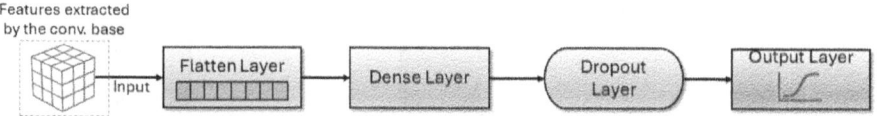

Fig. 7. Architecture of the classifier.

Table 3. Summary of hypermarameters employed for each model.

Hyperparameters	Pre-trained model		
	VGG16	ResNet50	EfficientNetB7
Learning Rate	1×10^{-4}	1×10^{-4}	1×10^{-4}
Batch Size	32	32	32
Epochs	30	30	30
Loss function (classifier)	Binary cross-entropy	Binary cross-entropy	Binary cross-entropy
Optimizer (classifier)	rmsprop	rmsprop	rmsprop

4.6 Performance Metrics

After the pre-trained convolutional base has extracted the features from the training images and the fully connected layer classifier has been trained using these features, the performance of the proposed scheme is evaluated using the test images to predict the classification of previously unseen data. In this study, we assessed the performance of the three pre-trained models using Accuracy (Acc), Sensitivity (Se), Precision (Pr), and F1-Score. These metrics are defined using TP (True Positive), TN (True Negative), FP (False Positive), and FN (False Negative) obtained from the confusion matrix and expressed by the set Eqs. 1 to 4.

$$Acc = \frac{TP}{TP + FP} \tag{1}$$

$$Se = \frac{TP}{TP + FN} \tag{2}$$

$$Pr = \frac{TP}{TP + FP} \tag{3}$$

$$F1 - Score = 2\,\frac{Pe\,Se}{Pe + Se} \tag{4}$$

5   Experimental Results

5.1   Comparative Analysis Between Evaluated Models

Table 4 shows the pre-trained models employed in this work with their respective number of parameters after feature extraction was performed on the MESSIDOR dataset's train data.

Table 4. Number of parameters of each model.

Pre-trained convolutional base	Trained Parameters	Untrained	Overall
VGG16	14714688	0	14714688
ResNet50	23534592	53120	23587712
EfficienNetB7	63786960	310727	64097687

The previous table shows that de EfficienNetB7 is the pre-trained model with the most parameters. Meanwhile, the VGG16 is the pre-trained model with fewer parameters. For each pre-trained convolutional base, the extracted features were classified by the personalized fully connected classifier, and the performance was assessed through the metrics presented previously. Figure 8 shows the evolution

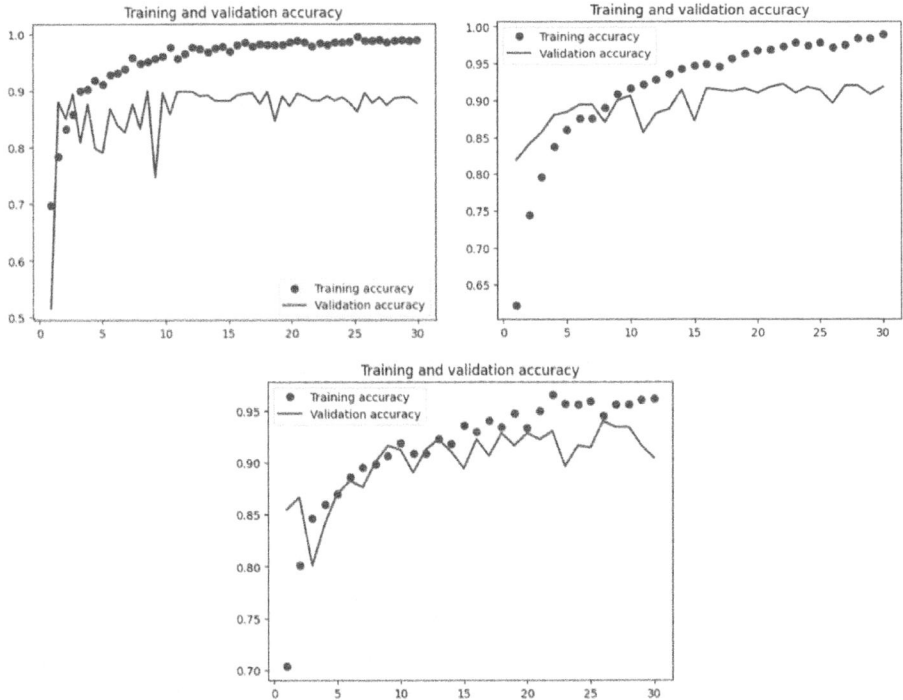

Fig. 8. Training and validation evolution Acc for (a) VGG16, (b) ResNet50, and (c) EfficienNetB7.

of training and validation accuracy through the 30 training epochs for each model, and Table 5 summarizes the comparison of the results obtained.

Table 5. Comparison analysis of the models evaluated for detecting DME with the set of MESSIDOR images.

Pre-trained model	Training Acc	Validation	Testing	Se (macro avg.)	Pr	F1-Score
VGG16	0.98	0.89	0.90	0.78	0.76	0.77
ResNet50	0.99	0.92	0.91	0.79	0.80	0.80
EfficienNetB7	**0.96**	**0.94**	**0.95**	**0.89**	**0.87**	**0.88**

As shown in Table 5, all models achieved an accuracy of > 90% with the test dataset; however, VGG16 and ResNet50 have poor Se and Pr compared to EfficienNetB7, which has an Acc of 95%, a Se of 89%, a Pr of 89% and a F1-Score of 88%. Therefore, EfficienNetB7 is the model selected in our proposed method because its performance is superior to that of the other models evaluated. Figure 9 shows the confusion matrix with the testing images for the best performance model, from which we can see that the model correctly classifies 33 of 41 test images with the possible presence of DME. In contrast, the model correctly classifies 283 of 293 test images in the possible absence of DME.

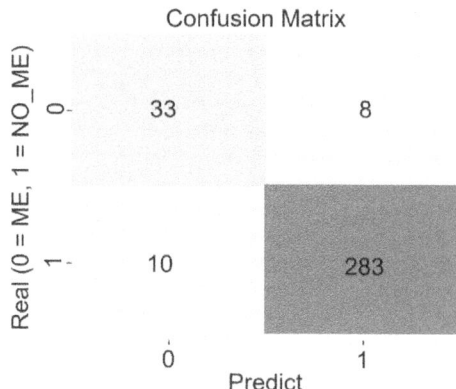

Fig. 9. Confusion matrix for DME detection using the EfficienNetB7 model.

5.2 Comparative with Related Works

In this section, we will compare the results of our proposed method based on the EfficienNetB7 pre-trained convolutional base and the fully connected classification layer with some of the related work that we studied in Sect. 3 to determine how close we are to their metrics and to have a notion of the efficiency of our best model (EfficienNetB7). Table 6 summarizes the comparison results.

Table 6. Comparison between related works and proposed method.

Model	Acc (%)	Re (%)	Pr (%)	F1-score (%)
Personalized CNN [15]	88.8	74.7	–	–
HE-CNN [16]	96.1	96.32	–	96.0
ResNet50 and SENet [11]	97.0	88.6	–	91.0
Our proposed method	**95.0**	**89.0**	**87.0**	**88.0**

Table 6 shows that the best Acc was obtained in [11], but its Se was low compared to our proposed method. In contrast, the work presented in [23] has outstanding Acc (96.1%), Se (96.32%), and F1-score (96.0%). In general terms, the results of our study are comparable with those of the related work, achieving an Acc (95. 0%) over the values obtained by [13] and an Acc slightly less than reported by [11, 23].

6   Conclusions

DME is a complication of diabetic retinopathy and is a disease that affects mainly people who have not been treated correctly, resulting in possible vision impairment or loss of vision if it is not detected in time. Early diagnosis of DME is essential to prevent such vision complications. Therefore, we have proposed an alternative solution to this problem using different pre-trained DCNN models designed for image detection and classification tasks.

This work used three pre-trained CNN models: VGG16, ResNet50, and EfficientNetB7 to evaluate the possibility of the presence of DME in fundus color images. The convolutional base of each model was employed for feature extraction, and a fully connected CNN classifier was used at the top of DCNN to classify the MESSIDOR images into two classes. DME risk and not DME risk.

This study employed different pre-processing techniques, hyperparameters, balancing, and augmentation data. These techniques help prevent overfitting and improve the training time, loss, and accuracy of the evaluated models. Our best result with the convolutional base of EfficientNet B7 and the fully connected CNN classifier achieves an accuracy of 95%, comparable with the state-of-the-art and suitable for the task of DME detection.

7   Future Research Lines

This study explores the implementation of pre-trained models to detect the risk of DME without the need for segmentation tasks. Although the proposed method has been demonstrated to be suitable for helping ophthalmologists diagnose DME, there are several aspects to consider for future work.

- Diagnosis of DME through fundus color images is not a conclusive method because the absence of visible hard exudates does not imply that it does not have a risk of DME. It is important to explore the detection of other signs, such as macular thickening and microaneurysms, using segmentation techniques and a combination of other imaging modalities, such as optical coherence tomography (OCT).
- Since visible hard exudates in DME are present near the macula center, the processing cost of the models can be reduced by obtaining a region of interest around the fovea, which allows the extraction of a sub-image with reduced size relative to the original. This can be done by applying segmentation techniques to locate the optic disc (OD) and finding the location of the fovea relative to the OD. With the approximate location of the fovea, we can extract an ROI around this point with a diameter of 2 OD.
- The proposed method was trained and evaluated only on the MESSIDOR dataset; therefore, it must be evaluated on other publicly available image sets, such as IDRiD, Eoptha, and EyePACS datasets.
- In this work, the focus is on determining the presence or absence of DME in fundus color images, addressing a binary classification problem. In future work, the aim will be to grade the severity of the disease, thus tackling a multiclass classification problem.
- In this work, only three pre-trained models were employed. Nowadays, there is a wide variety of models to pick from, so it is attractive to test and compare other models such as MobileNet, DenseNet, and ConvNeXt.
- An interesting future challenge is developing an application based on cloud computing to support healthcare personnel in diagnosing DME.

References

1. World Health Organization: Diabetes. https://www.who.int/es/news-room/fact-sheets/detail/diabetes. Accessed 27 Sep 2023
2. Romero-Aroca, P.: Ocular complications of diabetes and therapeutic approaches. J. Clin. Med. **11**, 5170 (2022)
3. Romero-Aroca, P., et al.: Referable diabetic retinopathy prediction algorithm applied to a population of 120,389 type 2 diabetics over 11 years follow-up. Diagnostics **14**, 833 (2024)
4. International Agency for the Prevention of Blindness: Diabetic retinopathy. https://www.iapb.org/learn/knowledge-hub/eye-conditions/diabetic-retinopathy/. Accessed 27 Sep 2023
5. Sánchez-Thorin, J., et al.: Panel de expertos sobre la atención inicial de la retinopatía diabética colombia. ALAD **8**, 566 (2018)
6. Das, A., McGuire, P., Rangasamy, S.: Diabetic macular edema: pathophysiology and novel therapeutic targets. Ophthalmology **122**, 1375–1394 (2015)
7. Ji, L., Chen, T.-Y., Liang, Y.: Early diagnosis and treatment of diabetic macular edema. Int. Eye Sci. **14**, 1809–1811 (2014)
8. Decenciére, E., et al.: Feedback on a publicly distributed image database: the messidor database. Image Anal. Stereology **33**, 231–234 (2014)

9. Escorcia-Gutierrez, J., et al: Analysis of pre-trained convolutional neural network models in diabetic retinopathy detection through retinal fundus images. In: International Conference on Computer Information Systems and Industrial Management, pp. 202–213. Springer (2022)
10. Escorcia-Gutierrez, J., et al.: Grading diabetic retinopathy using transfer learning-based convolutional neural networks. In: International Conference on Computer Information Systems and Industrial Management, pp. 240–252. Springer (2023)
11. Fu, Y., Lu, X., Zhang, G., Lu, Q., Wang, C., Zhang, D.: Automatic grading of diabetic macular edema based on end-to-end network. Expert Syst. Appl. **213**, 118835 (2023)
12. Chalakkal, R., Hafiz, F., Abdulla, W., Swain, A.: An efficient framework for automated screening of clinically significant macular edema. arXiv:2001.07002 (2020)
13. Zubair, M., et al.: Automated grading of diabetic macular edema using color retinal photographs. In: 2022 2nd International Conference of Smart Systems and Emerging Technologies (SMARTTECH), pp. 1–6 (2022)
14. Zubair, M., Umair, M., Ali Naqvi, R., Hussain, D., Owais, M., Werghi, N.: A comprehensive computer-aided system for an early-stage diagnosis and classification of diabetic macular edema," J. King Saud Univ. Comput. Inf. Sci. **35**, 101719 (2023)
15. Al-Bander, B., Al-Nuaimy, W., Al-Taee, M., Williams, B., Zheng, Y.: Diabetic macular edema grading based on deep neural networks. In: Proceedings of the Ophthalmic Medical Image Analysis Third International Workshop, (Athens, Greece), pp. 121–128. University of Iowa (2016)
16. Singh, R., Gorantla, R.: DMENet: diabetic macular edema diagnosis using hierarchical ensemble of CNNs. PLoS ONE **15**, e0220677–e0220677 (2020)
17. Simonyan, K., Zisserman, A.: Very deep convolutional networks for large-scale image recognition. arXiv:1409.1556 (2015)
18. He, K., Zhang, X., Ren, S., Sun, J.: Deep residual learning for image recognition. arXiv:1512.03385 (2015)
19. Tan, M., Le, Q.: EfficientNet: rethinking model scaling for convolutional neural networks. arXiv:1905.11946 (2020)
20. Thaker, T.: VGG 16 easiest explanation. VGG16: it is a convolutional neural. https://medium.com/nerd-for-tech/vgg-16-easiest-explanation-12453b599526. Accessed 30 May 2024
21. Pead, E., Megaw, R., Cameron, J., Fleming, A., Dhillon, B., Trucco, E., MacGillvray, T.: Automated detection of age-related macular degeneration in color fundus photography: a systematic review. Surv. Ophthalmol. **64**, 498–511 (2019)
22. Deep Learning with Python. Shelter island : Manning Publication (2021)
23. Wang, T.-Y., et al.: Diabetic macular edema detection using end-to-end deep fusion model and anatomical landmark visualization on an edge computing device. Front. Med. **9** (2022)

Gaussian Mixture Connectivity with α-Renyi Regularization for EEG-Based MI Classification

D. V. Salazar-Dubois$^{(\boxtimes)}$, A. M. Alvarez-Meza, and G. Castellanos-Dominguez

Signal Processing and Recognition Group, Universidad Nacional de Colombia,
170003 Manizales, Colombia
{dsalazard,amalvarezme,cgcastellanosd}@unal.edu.co

Abstract. Brain-computer interfaces (BCIs) have unveiled a transformative avenue for human computer interaction. So, motor imagery (MI) emerges as a compelling paradigm to mentally simulate movements without any overt physical execution. However, the inherent variability in brain signals, both inter and intra-subject, poses significant challenges to achieving robust classification. Here, we present the Gaussian Mixture Functional Connectivity Network with α-Renyi Regularization (GMRRNet) for electroencephalography-based MI classification. GMRRNet employs a Gaussian mixture block made up of kernel-based functional connectivities with different bandwidths to extract discriminant spatial features. Also, it uses an α-Renyi regularization to cut down on the mutual information between Gaussian-based channel matrices. Our method was evaluated on the GigaScience MI-EEG dataset, comprising recordings from 52 subjects performing left and right-hand MI tasks. The experimental results demonstrate that GMRRNet outperforms traditional models like EEGNet in terms of both accuracy and stability, particularly for subjects with medium to low MI skills. The model achieves an average accuracy of 71.7%, with a lower standard deviation, indicating improved consistency across trials. Finally, the spatial interpretability analysis highlights the GMRRNet capability to capture significant localized and global connections within the brain.

Keywords: Brain Computer Interfaces · Motor Imagery · Functional Connectivity · Regularizer · Explainable AI

1 Introduction

Nowadays, brain-computer interfaces (BCIs) have unveiled a transformative avenue for interfacing human thoughts with technology [13]. Within the BCI framework, motor imagery (MI) emerges as a compelling paradigm that capitalizes on the remarkable human ability to mentally simulate movements without any overt physical execution [3]. So, neuroimaging techniques are essential for the acquisition of brain signals. Electroencephalography (EEG) and magnetoencephalography respectively record electrical activity and magnetic fields

N. D. Duque-Méndez et al. (Eds.): CCC 2024, CCIS 2208, pp. 132–147, 2024.
https://doi.org/10.1007/978-3-031-75233-9_10

in the brain [23]. Functional magnetic resonance imaging detects changes in blood flow, indicating brain activity. Computed tomography and magnetic resonance imaging provide detailed structural images, useful in pathology identification. Moreover, positron emission tomography and single-photon emission computed tomography use radioactive materials to map how the brain works [33]. Although spatial and temporal resolution are essential for selecting the proper neuroimaging method, the cost factor is equally significant for decision-making in clinical and research contexts [25]. In particular, MI necessitates superior temporal resolution due to the fast-evolving cognitive processes it entails. Therefore, EEG and MEG are effective methods for documenting patterns of neural activity over time [2]. Additionally, EEG is commonly used by researchers and budget-constrained institutions that specialize in MI-BCI applications, both due to its portability and cost-effectiveness [11].

As a result, MI-EEG classification holds immense potential for BCI, including prosthetic control, neurorehabilitation therapies, and neuromarketing [8]. Yet, constant challenges of inter-subject variability, which involves the variation of brain patterns across individuals, and intra-subject variability, which involves the fluctuation of a person's own signals over time, are both present [16]. The subject's mental state, attention, and fatigue can also have a significant impact in this regard. Also, the quality of electrical activity patterns produced by the brain is essential for the regulation of external devices [22]. Nevertheless, these patterns demonstrate significant variation among subjects, even when presented with identical stimuli or conditions [29]. This variability is influenced by factors, such as gender, age, lifestyle, neurophysiological and psychological parameters, genetic differences, and cognitive processes [26]. Furthermore, traditional machine and deep learning methods dedicated to MI-BCI frequently struggle with interpretability, making it difficult to pinpoint which aspects of the EEG signal contribute most to successful classification. Further, it is difficult to strike the right balance between local features that are specific to a brain region and global patterns that show how the brain is working overall [16].

In this sense, feature extraction strategies are designed to convert the raw EEG into discriminant MI patterns. The latter can be broadly classified into spatial, time-frequency, and time approaches. In the time domain, amplitude modulation and variance-based measures [27]. Functional connectivity (FC) and common spatial patterns (CSP) are conventional techniques for feature extraction in the spatial domain. CSP employs a collection of spatial filters that have been learned to project the EEG signals into a reduced dimensional space [34]. FC records the similarity between channels, which can be used to determine relevant brain regions interactions [9]. Nevertheless, the selection of the most suitable representation method for the MI task is a difficult undertaking that necessitates a high level of subject-matter expertise and a prior understanding of the expected EEG.

On the other hand, the FBCSP method is the standard for MI-based BCI classification [35], using band-pass filters to extract features, although it is sensitive to noise and overfitting. To improve, regularized variants of CSP have been created, including covariance regularization, cost functions, and L1 norms for outlier data. In deep learning, various architectures have been tested, such as deep belief networks [5], convolutional neural networks [21] and recurrent neural networks [32], each with its own challenges, especially overfitting and the need for large datasets. Models like TCFussionnet aim to simplify and improve generalization [20]. Despite advances, deep learning approaches remain less interpretable compared to classical covariance-based methods.

Recently, deep learning methods have been applied to the classification of MI-based EEG signals (EEG-MI). A seminal approach, EEGNet [14], employs temporal and spatial filters with convolutional layers to process raw EEG data and enhance interpretability by displaying neurophysiologically understandable features. However, its performance varies significantly between subjects, especially in cross-subject classification due to data quality [14]. DeepConvNet is notable for recognizing spatial and temporal patterns, though it requires a lot of training data to avoid overfitting. A convolutional network proposed in [36] also simultaneously extracts temporal and spatial features, but its high computational complexity may not always identify distinct patterns. Other methods, such as ShallowConvNet, Graph Convolution Neural Networks, and EEG-transformer [31], also address EEG-MI issues, though they face similar problems of subject variability and overfitting.

Here, we present GMRRNet, a Gaussian mixture functional connectivity network with α-Renyi regularization that can be used to classify MI based on EEG signals. Our method tries to keep the learning process understandable by adding a Gaussian mixture approach made up of kernel-based FC with various bandwidths. These kernels enable the identification of various similarities among EEG channels, thereby coding the local and global connections between the various brain regions involved. Moreover, a custom loss based on α Renyi is also suggested. This loss takes classification accuracy into account and tries to reduce the mutual information among Guassian-based channel matrices to avoid overfitting and enhance spatial interpretabilty. Results obtained on a well-known MI dataset containing 50 subjects demonstrate that GMRRNet is capable of dealing with inter- and intra-subject variability, as well as establishing spatial explainability, compared to baseline deep learning methods for EEG-based MI.

The rest of the document is structured as follows: Sect. 2 covers the materials and methods. Sections 3 and 4 detail the experiments and discuss the results. Lastly, Sect. 5 presents the concluding remarks.

2   Materials and Methods

2.1   Gaussian Mixture Functional Connectivity Fundamentals

A real-valued auto-correlation function, $R_x(\tau)$, with $\tau \in \mathbb{R}$, of a weak-sense stationary stochastic process x can be defined as: $R_x(\tau) = \int_{\mathbb{R}} \exp(j2\pi\tau f)dP_x(f)$, where $P_x(f) \in \mathbb{R}$ is a monotonic spectral distribution function over frequency $f \in \mathbb{R}$. Bochner's theorem states that a stationary positive-definite kernel can be written from the pairwise correlation between random vectors $\boldsymbol{x}, \boldsymbol{x}' \in \mathbb{R}^T$, i.e., EEG channels, through a generalized, kernel-based covariance, if and only if the following assumption holds [4]:

$$\kappa(\boldsymbol{x} - \boldsymbol{x}') = \int_{\mathbb{R}} \exp\left(j2\pi(\boldsymbol{x} - \boldsymbol{x}')^{\top}\boldsymbol{f}\right) S_{\boldsymbol{xx}'}(\boldsymbol{f})d\boldsymbol{f}, \tag{1}$$

where $\boldsymbol{f} \subseteq \Omega$ is the frequency domain within the set Ω and $S_{\boldsymbol{xx}'}(\boldsymbol{f}) \in \mathbb{C}$ is the cross-spectral density preserving: $S_{\boldsymbol{xx}'}(\boldsymbol{f}) = dP_{\boldsymbol{xx}'}(\boldsymbol{f})/d\boldsymbol{f}$, with $P_{\boldsymbol{xx}'}(\boldsymbol{f}) \in [0,1]$ being the cross-spectral distribution from the mapping kernel, $\kappa \colon :\mathbb{R}^T \times \mathbb{R}^T \to \mathbb{R}$. Then, $P_{\boldsymbol{xx}'}(\boldsymbol{f})$ can be written as below [10]:

$$P_{\boldsymbol{xx}'}(\boldsymbol{f}) = 2 \int_{\boldsymbol{f} \in \Omega} S_{\boldsymbol{xx}'}(\boldsymbol{f})d\boldsymbol{f} = 2 \int_{\boldsymbol{f} \in \Omega} \mathscr{F}\left\{\kappa(\boldsymbol{x}, \boldsymbol{x}')\right\} d\boldsymbol{f}. \tag{2}$$

Notation $\mathscr{F}\{\cdot\}$ stands for the Fourier transform.

Consequently, the frequency-based interpretation of the kernel-based pairwise dependencies that were estimated between vectors of random functions is preserved by Eq. 2. Additionally, the imposed stationary kernel facilitates the extraction of nonlinear data dependencies. Here, to code local and global EEG-based channel similarities, we propose to use a mixture of $G \in \mathbb{N}$ Gaussian kernels to rule the cross-spectral distribution, yielding:

$$\kappa(\boldsymbol{x} - \boldsymbol{x}') = \mathcal{M}\{\kappa_{\sigma_g}(\boldsymbol{x} - \boldsymbol{x}') : g \in G\}, \tag{3}$$

where $\mathcal{M}\{\cdot\}$ stands for a deep learning model and:

$$\kappa_{\sigma_g}(\boldsymbol{x} - \boldsymbol{x}') = \exp\left(\frac{-\|\boldsymbol{x} - \boldsymbol{x}'\|_2^2}{2\sigma_g^2}\right), \tag{4}$$

being $\sigma_g \in \mathbb{R}^+$ the g-th bandwidth. Notably, pattern classification prefers the Gaussian kernel due to its universal approximation ability and its straightforward explainability regarding the used length scale.

2.2 Gaussian Mixture Connectivity Network with α-Renyi Regularization (GMRRNet)

Our FC in Eq. 3 can be coupled with an EEG-MI-devoted network to favor data classification and spatial interpretability. Nonetheless, to avoid overfitting, we also incorporate an α-Renyi regularizer from learned kernels. Thus, a kernel-based estimator of Renyi's entropy can be defined as [37]:

$$S_\alpha(\tilde{\mathbf{K}}) = \frac{1}{1-\alpha} \log\left(\tilde{\mathbf{K}}^\alpha\right), \tag{5}$$

where $\alpha > 0$, $\alpha \neq 1$, $\tilde{\mathbf{K}} \in \mathbb{R}^{C \times C}$ gathers elements from a given kernel function, and $\mathrm{tr}(\tilde{\mathbf{K}}) = 1$. Alike, the α-order joint entropy can be written as:

$$S_\alpha(\tilde{\mathbf{K}}_g, \tilde{\mathbf{K}}_{g'}) = S_\alpha\left(\frac{\tilde{\mathbf{K}}_g \circ \tilde{\mathbf{K}}_{g'}}{\mathrm{tr}(\tilde{\mathbf{K}}_g \circ \tilde{\mathbf{K}}_{g'})}\right). \tag{6}$$

∘ stands for the Hadamard product. Afterward, given a set of Gaussian kernels $\{\tilde{\mathbf{K}}_g\}_{g=1}^G$, their α-Renyi mutual information can be written as:

$$I_\alpha\left(\{\tilde{\mathbf{K}}_g\}_{g=1}^G\right) = \sum_{g=1}^G S_\alpha(\tilde{\mathbf{K}}_g) - S_\alpha\left(\frac{\prod_{g=1}^G \tilde{\mathbf{K}}_g}{\operatorname{tr}\left(\prod_{g=1}^G \tilde{\mathbf{K}}_g\right)}\right). \tag{7}$$

Lastly, our GMRRNet loss computes a trade-off between MI accuracy, e.g., ruled by a conventional cross-entropy classification loss, and the kernel-based α-Renyi regularizer in Eq. 7 as:

$$\mathcal{L}(\boldsymbol{y}, \hat{\boldsymbol{y}} | \mathcal{M}) = -\lambda \sum_{q=1}^Q y_q \log(\hat{y}_q(\mathcal{M})) + (1 - \lambda) I_\alpha\left(\{\tilde{\mathbf{K}}_g(\mathcal{M})\}_{g=1}^G\right), \tag{8}$$

where $\lambda \in [0, 1]$ is a regularization hyperparameter and $\boldsymbol{y} \in \{0, 1\}^Q$ are the $1 - Q$ target MI class-membership labels, while $\hat{\boldsymbol{y}} \in [0, 1]^Q$ are the predicted MI class-membership probability labels. The notations $\hat{y}_q(\mathcal{M})$ and $\tilde{\mathbf{K}}_g(\mathcal{M})$ highlight the predicted label probability and Gaussian kernel dependency on the deep learning model.

Besides, the deep learning loss normalization strategy suggested in [17] is used on each term in Eq. 8 to help with hyperparameter tuning. Finally, model parameters, stacking the weight matrices and bias vectors, are optimized using a gradient descent-based framework with back-propagation [19]. Our GMRRNet sketch for EEG-based MI classification is summarized in Fig. 1.

3   Experimental Set-Up

We test our GMRRNet for EEG-based MI classification. We will provide the description of the studied dataset, training scheme and method comparison.

3.1   GigaScience Dataset

The GigaScience database includes extensive MI-EEG recordings from 52 subjects and employs a 10-10 electrode placement system comprising 64 channels, a standard configuration for recording MI-EEG data. This electrode setup is designed to capture brain activity throughout each trial. The collection, publicly available at http://gigadb.org/dataset/100295 and produced by researchers in [7], holds 19 females and 33 males, performing two MI tasks: left and right hand. Of note, 50 subjects are right-handed. Data were gathered in a laboratory with a noise level between 37 and 39 decibels during one of the next time slots: T1 (9:30-12:00), T2 (12:30-15:00), T3 (15:30-18:00), or T4 (19:00-21:30). One hundred trials of each label per subject were recorded, where each channel was sampled at 512 Hz. Furthermore, EMG and EEG signals were recorded simultaneously with the same system and sampling rate to check actual hand

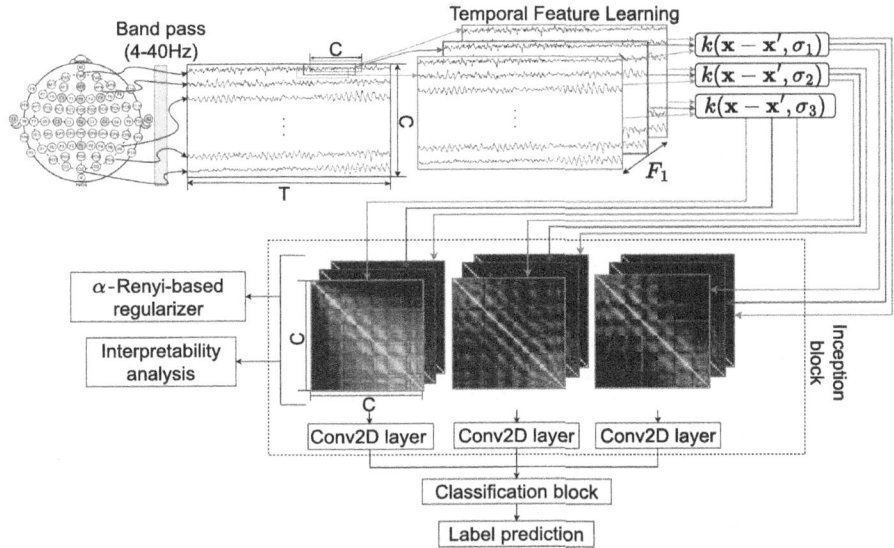

Fig. 1. GMRRNet's EEG-based MI discrimination pipeline. For illustrative purposes, we consider three Gaussians $(G = 3)$ and add a temporal block for feature enhancement.

movements. For concrete testing, our study focuses specifically on a subset of 50 subjects who meet the criterion of having at least 100 recorded EEG trials.

The experiment involved non-motor and MI tasks, starting with data collected under six different types of noise: eye blinking, eye movements, head movement, chewing, and resting state. Each noise type was recorded twice for five seconds, except for the resting state condition, which was recorded for 60 s. Figure 2 depicts the main protocol for MI-EEG data acquisition. Each trial commenced with a two-second fixation on a cross on a black screen. There was a cue that followed the task to indicate whether it was a right- or left-hand MI. The subjects were then instructed to imagine the sensation of touching each digit of the designated hand with the thumb, beginning with the index finger and progressing sequentially to the little finger, as part of the task. This emphasis on the kinesthetic rather than the visual aspect guaranteed that the subjects were indeed envisioning the motor task. The MI task persisted for three seconds, following which a blank screen indicated a rest period. The duration of this recovery interval was randomly assigned to 2.1–2.8 s.

The task, break, and recovery were repeated 20 times to gather a run. Each subject performed between five and six runs. After each run, a cognitive questionnaire was given as an additional data collection point. Additionally, labeling and excluding trials with high voltage magnitudes helped ensure data quality. Lastly, to maintain the engagement and motivation of the subjects, they were provided with feedback on the accuracy of their performance after each run.

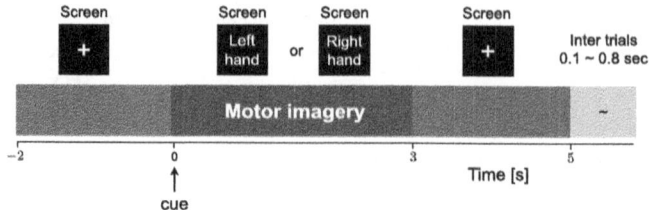

Fig. 2. GigaScience dataset experimental protocol involving MI tasks (single trial illustration). Adapted from [7]

3.2 Training Details and Method Comparison

Initially, subject recordings are loaded using a custom database loader module available at https://github.com/UN-GCPDS/python-gcpds.databases. Subsequently, each signal is downsampled from 512 Hz to 128 Hz using the Fourier method provided by https://docs.scipy.org/doc/scipy/reference/generated/scipy.signal.resample.html. A fifth-order Butterworth bandpass then filters each time series trial, restricting frequencies to the range of 4 to 40 Hz. Additionally, we clip records from 0.5 s to 2.5 s post-cue onset to exclusively focus on MI task-related information. The primary goal of this straightforward preprocessing is to explore the brain rhythms including theta, alpha, and three types of beta waves. Theta waves (4–8 Hz), originating in the hippocampus and various cortical structures, are associated with an online state and are involved in sensorimotor and mnemonic functions. Contrarily, sensory stimulation and movements suppress alpha-band (8–13 Hz), which attention, working memory, and mental tasks modulate, potentially serving as an indicator of higher motor control functions. The preprocessing also investigates three categories of beta waves: Low beta waves (12–15 Hz), also known as beta one, are primarily linked to focused and introverted concentration; mid-range beta waves (15–20 Hz), or beta two, are associated with increased energy, anxiety, and performance; and high beta waves (18–40 Hz), or beta three, are correlated with significant stress, anxiety, paranoia, high energy, and arousal [12].

We divided the trials of each subject's data using the standard 5-fold cross-validation scheme with an 80-20 split. This entails shuffling the data, assigning 80% for training and reserving the remaining 20% for testing, repeating this process five times. We used the RandomSearch approach from SKlearn to find the best combination of hyperparameters for our GMRRNet. Specifically, we tuned the standard deviations of three Gaussian kernels within the range $[0, 5]$, and the number of filters used, F_1 and F_3, were searched within the set $\{2, 3, \ldots, 10\}$, while F_2 within the range $\{2, 4, \ldots, 20\}$. We employed the Adam optimizer with a learning rate of 0.001. Regarding the loss function, we combined the minimization of the Normalized Binary Crossentropy between the actual and predicted labels with the minimization of the α-Renyi mutual information between the outputs of the Gaussian kernels, assigning weights of 0.8 and 0.2, respectively, to impose a trade-off between MI classification and automatic functional connectivity extraction ($\lambda = 0.8$). All tests were carried out over 150 epochs with a batch size of 32. The GMRRNet architecture is presented in Table 1.

Table 1. GMRRNet architecture for MI classification. C: # of channels. F.: # of filters. GK.: Gaussian-based functional connectivity. σ.: Gaussian bandwidth. Three Gaussians are used ($G = 3$).

Layer	Output Dimension	Parameters	Connected to
Input	$(C, T, 1)$	$--$	$--$
Conv1	(C, T, F_1)	kerSize(1×64)	Input
BatchNorm1	(C, T, F_1)	$--$	Conv1
GK1	(C, C, F_1)	$\sigma = 0.8$	BatchNorm1
GK2	(C, C, F_1)	$\sigma = 0.2$	BatchNorm1
GK3	(C, C, F_1)	$\sigma = 4.8$	BatchNorm1
ConvGK1	(C, C, F_2)	kerSize(3×3)-ReLU	GK1
ConvGK2	(C, C, F_2)	kerSize(3×3)-ReLU	GK2
ConvGK3	(C, C, F_2)	kerSize(3×3)-ReLU	GK3
GMConcatenation	$(C, C, 3 * F_2)$	axis$=-1$	ConvGK1, ConvGK2, ConvGK3
Conv2	(C, C, F_3)	kerSize(3×3)	GMConcatenation
BatchNorm2	(C, T, F_1)	$--$	Conv2
flatten	$(C * C * F_3)$	$--$	BatchNorm2
ConcGK	$(C, C, 3 * F_1)$	axis$=-1$	GK1, GK2, GK3
Dense1	(64)	ReLU	flatten
Transpose	$(3 * F_1, C, C)$	$--$	ConcGK
Output	(K)	ReLU	Dense1
Softmax	(K)	$--$	Output
Entropy	$(3 * F_1)$	$\alpha = 2$	Transpose
JointEntropy	(1)	$\alpha = 2$	Transpose
ConcEntropies	$(3 * F_1 + 1)$	axis$=-1$	Entropy, JointEntropy

We categorized subjects into groups based on their classification performance to gain insights about the inter-subject variability issue. The baseline EEG-Net [14], is used for method comparison. Additionally, we test the following networks for average classification performance: The straightforward Shallow-convnet [28], the Deepconvnet [28], and the TCFussionnet [20]. As quantitative assessment, we compute the well-known accuracy measure for the testing set as:

$$ACC = \frac{T_P + T_N}{F_N + F_P + T_N + T_P} \cdot 100[\%], \qquad (9)$$

where $F_N, F_P, T_N, T_P \in \mathbb{N}$ correspond to the number of false negatives, false positives, true positives, and true negatives. All experiments were conducted using Python 3.8 and the Tensorflow 2.15.0 API within the Kaggle environment.

4   Results and Discussion

4.1   Inter and Intra-subject Variability Classification Results

Figure 3 depicts the subject-dependent mean classification accuracy concerning the EEGNet and our GMRRNet. Three MI performance groups are considered: high-G1 (green-MI accuracy greater than 75 [%]), medium-G2 (yellow-MI accuracy between 75 [%] and 55 [%]), and low-G3 (orange-MI accuracy lower than 55 [%]). EEGNet arranges the subjects in descending order of accuracy. Moreover, the blue bars indicate the degree to which GMRRNet outperformed EEGNet in terms of accuracy, while the red bars indicate a decrease in accuracy. The average accuracy of GMRRNet and EEGNet was 69.47% and 71.71%, respectively, showing an improvement in accuracy of 2.24%.

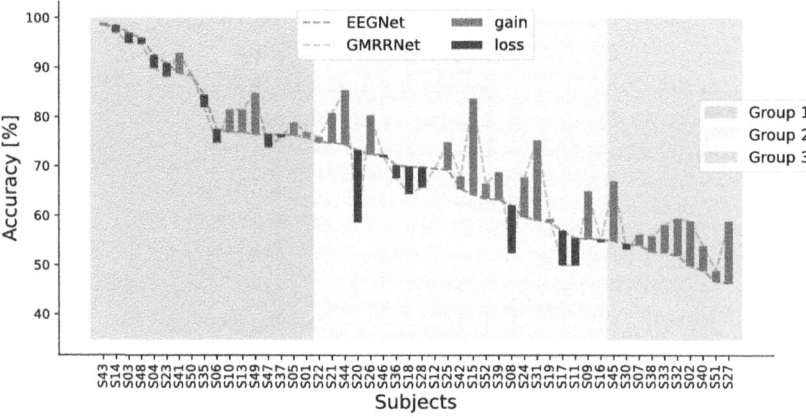

Fig. 3. Subject-dependent MI classification results (inter-subject variability analysis). The mean accuracy is presented for the baseline EEGNet vs. our GMRRNet. The background code indicates the group membership (high-G1, medium-G2, and low-G3), while the blue and red bars indicate outperformed or decreased performance. (Color figure online)

As shown in Fig. 4, though some subjects reveal a decrease in accuracy (G1), possibly due to overfitting or a lack of more exhaustive bandwidth tuning, on average, our approach achieves higher performance than EEGNet within a shallow network. It is worth noting that, there is a significant improvement for G2 and G3. There is also a noticeable overall improvement in the standard deviation. Therefore, our approach not only outperforms EEGNet in terms of accuracy but also reduces variability across all groups. The latter corroborates the GMMR-Net benefits in dealing with the intra-subject variability issue from an automatic Gaussian functional connectivity-based feature enhancement.

Next, Fig. 5 shows the intra-subject variability analysis for the 100 [%] minus the standard deviation MI accuracy. Positive values indicate that GMRRNet

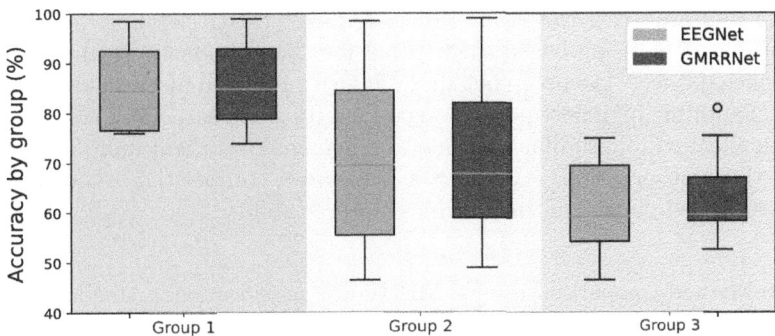

Fig. 4. Inter-subject variability results (group analysis). EEGNet (blue) vs. our GMR-RNet (orange) accuracy boxplots. The background codes the group membership (high–G1, medium–G2, and low–G3). (Color figure online)

exhibits less data dispersion, reflecting greater consistency compared to EEGNet. As seen, 16 subjects present higher variability in GMRRNet, which translates to loss in our plot. Among these, seven subjects show a decrease of more than two points. The remaining subjects demonstrated a gain, with 17 of them exceeding three points. This suggests that, for the majority of subjects, GMRRNet significantly improves data consistency compared to EEGNet.

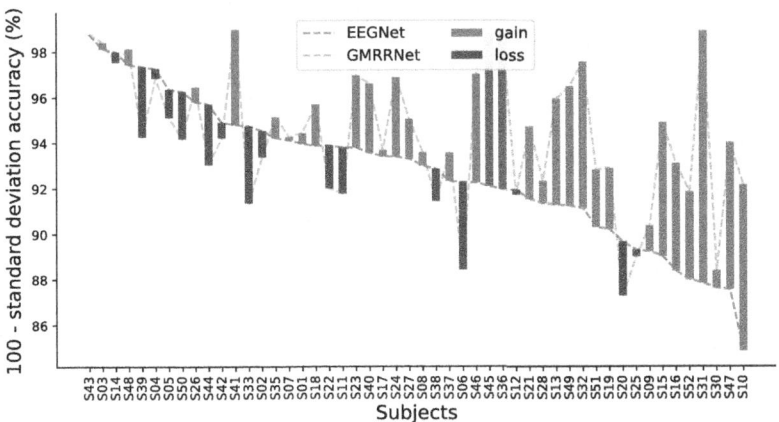

Fig. 5. Intra-subject variability results regarding the 100-standard deviation [%] along the 5-cross-validation scheme. Subjects are sorted concerning the EEGNet standard deviation-based classification performance, while the blue and red bars indicate outperformed or decreased performance regarding the prediction variability along folds. (Color figure online)

Table 2 summarizes the results obtained by GMRRNet and several state-of-the-art deep learning-based algorithms for MI classification applied to the GigaSicence dataset. Deepconvnet shows inferior performance, making it unsuitable for handling high intra-class variability. In contrast, Shallowconvnet and TCFussionnet exhibit similar values of quality, making them more competitive. Despite these observations, GMRRNet achieves a competitive score and shows a lower standard deviation, indicating greater stability.

Table 2. Method comparison results. MI average classification $\pm$ standard deviation along subject performances on the GigaScience dataset.

Approach	Accuracy
Deepconvnet [28]	62.5 ± 13.0
EEGNet [14]	69.5 ± 14.3
TCFussionnet [20]	72.7 ± 14.0
Shallowconvnet [28]	74.9 ± 13.9
GMRRNet (ours)	71.7 ± 13.7

4.2   GMRRNet Spatial Interpretability Results

Figure 6 represents the average topomap difference of the class membership trials belonging to left and right MI for each provided Gaussian bandwidth. Likewise, Fig. 7 depicts the GMRRNet-based FC for subject 43. The latter highlights the channels that have the strongest connections for discriminating MI patterns. The topoplots of the classes are normalized by the highest overall value. As seen for subject 43, as a representative sample of G1, we can observe that the features extracted from the first Gaussian with $\sigma = 0.8$ are minimal compared to the others. This is because, with a smaller deviation, the connections are more localized, resulting in a much lower average. The connectivity matrices obtained with $\sigma = 2.2$ and $\sigma = 4.8$, respectively, reveal more prominent areas. Both classes highlight the differences around the C3 and C4 areas, while the third kernel also emphasizes differences near the C5 region, despite some similarities between them. Subjects representing G2 (subject 25) and G3 (subject 27) highlight less localized differences in the sensorimotor area, corroborating the inter-subject variability issue in the MI task. Still, our GMRRNet can code a trade-off between performance in discrimination and localized features from Gaussian-based connections, based on how each subject MI-skill.

4.3   Limitations

Despite the promising results of the GMRRNet for EEG-based MI classification, there are several limitations and drawbacks that need to be addressed.

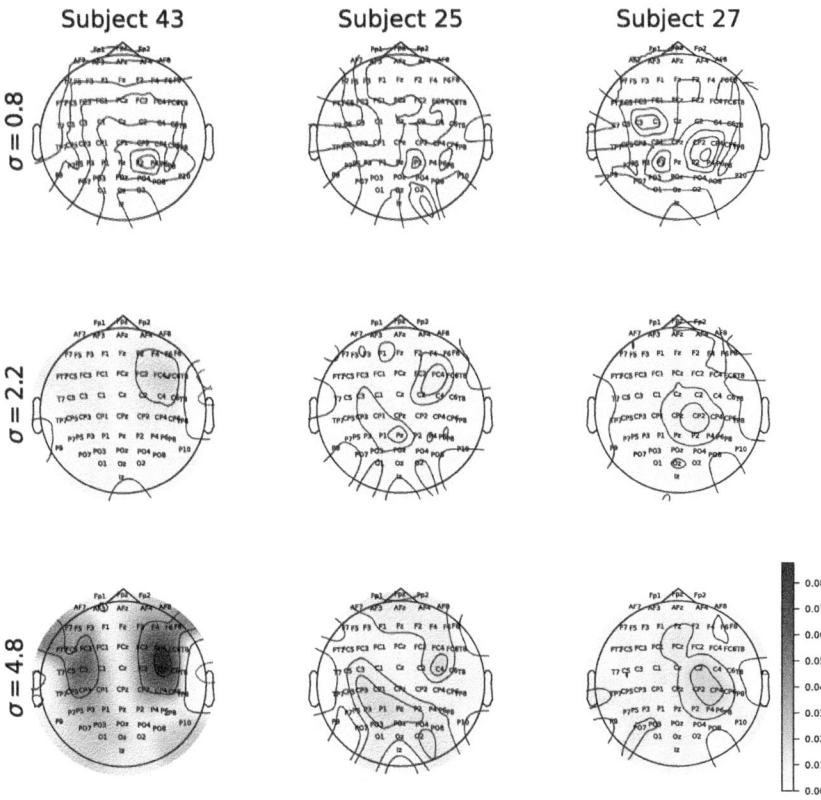

Fig. 6. Spatial interpretability results. Our GMRRNet learns the EEG channel features of the Gaussian mixture-based FC matrices, which we then use to create the brain topomap of the differences between left and right hand MI trials.

Firstly, while GMRRNet shows improved performance and stability compared to baseline models, e.g., EEGNet, it still exhibits variability in classification accuracy across different subjects. This indicates that the model may require further refinement to handle the diverse range of EEG signal characteristics more effectively. Additionally, the computational complexity of the model, especially due to the use of multiple Gaussian kernels and convolutional layers, can be a significant issue. This complexity can lead to longer training times and higher resource requirements, which may not be feasible for all applications, particularly in real-time or resource-constrained environments. Furthermore, although GMRRNet enhances interpretability by extracting spatial features through Gaussian kernels, the quantitative interpretability of the model's deeper layers and the specific contributions of these features to the classification decision remain somewhat opaque. Future research should focus on optimizing the model's architecture to reduce complexity, improving its adaptability to various EEG signal character-

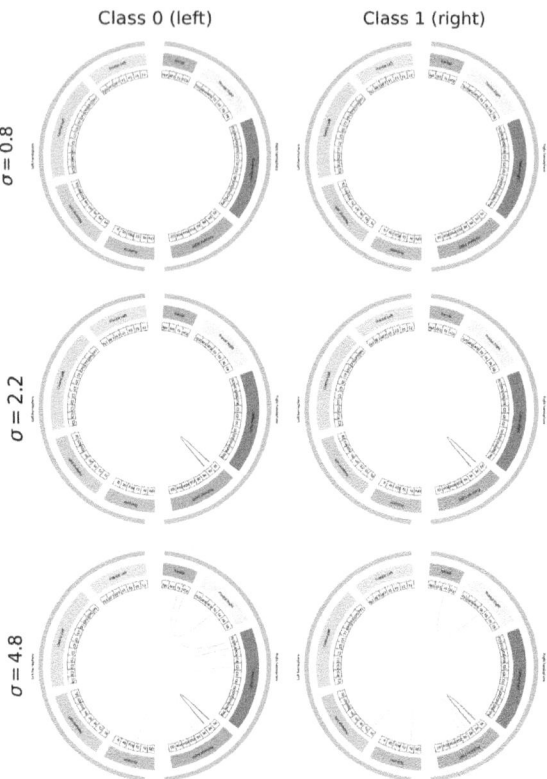

Fig. 7. GMRRNet-based average FC are presented for subject 43. Besides, it highlights in color the main parts of the brain: frontal left, frontal, frontal right, central right, posterior right, posterior, posterior left, central left. FC values higher than the 0.75 percentile are shown.

istics, and enhancing the transparency of its decision-making process to fully leverage its potential in practical BCI applications.

5   Conclusions

We proposed a Gaussian Mixture Functional Connectivity deep learning app-roach with α-Renyi Regularization enhancement, termed GMRRNet, which demonstrates significant potential for EEG-based motor MI classification. Our approach addresses key challenges in the field, including inter-subject and intra-subject variability, by effectively extracting both temporal and spatial fea-tures through a novel combination of convolutional layers and Gaussian ker-nels. Namely, we automatically extracts channel-based spatial features within a deep learning framework grounded on the Bockner's theorem for stationary ker-nels. Thereby, subjects were grouped based on their classification performance into high, medium, and low categories. The results indicated that GMRRNet generally outperformed the baseline EEGNet, with notable improvements in accuracy, particularly for subjects with medium and low MI skills. Besides, the intra-subject variability analysis showed that GMRRNet achieved greater consistency in classification accuracy across multiple trials compared to EEGNet. Specifically, GMRRNet exhibited lower standard deviation values, indicating reduced data dispersion and enhanced reliability in EEG signal interpretation. In turn, the spatial interpretability results of the GMRRNet model are highlighted through the analysis of EEG channel features, specifically examining the differences between left and right hand MI trials. Our study utilized Gaussian kernels with varying standard deviations to explore localized and global connections within the brain. For subjects with high MI performance, the results showed that features extracted with smaller deviations were minimal and highly localized, whereas larger deviations revealed more pronounced areas, particularly around the C3 and C4 regions. Hence, our Gaussian mixture can capture more significant spatial features.

GMRRNet effectively balances performance in MI classification with the extraction of interpretable spatial features, adapting to the individual skill levels and noise conditions present in the EEG data. Then, GMRRNet not only achieves competitive classification accuracy compared to existing state-of-the-art methods but also offers enhanced stability and interpretability. This makes it a promising tool for various BCI applications, including neurorehabilitation and prosthetic control, where robust and reliable performance is crucial. Future work will focus on further refining the model regarding the number of Gaussians, the bandwidth hyperparameter tuning, and the enhancement of the spatio-temporal representation. Then, Bayesian hyperparameter optimization [18] and variational approaches [38] will be considered. Besides, quantitative measures for network explainability will be studied based on class activation maps [24]. Evaluating the method's potential and applicability in different scenarios, such as serious games [8], emotion recognition [15], educational technology [30], cognitive workload assessment [6], and mental health monitoring [1], remains an essential aspect of its ongoing development.

Acknowledgements. Under grants provided by the program: "Alianza científica con enfoque comunitario para mitigar brechas de atención y manejo de trastornos mentales relacionados con impulsividad en Colombia (ACEMATE)-91908"—Project: "Sistema multimodal apoyado en juegos serios orientado a la evaluación e intervención neurocognitiva personalizada en trastornos de impulsividad asociados a TDAH como soporte a la intervención presencial y remota en entornos clínicos, educativos y comunitarios-790-2023", funded by Mincienicas. Also, authors thank to the project: "Sistema de visión artificial para el monitoreo y seguimiento de efectos analgésicos y anestésicos administrados vía neuroaxial epidural en población obstétrica durante labores de parto para el fortalecimiento de servicios de salud materna del Hospital Universitario de Caldas—SES HUC-(HERMES-57661)", funded by Universidad Nacional de Colombia.

References

1. Ali, A., Afridi, R., Soomro, T.A., Khan, S.A., Khan, M.Y.A., Chowdhry, B.S.: A single-channel wireless EEG headset enabled neural activities analysis for mental healthcare applications. Wirel. Pers. Commun. **125**(4), 3699–3713 (2022)
2. Alsharif, A., Salleh, N., Baharun, R., Safaei, M.: Neuromarketing approach: an overview and future research directions. JATIT **98**(7), 991–1001 (2020)
3. Altaheri, H., et al.: Deep learning techniques for classification of EEG MI signals: a review. Neural Comput. Appl. **35**(1), 14681–14722 (2023)
4. Bochner, S.: Harmonic Analysis and the Theory of Probability. UC Press (2020)
5. Chao, H., Liu, Y.: Emotion recognition from multi-channel EEG signals by exploiting the deep belief-conditional random field framework. IEEE Access **8**, 33002–33012 (2020)
6. Chikhi, S., Matton, N., Blanchet, S.: EEG power spectral measures of cognitive workload: a meta-analysis. Psychophysiology **59**(6), e14009 (2022)
7. Cho, H., Ahn, M., Ahn, S., Kwon, M., Jun, S.: EEG datasets for MI brain-computer interface. GigaScience **6**(7), gix034 (2017)
8. Delisle, D., Luiz, H., Carvalho, J.: authors: Multi-channel EEG-based BCI using regression and classification methods for attention training by serious game. Biomed. Signal Process. Control **85**, 104937 (2023)
9. García, D., Alvarez, A., Castellanos, G.: Single-trial kernel-based functional connectivity for enhanced feature extraction in motor-related tasks. Sensors **21**(8), 2750 (2021)
10. García, D., Álvarez, A., Castellanos, G.: KCS-FCnet: kernel cross-spectral functional connectivity network for EEG-based MI classification. Diagnostics **13**(6), 1122 (2023)
11. Janapati, R., Dalal, V., Sengupta, R.: Advances in modern EEG-BCI signal processing: a review. Mater. Today Proc. **80**, 2563–2566 (2023)
12. Kropotov, J.D.: Chapter 2.3 - beta and gamma rhythms. In: Kropotov, J.D. (ed.) Functional Neuromarkers for Psychiatry, pp. 107–119. Academic Press (2016)
13. Kumar, M.K., Parameshachari, B., et al.: Comparative analysis to identify efficient technique for interfacing BCI system. IOP Conf. Ser. Mater. Sci. Eng. **925**(1), 012062 (2020)
14. Lawhern, V.J., Solon, A.J., Waytowich, N.R., Gordon, S.M., Hung, C.P., Lance, B.J.: EEGNet: a compact convolutional neural network for EEG-based brain-computer interfaces. J. Neural Eng. **15**(5), 056013 (2018)
15. Li, X., et al.: EEG based emotion recognition: a tutorial and review. ACM Comput. Surv. **55**(4), 1–57 (2022)
16. Lionakis, E., Karampidis, K., Papadourakis, G.: Current trends, challenges, and future research directions of hybrid and deep learning techniques for BCI-MI. Multimodal Technol. Interact. **7**(10) (2023)
17. Ma, X., Huang, H., other: Normalized loss functions for deep learning with noisy labels. In: III, H.D., Singh, A. (eds.) ICML, vol. 119, pp. 6543–6553. PMLR (2020)
18. Miah, M.O., Habiba, U., Kabir, M.F.: ODL-BCI: optimal deep learning model for BCI to classify students confusion via hyperparameter tuning. Brain Disord. **13**, 100121 (2024)
19. Murphy, K.P.: Probabilistic Machine Learning: An Introduction. MIT press (2022)
20. Musallam, Y.K., et al.: Electroencephalography-based MI classification using temporal convolutional network fusion. Biomed. Signal Process. Control **69**, 102826 (2021)

21. Pandey, P., Miyapuram, K.P.: BRAIN2DEPTH: lightweight CNN model for classification of cognitive states from EEG recordings. arXiv:2106.06688 (2021)
22. Pérez, S., Santamaria, E., et al.: EEGSym: overcoming inter-subject variability in MI based BCIs with deep learning. IEEE Trans. Neural Syst. Rehabil. Eng. **30**, 1766–1775 (2022)
23. Philip, B.S., Prasad, G., Hemanth, D.J.: Non-stationarity removal techniques in meg data: a review. Procedia Comput. Sci. **215**, 824–833 (2022)
24. Poppi, S., Cornia, M., Baraldi, L., Cucchiara, R.: Revisiting the evaluation of class activation mapping for explainability: a novel metric and experimental analysis. In: Proceedings of the IEEE/CVF, pp. 2299–2304 (2021)
25. Rashid, M., et al.: Current status, challenges, and possible solutions of EEG-based brain-computer interface: a comprehensive review. Front. Neurorobotics **14**, 25 (2020)
26. Sannelli, C., Vidaurre, C., Müller, K.R., Blankertz, B.: A large scale screening study with a SMR-based BCI: Categorization of BCI users and differences in their SMR activity. PLoS ONE **14**(1), e0207351 (2019)
27. dos Santos, E.M., Cassani, R., Falk, T.H., Fraga, F.J.: Improved MI brain-computer interface performance via adaptive modulation filtering and two-stage classification. Biomed. Signal Process. Control **57**, 101812 (2020)
28. Schirrmeister, R., et al.: Deep learning with convolutional neural networks for EEG decoding and visualization. Hum. Brain Mapp. **38**(11), 5391–5420 (2017)
29. Seghier, M.L., Price, C.J.: Interpreting and utilising intersubject variability in brain function. Trends Cogn. Sci. **22**(6), 517–530 (2018)
30. Tang, H., Dai, M., Du, X., Hung, J.L., Li, H.: An EEG study on college students' attention levels in a blended computer science class. IET **61**(4), 789–801 (2024)
31. Tobón, M., Álvarez, A., Castellanos, G.: Subject-dependent artifact removal for enhancing MI classifier performance under poor skills. Sensors **22**(15) (2022)
32. Tortora, S., Ghidoni, S., Chisari, C., Micera, S., Artoni, F.: Deep learning-based BCI for gait decoding from EEG with LSTM recurrent neural network. J. Neural Eng. **17**(4), 046011 (2020)
33. Veena, N., Anitha, N.: A review of non-invasive BCI devices. Int. J. Biomed. Eng. Technol. **34**(3), 205–233 (2020)
34. Wang, H., et al.: Diverse feature blend based on filter-bank common spatial pattern and brain functional connectivity for multiple MI detection. IEEE Access **8**, 155590–155601 (2020)
35. Wei, X., Dong, E., Zhu, L.: Multi-class MI-EEG classification: using FBCSP and ensemble learning based on majority voting. In: Proceedings of the 2021 China Automation Congress (CAC), pp. 872–876. IEEE (2021)
36. Xie, Y., Oniga, S.: Classification of MI EEG signals based on data augmentation and convolutional neural networks. Sensors **23**(4) (2023)
37. Yu, S., Giraldo, L.G.S., Jenssen, R., Principe, J.C.: Multivariate extension of matrix-based rényi's α-order entropy functional. IEEE Trans. Pattern Anal. Mach. Intell. **42**(11), 2960–2966 (2019)
38. Zancanaro, A., Zoppis, I., Manzoni, S., Cisotto, G.: veegnet: a new deep learning model to classify and generate EEG. In: ICT4AWE, vol. 2023, pp. 245–252. Science and Technology Publications (2023)

Optimisation of User Interaction for the Calculation of the Cobb Angle in Lumbar Spine MRI Images

William A. Romero R.[1]([envelope])[iD], Susana Uribe Velásquez[2],
Daniel Restrepo Quiñones[2], and Duván A. Gómez[2][iD]

[1] University of Montpellier, CARTIGEN, CHU de Montpellier, Montpellier, France
william.romero@umontpellier.fr
[2] Universidad EIA, Medellin, Colombia
{susana.uribe7,daniel.restrepo53,duvan.gomez36}@eia.edu.co

Abstract. Clinical studies conducted on large populations of patients require computer-based tools to efficiently and accurately process and interpret large amounts of imaging data. However, certain tasks are still performed manually. As a result, when a manual task needs to be repeated across a cohort of hundreds of patients, it can become a bottleneck in the analysis workflow. One example of these manual tasks is the computation of the Cobb angle. To compute the Cobb angle, users typically identify four points on spine images: two points for the tangent line to the upper end-plate of the upper vertebra and two points for the tangent line to the lower end-plate of the lower vertebra. This process requires four mouse clicks (a 4-point definition) for user interaction with the graphical interface. This article introduces a method where users need to specify only two points on the image for Cobb angle calculation, significantly streamlining the process. The main stages include: selecting vertebrae using two mouse click; extracting sub-images, segmenting vertebrae, and calculating bounding boxes; and finally, calculating tangent lines and determining the Cobb angle. Such efficiency is vital when managing images from a large patient cohort in a clinical study. The computational method has been published in an extension module for 3D Slicer. The evaluation of the method was performed on magnetic resonance images of the lumbar spine. The proposed method reduced the required interaction time by about 50% while maintaining an accuracy of 90%.

Keywords: Cobb Angle · Lumbar MRI · Semi-Automatic Segmentation

1 Introduction

In 2020, 619 million people worldwide were affected by Low Back Pain (LBP), and projections estimate this number will rise to 843 million by 2050. This

N. D. Duque-Méndez et al. (Eds.): CCC 2024, CCIS 2208, pp. 148–155, 2024.
https://doi.org/10.1007/978-3-031-75233-9_11

increase is primarily due to population growth and aging [5]. According to the World Health Organization (WHO), LBP is the leading cause of disability globally [16]. Consequently, many individuals could benefit from novel physical assistance devices and treatments for improving their motor skills.

In addition, Magnetic Resonance Imaging (MRI) is a valuable tool for lower spinal imaging, enabling detailed assessment of soft tissues, including intervertebral discs, nerves, and spinal cord abnormalities. The use of low-field MRI scanners (0.1–0.3T) is becoming increasingly common, offering several advantages compared to high-field MRI systems (1.0–3.0T). These advantages include reduced costs and maintenance, making low-field MRI scanners more accessible and affordable for smaller clinics and healthcare facilities [2,13]. By analysing MRI images, clinical researchers can gain insights into the pathophysiology of diseases, refine diagnostic criteria, and improve therapeutic strategies.

In clinical studies involving large patient populations, image analysis becomes critically important. This has led to a growing interest in automated segmentation methods, particularly those based on deep learning [8,12], to efficiently and accurately process vast amounts of imaging data. However, some tasks still require manual intervention. When a manual task must be repeated across a cohort of hundreds of patients, it can become a bottleneck in the analysis workflow. An example of such a manual task is the computation of the Cobb angle, which is used to assess the degree of spinal curvature, especially in scoliosis cases. The Cobb angle is determined by drawing lines parallel to the upper and lower end-plates of the vertebrae of interest, with the angle between these lines indicating the extent of spinal curvature (Fig. 1). Most medical image analysis software applications require the user to identify four points (i.e. Weasis DICOM viewer [14]): two for the tangent line to the upper end-plate of the upper vertebra and two for the tangent line to the lower end-plate of the lower vertebra (Fig. 1).

This paper presents a semi-automatic method for vertebrae segmentation and automatic Cobb angle calculation between two user-selected vertebrae. The primary objective of this study is not vertebral segmentation itself; rather, vertebral segmentation is utilised as an intermediate step to geometrically define the parameters necessary for calculating the Cobb angle between two selected vertebrae within the L1-L5 region, using MRI images obtained from a low-field scanner (0.25T). The method requires the user to identify only two points on the image: one for the upper vertebra and one for the lower vertebra as presented in Fig. 1. Vertebrae segmentation is performed by a U-Net neural network [15] trained to predict vertebrae within a sub-image and simplify the problem to a single class segmentation. This solution has been released as a module extension for 3D Slicer [4].

2   Materials and Methods

2.1   Magnetic Resonance Images of the Lumbar Spine

In vivo, sagittal fast spin-echo (FSE) images of the lumbar spine were used to assess the method. These images were acquired using a low-field scanner (Esaote

Baseline algorithm   **Proposed algorithm**

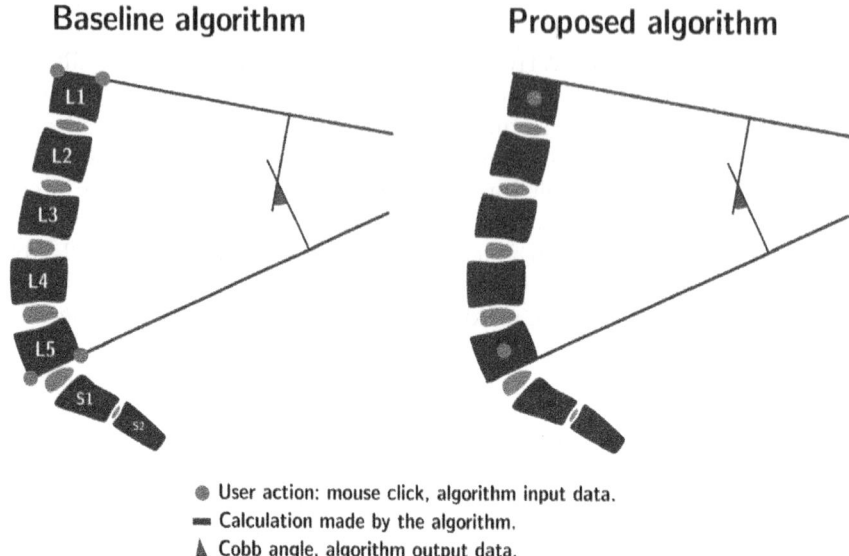

● User action: mouse click, algorithm input data.
▬ Calculation made by the algorithm.
▲ Cobb angle, algorithm output data.

Fig. 1. Schematic representation (on the sagittal plane) of user actions (input data represented by orange dots), calculations performed by the algorithm (blue lines and segmentation), and the Cobb angle (algorithm output represented by a green angle). The baseline algorithm is the default pathway in most image analysis tools. The proposed algorithm reduces the number the user actions. (Color figure online)

G-scan Brio 0.25T). Relevant imaging parameters: TE = 125 ms, TR = 3000 ms, echo train lenght = 6, in-plane resolution = 0.6 mm, slice thickness = 4 mm.

2.2 Semi-automatic Vertebra Segmentation

Vertebrae segmentation was performed using a U-Net, convolutional neural network [15]. In this process, a sub-image of size 96 × 96 pixels was fed into the U-Net model to narrow the problem down to a single class segmentation. The extraction of the sub-image is based on the localisation in the image space made by the user.

Sub-images were normalised by rescaling pixel values to a specific range [0, 1] using the min-max normalisation technique. To expand the training data, data augmentation techniques, such as random pixel dropout, rotation, flipping, and elastic transformation, were applied. The feature maps from five scales were then aggregated to merge shallow and deep semantic information. This was achieved by applying 90 filters of size 3 × 3.

2.3 Cobb Angle

The Cobb angle is a measurement used to quantify the degree of spinal curvature, particularly in cases of scoliosis [3]. It is defined as the angle formed between

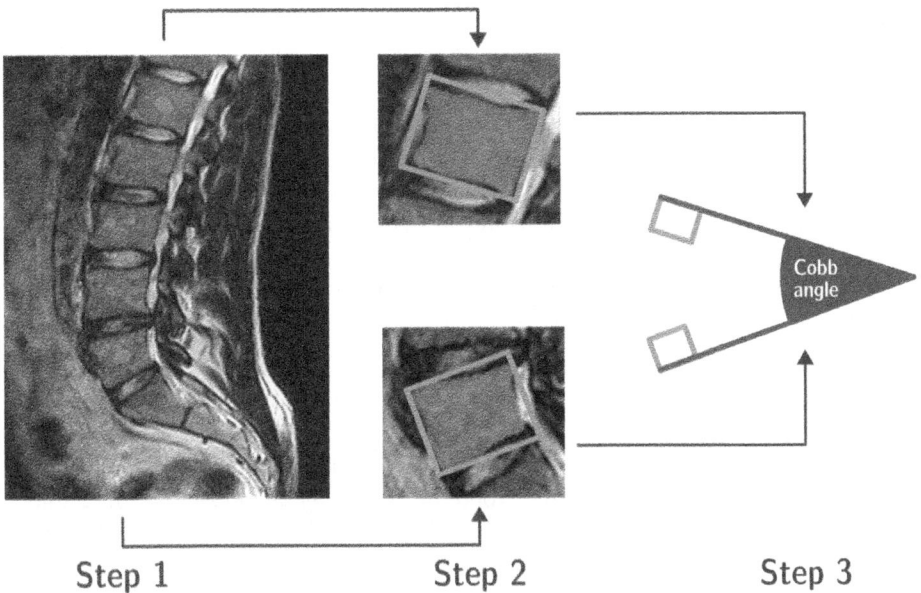

Fig. 2. Cobb angle calculation workflow. Stage 1: select vertebrae using a mouse click. Stage 2: extract sub-images, segment vertebrae, and calculate bounding boxes. Stage 3: calculate tangent lines and determine the Cobb angle.

vertebrae at the upper and lower ends of the curve. The Cob angle θ can be calculated as

$$\theta = \arctan\left(\frac{|m_1 - m_2|}{1 + m_1 \cdot m_2}\right) \qquad (1)$$

where m_1 and m_2 are the slopes of the lines tangent to the upper endplate of the upper vertebra and the lower endplate of the lower vertebra, respectively (Fig. 2). There are different models for the calculation of the Cobb angle [6] from the original model introduced in 1948 [3].

2.4 Measurement of User Interaction Time to Calculate the Cobb Angle

The time spent by an experienced user calculating the Cobb angle was measured sequentially across 10 different sagittal lumbar MR images. The user performed these measurements using three different tools provided by the Weasis DICOM viewer [14]: *Cobb's angle* (a 4-point definition), *Open angle* (a 3-point definition), and *Angle* (a 3-point definition).

The Cobb angle calculated with the 4-point method is taken as a reference ($\theta_{\text{reference}}$) for calculating the relative error (RE) as

$$RE = \frac{\theta_{\text{target}} - \theta_{\text{reference}}}{\theta_{\text{reference}}} \times 100 \tag{2}$$

where θ_{target} is the Cobb angle calculated with the method for evaluation. A module extension for 3D Slicer (LS-SEG) has been developed in python using SimpleITK [11] and TensorFlow [1]. This module includes the proposed method for vertebrae segmentation and automatic Cobb angle calculation between two user-selected vertebrae (2 points). The time spent by an experienced user calculating the Cobb angle with LS-SEG was measured on the same set of images used on Weasis DICOM viewer.

3   Results

The average time of user interaction for calculating the Cobb angle across different interaction tools (as described in Sect. 2.4) is presented in Table 1. When using the *Angle* tool, after the user defines the three base points, they must readjust the angle (by dragging with the mouse) to ensure the lines are tangent to the upper and lower end-plates of the vertebrae of interest. This adjustment adds approximately 5 s.

The Fig. 3 presents the measurement of user interaction time to calculate the Cobb angle across images. A slight increase in user interaction time is observed in the 4-Point method from image 7 out of 10. This may be due to cognitive or visual fatigue of the user. The proposed method reduced the required interaction time by about 50% while maintaining an accuracy of 90%.

Table 1. Measurement of the average user interaction time to calculate the Cobb angle: relative error was calculated using Eq. 2. Time measurements are presented in seconds (s) and milliseconds (ms).

Method	Average time (s:ms) ± standard deviation (ms)	Relative error (%)
Cobb's angle (4-point)	4:600 ± 500	Reference
Open angle (3-point)	1:860 ± 100 + 3:400 adjustment	0.9
Angle (3-point)	1:600 ± 100 + 5:000 adjustment	1.1
Proposed method (2-point)	1:810 ± 91	0.9

4   Discussion

The Cobb angle is the standard method for quantifying the degree of scoliosis. The clinical standard imaging protocol for scoliosis assessment primarily involves the use of standing posteroanterior (PA) and lateral radiographs of the spine including: Cervical Spine (C1-C7), Thoracic Spine (T1-T12) and Lumbar Spine

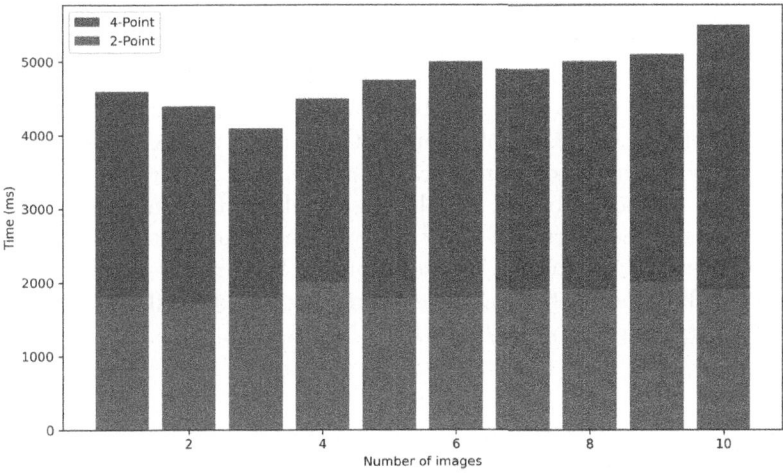

Fig. 3. Measurement of user interaction time (4-Point vs 2-Point) across images. Average user interaction time presented in milliseconds (ms).

(L1-L5) [7]. This study was conducted using MRI images obtained from a low-field scanner (0.25T), focusing on the lumbar spine, specifically from L1 to L5.

While several deep-learning (DL) methods for automatic vertebral segmentation and spinal curvature calculation exist [8,12]; this work aims to improve a manual task frequently performed in Quality Assurance and Quality Control (QA/QC) procedures on lumbar MR images (L1-L5). The aim of this work is not vertebral segmentation itself; rather, vertebral segmentation serves as an intermediate step in the geometric definition required for calculating the Cobb angle between two arbitrary vertebrae within L1-L5.

This study measured the time required to define the geometric coordinates necessary for the Cobb angle algorithm on an image. In practice, the image analysis time encompasses selecting the DICOM series in the corresponding patient study, viewing the image of interest (loading it into the DICOM viewer), and executing the task. When this procedure is performed on a cohort of over 100 patients, the marginal time saved on a single patient becomes significant across the entire study.

It is well known that diagnostic accuracy significantly decreases due to visual fatigue and cognitive load after a working day [9,10]. The proposed method has the potential to reduce both cognitive and visual load. Rather than focusing on four specific points, it is much easier to indicate the area corresponding to the vertebrae of interest. This presents an interesting hypothesis to be validated in a future study.

In this study, participants had prior knowledge of the experiment, which may have biased them to define the angles as quickly as possible. Therefore, we consider that the measured times cannot be generalised. However, the measured times do indicate the extent to which this process can be accelerated.

5 Conclusion

A method for reducing the number of actions (mouse clicks) required by the user to calculate the Cobb angle has been described. This method only requires the selection of the vertebrae of interest. Once the user has selected the vertebrae of interest, the algorithm extracts sub-images, segments the vertebrae, and calculates bounding boxes. Finally, it calculates tangent lines (the upper end-plate of the upper vertebra and the lower end-plate of the lower vertebra) and determines the Cobb angle. The solution has been released within a module extension for 3D Slicer: LS-SEG. The source code and production version are available at https://github.com/CARTIGEN/LS-SEG.

References

1. Abadi, M., et al.: Tensorflow: large-scale machine learning on heterogeneous distributed systems. arXiv preprint arXiv:1603.04467 (2016)
2. Arnold, T.C., Freeman, C.W., Litt, B., Stein, J.M.: Low-field MRI: clinical promise and challenges. J. Magn. Reson. Imaging **57**(1), 25–44 (2023)
3. Cobb, J.: Outline for the study of scoliosis. Instructional course lecture (1948)
4. Fedorov, A., et al.: 3D slicer as an image computing platform for the quantitative imaging network. Magn. Reson. Imaging **30**(9), 1323–1341 (2012)
5. Ferreira, M.L., et al.: Global, regional, and national burden of low back pain, 1990–2020, its attributable risk factors, and projections to 2050: a systematic analysis of the global burden of disease study 2021. Lancet Rheumatol. **5**(6), e316–e329 (2023)
6. Hellen, D.W., Teusner, M.: Two new improved methods for determination of cobb's angle of scoliosis. medRxiv, pp. 2020–06 (2020)
7. Kim, H., et al.: Scoliosis imaging: what radiologists should know. Radiographics **30**(7), 1823–1842 (2010)
8. Kim, K.C., Cho, H.C., Jang, T.J., Choi, J.M., Seo, J.K.: Automatic detection and segmentation of lumbar vertebrae from X-ray images for compression fracture evaluation. Comput. Methods Programs Biomed. **200**, 105833 (2021)
9. Krupinski, E.A., Berbaum, K.S., Caldwell, R.T., Schartz, K.M., Kim, J.: Long radiology workdays reduce detection and accommodation accuracy. J. Am. Coll. Radiol. **7**(9), 698–704 (2010)
10. Lee, C.S., Nagy, P.G., Weaver, S.J., Newman-Toker, D.E.: Cognitive and system factors contributing to diagnostic errors in radiology. Am. J. Roentgenol. **201**(3), 611–617 (2013)
11. Lowekamp, B.C., Chen, D.T., Ibáñez, L., Blezek, D.: The design of simpleITK. Front. Neuroinform. **7**, 45 (2013)
12. Lu, J.T., et al.: Deep spine: automated lumbar vertebral segmentation, disc-level designation, and spinal stenosis grading using deep learning. In: Machine Learning for Healthcare Conference, pp. 403–419. PMLR (2018)
13. Marques, J.P., Simonis, F.F., Webb, A.G.: Low-field MRI: an MR physics perspective. J. Magn. Reson. Imaging **49**(6), 1528–1542 (2019)

14. Roduit, N: Weasis dicom viewer. version 4.4.0 (nd). https://github.com/nroduit/Weasis. Accessed 10 June 2024
15. Ronneberger, O., Fischer, P., Brox, T.: U-net: convolutional networks for biomedical image segmentation. In: MICCAI 2015, Part III, pp. 234–241. Springer (2015)
16. World Health Organization: Low back pain fact sheets (nd). https://www.who.int/news-room/fact-sheets/detail/low-back-pain. Accessed 10 June 2024

Implementation of Convolutional Neural Networks for Automated Disease Detection in Cucumber Crops

Andrea Menco Tovar[(✉)][iD], Edwin Puertas[(✉)][iD],
and Juan Carlos Martinez-Santos[iD]

Universidad Tecnológica de Bolívar, Cartagena, Colombia
{amenco,epuerta,jcmartinezs}@utb.edu.co

Abstract. This study presents an automated system for detecting diseases in cucumber crops using Convolutional Neural Networks (CNN). Given the importance of agriculture and the challenges crops face due to pests and diseases, a vision-based approach is proposed to improve the early and accurate identification of diseases in cucumbers. Three CNN architectures were evaluated: Xception, VGG16, and ResNet50, using a balanced dataset of images of healthy and diseased cucumber leaves. The Xception model showed the best performance with an accuracy of 93.45% and a loss of 0.4842, surpassing the other models. Image preprocessing and transfer learning were key to achieving these results. Despite the good results, challenges were identified in accurately classifying some images, suggesting areas for future improvement. This system provides a valuable tool for farmers, enabling early detection and rapid decision-making to control diseases, which can significantly improve crop quality and yield. Future research could integrate this system with mobile technologies and drones for more efficient real-time monitoring.

Keywords: Disease detection · Transfer learning · Artificial intelligence in agriculture · Convolutional Neural Networks (CNN) · Cucumber crops

1 Introduction

Agriculture plays a vital role in humanity; without it, it would be impossible to feed the world's population. It is also considered the backbone of the economy of countries, especially developing countries [5,6]. Agricultural development is one of the most important means to end extreme poverty, boost shared prosperity, and feed the entire population [3]. However, this food security is affected by diseases and pests, causing crop yield losses [19].

Plant diseases affect the quality and yield of agriculture [6]. Cucumber *(Cucumis sativus L.)* is a vegetable belonging to the cucurbit family exceptionally susceptible to several formidable pests, bacterial and fungal diseases that can rapidly devastate entire fields if left unchecked [18]. By 2020, the global cucumber

N. D. Duque-Méndez et al. (Eds.): CCC 2024, CCIS 2208, pp. 156–167, 2024.
https://doi.org/10.1007/978-3-031-75233-9_12

planting area was estimated to be about 2.25 million ha and the global yield was 90.35 million tons [8]. Several potential diseases threaten cucumber crops, and lack of identification and prevention causes a significant reduction in vegetable yield and quality [8]. Frequently described and known phytopathogens for this crop worldwide are anthracnose caused by the fungus *Colletotrichum orbicularis*, powdery mildew, leaf spot, downy mildew, and cucumber mosaic [17].

Generally, traditional manual early diagnostic methodologies are usually laborious, time-consuming, and subjective, resulting in low efficiency and misdiagnosis. On the other hand, techniques based on laboratory tests and biosensors are difficult for farmers to implement due to lack of experience and high cost [2,8,17]. The complexity of some recognition methods and resource availability make monitoring and control challenging. Currently, due to technological advances, there are alternative methods such as machine vision systems for automated detection and classification of crop diseases from color transformations, support vector machines, artificial neural networks, and probabilistic techniques that together with Deep Learning are successfully used for automatic recognition of plant diseases [6,19]. Deep learning techniques, in particular Convolutional Neural Networks (CNN), are a suitable method in terms of accuracy, well-defined results, and precision for detecting pests and diseases given their strong ability to identify objects and establish patterns in images, thus optimizing crop management [10].

In our research, we utilized the database available in the Kaggle competition Cucumber Plant Diseases Dataset to develop a computer vision system for the early and reliable detection and identification of cucumber diseases. This system aims to provide an effective and accessible tool for farmers to improve the management and control of pests and diseases in these crops, thus optimizing the yield and quality of vegetables. To this end, we will apply advanced deep learning techniques to achieve an automated and reliable diagnosis. The remainder of this article is organized as follows: Sect. 2 presents the conceptual framework. Section 3 presents the state of the art. Section 4 includes the methodology, which is broken down into three parts: Dataset, Preprocessing, and Technique. The results of this proposed work are presented in Sect. 5. Finally, Sect. 6 concludes the paper.

2   Conceptual Framework

Plant Disease
A plant disease is defined as an abnormal condition that affects the growth, development, or function of a plant. These diseases are typically caused by pathogenic organisms, including fungi, bacteria, viruses, and nematodes, as well as by adverse environmental factors. The manifestations of plant diseases may be visible or non-visible. Visible symptoms include leaf spots, discoloration, wilting, abnormal growth, necrosis, and premature leaf or fruit drop. These symptoms are readily discernible and serve as an alert to the grower that a disease may be present. However, some symptoms are not visible to the naked eye, such as loss of

vigor, which may be reflected in a decrease in flower or fruit production. In more severe cases, the disease can result in the death of the plant. It is important to note that both visible and non-visible symptoms are equally important, as they can indicate a problem plant. Diseases represent a significant challenge for agriculture and horticulture, as they can seriously compromise crop productivity and health if not properly managed [5, 21].

Common Cucumber Plant Diseases
Anthracnose
Anthracnose is a disease that can affect various parts and aerial parts of plants. Its symptoms vary depending on the cucurbit infected. On cucumber leaves, irregular brown spots develop, with a tendency for the center of the spot to be lost. The latter usually shows yellow halos around the spots. Infections on elongated and sunken stems can occur on cucumbers, being characterized by the exudation of a reddish gum. On fruit, they manifest as black sunken spots measuring between one-quarter and one-half inch in width and depth, with white mycelia and salmon-colored sticky spores becoming visible during periods of high humidity [12].

Powdery Mildew
The disease caused by the pathogen *Pseudoperonospora cubensis* affects all Cucurbitaceae, both edible and inedible-skinned, both outdoors and in greenhouses. The pathogen requires free water to disperse spores and cause infection. Its mycelium is colorless, and the sporangia are gray and visible on the underside of leaves. Symptoms appear on the leaves, initially as light green spots that later turn yellow with angular shapes on the upper side, while the underside shows the development of mycelium and sporangia. Eventually, these spots become necrotic, and the leaves dry up completely [4].

Leaf Spot
Leaf spot of cucumber, symptoms begin as small, yellow, lunate spots on the leaves which gradually enlarge to about 1 cm (0.4 in.) becoming angular and occasionally attacking petioles and stems. The fungus *C. cassiicola* has a wide host range in tropical and subtropical climates, damaging both leaves and fruit [21].

Downy Mildew
Downy mildew can impact plants at all stages of development, although it usually manifests late in the season in many growing areas. Infection is primarily concentrated in the leaves, but the resulting reduction in photosynthetic capacity causes reduced plant growth, lower yields, and fruit damage after leaf drop. The first signs manifest as small chlorotic spots on the upper surface of leaves, usually starting on older leaves. These lesions expand and take on an angular shape as they follow the main veins of the leaves. Finally, the affected tissue necroses, causing leaf death and leaf drop. A gray to purple downy mildew develops on the underside of the leaves [15].

Cucumber Mosaic

Cucumber mosaic virus (CMV) is a member of the genus Cucumovirus and is widely distributed globally, occurring in both temperate and tropical climates. However, it is unable to survive in extremely dry conditions. Infection results in severe damage to the host plant. Symptoms induced by cucumber mosaics include light green to dark green mosaics, widespread chlorosis, stunting, leaf filiform, and local chlorotic lesions. These symptoms are host-specific. Cucumber mosaic is transmitted by aphids and parasitic weeds [1,13].

3   State of the Art

In related work, several techniques for detecting and classifying plant leaf diseases have been presented, providing evidence that the identification and diagnosis of cucumber leaf disease symptoms have made great advances, based on machine learning algorithms and computer vision.

Recent research has used CNN to detect cucumber diseases or pests from images. [16] conducted their work using a dataset built by the authors with a total of 4868 images that included diseases and pests such as spider mites, leaf miners, downy mildew, powdery mildew, a viral disease, and healthy class leaves. They developed a CNN built from scratch, achieving 98.19% recognition accuracy and 100% recognition accuracy with the public plant disease dataset.

Following the trend of existing solutions, [20] implemented an algorithm based on CNN called You Only Look Once-YOLO v4 to detect, identify, and classify objects in real-time, obtaining through this an accuracy of 80% considering these results very positive in the application of artificial intelligence in Vietnam. [9] used the YOLO v2 model combined with the ONNX (Open Neural Network Exchange) model to analyze deep features, obtaining a classification model with an accuracy of 96.4%.

Feature extraction is the essential step to obtain important information. Color, shape, geometric features, and texture are features that are extracted from images to determine diseased crops [10]. [7] applied various traditional machine learning algorithms using K-means and random forest-based image segmentation offering an accuracy of 89, 93%, on the other hand, they applied CNN-based transfer learning comparing models such as InceptionV3, MobileNetV2, and VGG16, where MobileNetV2 achieves the highest accuracy with 93.23 %.

Zhang et al. [14]demonstrated the recognition of seven types of diseases in cucumber plant leaves, for this they used the K-means clustering algorithm to extract shape and color features and then classify them by sparse representation, their method reached an overall accuracy of 85.7% [11]. In this opportunity they worked with images of greenhouse cucumber diseases, building a model from a state-of-the-art EfficientNet method obtaining an accuracy of 97% for the classification of powdery mildew, downy mildew, healthy leaves, and the combination of powdery mildew and downy mildew.

4   Methodology

4.1   Dataset

In this project, we utilized the database available on the Kaggle website, designated as the "Cucumber Plant Diseases Dataset". This dataset consists of three subfolders, each containing distinct data sets: training, test, and single prediction. The training data set comprises 242 images of healthy cucumber leaves and 227 images of diseased cucumber leaves. The test data set consists of 123 images of healthy cucumber leaves and 99 images of diseased cucumber leaves. The single prediction data set comprises two images of healthy cucumber leaves and two images of diseased cucumber leaves, resulting in a total storage requirement of 797 megabytes. It is important to note that only models with training and test data were evaluated, Fig. 1.

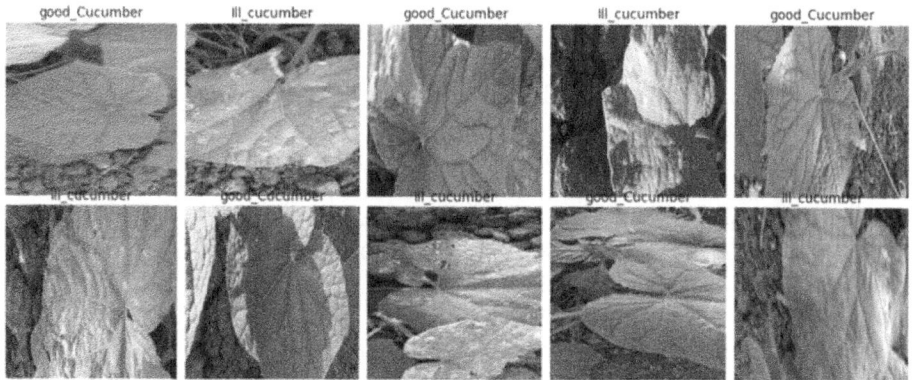

Fig. 1. Sample images from the cucumber leaf image dataset.

The data is divided only between diseased and healthy leaves, with no labeling to discriminate which type of disease is the subject of study. In light of the existing research on cucumber leaf diseases and the characteristics of the images under study, it can be reasonably concluded that the disease in question is Alternaria Leaf Spot. This is because infected leaves exhibit yellow spots, which may be large or small.

4.2   Preprocessing

First, features were extracted from the training and test images using the extract_features function. Next, data generators were created for each set of images using flow_from_directory from Keras. The images were then resized to a size of 224 × 224 pixels for VGG16 and ResNet50, 299 × 299 pixels for Xception. To ensure that the images have the same dimensions and fit the input format required by the model.

Subsequently, the images were normalized by dividing the pixel values by 255, thus ensuring that the values fall within the range [0,1]. For VGG16 and ResNet50, it was necessary to subtract specific mean RGB values from all images in the dataset. It should be noted that these mean values are provided by the model architecture documentation.

Finally, the images were grouped into batches for efficient processing during training. Batches of a fixed size, containing a specific number of images, were created. Finally, the base layer of the convolutional network was frozen to prevent the base layer weights from being trained and to capitalize on the knowledge stored in the base subnetwork.

4.3   Technique

Consider the advantages of CNN applied to image processing, this paper presents three CNN architectures such as VGG16, ResNet50 and Xception, with a transfer learning approach, the selection of these architectures was based on their good performance in similar plant disease classification problems [5,6] Fig. 2.

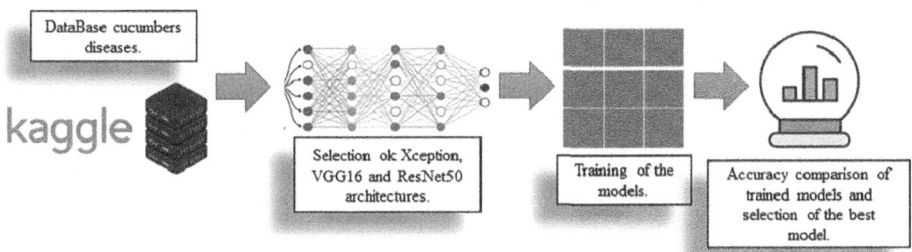

Fig. 2. Convolutional Neural Network. Proposed method for cucumber disease recognition.

Xception Model
The general structure of Xception comprises a separable depth convolution, followed by a depthwise convolution. Subsequently, a 3×3 convolution is performed independently for each input channel, whereby the filter is applied to each channel separately. Subsequently, a pointwise convolution, defined as a $1 \times 11 \times 11 \times 1$ convolution, is applied for the purpose of combining the features extracted from the depthwise convolution. This operation results in a transformation of the number of output channels. A $1 \times 11 \times 11 \times 1$ filter is applied to each of the output channels. The network is then constituted by blocks comprising a series of separable convolutional layers. Each block comprises a depthwise convolution followed by a pointwise convolution. The blocks are arranged in sequence to form the complete network. Additionally, Xception employs residual connections to facilitate training. Residual connections permit the original input of a block to be added to the block's output after passing through several layers, thereby mitigating the issue of vanishing gradients and accelerating the training process.

First, the image data generators for training and testing were defined, to apply specific transformations to the training data, such as resizing, 45° rotations, vertical and horizontal rotations, and resizing only to the test data. Then, the pre-trained Xception model was loaded and customized by adding additional layers to adapt it to the specific task. The model was compiled by specifying the Adam optimizer and the sparse categorical cross-entropy loss function. Stream data generators were created for the training datasets by loading images from specific directories and splitting them into batches for processing. Finally, cross-validation is performed using the K-fold method to evaluate model performance on different subsets of training and test data.

Model VGG16

The overall structure of VGG16 comprises 16 learning weight layers, 13 of which are convolutional. The aforementioned layers apply kernels to the input images to extract features. After each convolutional layer, a ReLU (Rectified Linear Unit) activation function is applied to introduce non-linearity. Subsequently, three fully connected layers perform the same function as in a traditional neural network, whereby each neuron is connected to all neurons in the preceding layer. Additionally, clustering layers reduce the spatial dimension of the extracted features, retaining only the most pertinent ones. Finally, a softmax layer for classification is employed, responsible for converting the outputs into probabilities.

It starts with the weights pre-trained in ImageNet. The last layers of the network are excluded, to allow customization of the model. Next, all layers of the VGG16 model are frozen to ensure that the weights of these layers are not updated during training. Then, a new top layer is added to the model, consisting of a global clustering layer, a dense layer with ReLU activation, and a dense output layer with Softmax activation for binary classification. The new model is compiled using the Adam optimizer, the categorical Cross Entropy loss function, and the accuracy metric. Subsequently, the test data is loaded, resized to the specified dimensions, and split into batches for processing.

Model ResNet50

The general structure of ResNet50 is composed of an input layer, which is in turn made up of a convolutional layer using filters of size $7 \times 77 \times 77 \times 7$ with a stride of 2. This layer has 64 filters and is followed by a batch normalization layer and a ReLU activation function. Subsequently, a max pooling layer with a size of $3 \times 33 \times 33 \times 3$ and a stride of 2 is applied, allowing the spatial dimensions of the image to be reduced and thus facilitating the preservation of the most significant features. Subsequently, the network is divided into four principal stages, each comprising multiple residual blocks. Each residual block is designed to learn a residual function in lieu of the direct mapping function, thereby facilitating the training of the deep network. The first stage contains three residual blocks, each of which has 64 filters in each convolutional layer. The second stage is composed of four residual blocks, each of which the third stage comprises six residual blocks, each with 256 filters in the convolutional layers. The fourth stage is constituted by three residual blocks, each with 512 filters in the convolutional layers. In the third stage, each residual block has 256 filters in the convolutional layers. Subsequently, a global average pooling layer is applied, whereby the spatial dimensions are reduced to 1×1 by averaging the values in each feature map. This results in a compact representation of the image. The network then concludes with a fully connected layer, which generates the output for classification.

First, the loss function is imported to measure the discrepancy between the true labels and the model predictions during training. Then, the model is compiled using the compile method. Here, the optimizer to be used to adjust the model weights during training is specified in this case, Adam. After compiling the model, we proceed to train it with training data using the fit method. The training data generator (train_generator), the number of training epochs (in this case, 10), and the validation data generator are specified. During training, the model will adjust its weights to minimize the specified loss function.

Once training is complete, the model is evaluated on the test data using the evaluate method. This involves passing the test data generator (test_generator) to the model, which will calculate the loss and accuracy of the model on the test data. Finally, the percentage loss and accuracy obtained from the model evaluation is printed out. This provides a quantitative measure of the model's performance on the test data, allowing further evaluation of its ability to generalize to unseen data.

For transfer learning applied to each model, Imagenet weights were used, where the last layer of each previously trained model is exchanged for a pooling layer to reduce dimensionality, a flattened layer that converts the output into a one-dimensional vector and a vector hidden dense layer with 1024 neurons, followed by a dense hidden layer with 512 neurons.

The tests were performed on a computer with 8 GB of Ram and a twelfth-generation IntelCore i5 processor. The programming language was Python 3.10.13 and the libraries used were Tensorflow version 2.15, Matplotlib, NumPy, Scikit-learn.

5   Results

Table 1 presents a comparative analysis of the three selected models, Xception, VGG16, and ResNet50, based on the results obtained by evaluating them with the test data set (469 images that were not used in training). The same dataset described in the previous section was employed for training each of the models. To identify the optimal model, the highest value of the accuracy metric was considered. This metric measures the percentage of correct measurements and is therefore highly reliable, given that the data set is not excessively imbalanced.

Table 1. Results of the evaluation of each model with the test set.

Model	Accuracy (%)	Loss
Model 1: Xception	93.45	0.4842
Model 2: VGG16	55.41	0.7032
Model 3: ResNet50	45.00	0.2662

It is observed that the Xception model presents an accuracy of 93, 45 % and a loss of 0, 4842 surpassing the other previously trained models. It is evidenced that the training loss decreases steadily during the first 5 epochs, then it stabilizes. The test loss increases significantly in epoch 8, t h e training accuracy increases steadily during the first 5 epochs and then stabilizes, and finally, the test accuracy also shows an irregular trend, with inconsistent performance Fig. 3.

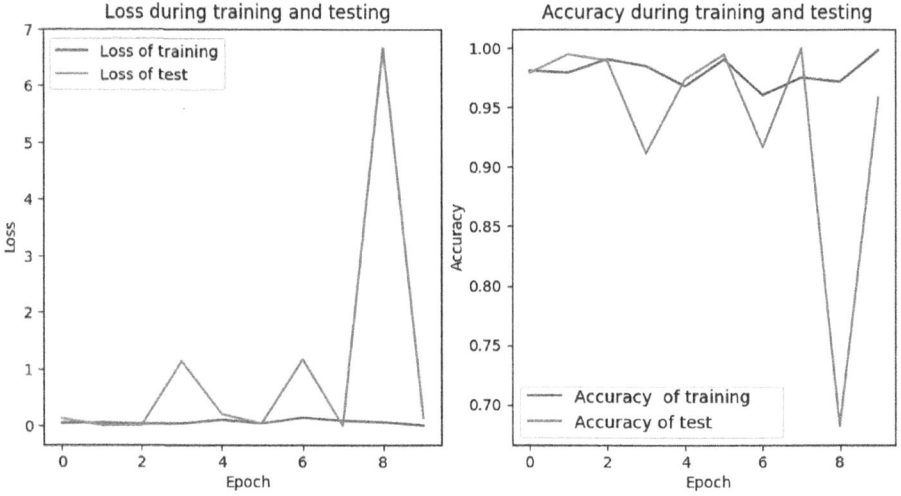

Fig. 3. Learning curve of the Xception model.

In the context of the confusion matrix, the true positives are 55 (cases where "Sick" was correctly predicted as "Sick") plus 77 (cases where "Sana" was correctly predicted as "Sana"), for a total of 132. Additionally, there were instances where the model incorrectly classified healthy leaves as sick, with 55 instances where "healthy" was incorrectly predicted as "Sick". The true negatives, represented by 77 cases, are instances where the model accurately identified "healthy" leaves. False negatives, with 46 cases, represent instances where "healthy" was incorrectly predicted as "healthy".

The evaluation metrics analysis revealed that the model exhibited an accuracy of 70%, a sensitivity (recall) of 54% for detecting "sick" cases, and a specificity of 58% for detecting "healthy" cases. This indicates that while the model is reasonably accurate in identifying healthy leaves, it has limitations in correctly identifying sick leaves. Although 222 images were used for the test set, this sample size may not be sufficient to ensure the results' generalizability to larger and more diverse datasets. With the inclusion of additional test images, the results may potentially diverge, as shown in Fig. 4.

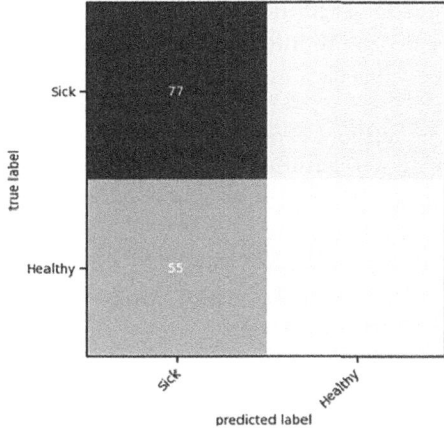

Fig. 4. Confusion matrix of Xception from 222 images.

It is crucial to highlight that this study primarily concentrates on the accuracy and loss metrics to assess the model's performance. Nevertheless, the incorporation of additional metrics, such as the F1-score and the area under the ROC curve, could provide a more comprehensive assessment of the model's performance. This should be considered in future research.

6 Conclusion

The Xception model showed superior performance compared to the other architectures evaluated (VGG16 and ResNet50), achieving an accuracy of 93.45% and a loss of 0.4842 on the test data set. This demonstrates its high effectiveness and accuracy for disease detection in cucumber leaves. The use of transfer learning allowed taking advantage of pretrained models on large datasets, significantly improving the accuracy of the models and reducing the required training time.

On the other hand, image preprocessing, which included resizing, normalization and data augmentation, was crucial to the performance of the system. These steps ensured that the images were properly adjusted to the format required by the models, improving the consistency and accuracy of the training.

Despite the high performance of the Xception model, challenges were observed in the accurate classification of some images, especially in the case of healthy leaves misclassified as diseased. This indicates the need to continue to optimize the models and possibly integrate additional data enhancement or post-processing techniques.

The development of this automated disease detection system provides a valuable tool for growers, facilitating the early identification of diseases and enabling rapid action to be taken to mitigate their impact. This can contribute significantly to improving the yield and quality of cucumber crops.

Future research could focus on integrating these systems with mobile applications or drones for real-time monitoring, as well as expanding the data set to include a wider variety of diseases and environmental conditions.

Acknowledgments. The authors express their gratitude to the Call 933 "Training in National Doctorates with a Territorial, Ethnic and Gender Focus in the Framework of the Mission Policy - 2023" of the Ministry of Science, Technology and Innovation (Minciencia). In addition, we thank the team of the Artificial Intelligence Laboratory VerbaNex (https://github.com/VerbaNexAI), affiliated with the UTB, for their contributions to this project.

References

1. Agrio: Cmv-12 (2018). https://agrio.app/library/CMV-12/
2. Alsaadi, I.A., Khodaparast, G.: Knowledge-based systems: an overview and future directions. https://philpapers.org/rec/ALSAKB
3. Bank, W.: Agriculture overview. https://www.bancomundial.org/es/topic/agriculture/overview
4. BASF: Mildiu del pepino (2021). https://www.agro.basf.es/es/Camposcopio/Secciones/Enfermedades-y-plagas/Mildiu-del-pepino/#:~:text=El%20mildiu%20del%20pepino%20es,por%20el%20pat%C3%B3geno%20Pseudoperonospora%20cubensis. Accessed 31 May 2024
5. Food, of the United Nations (FAO), A.O.: Organización de las naciones unidas para la alimentación y la agricultura. https://www.fao.org
6. Jia, W., et al.: Detection and recognition of tomato leaf diseases based on improved convolutional neural network. Comput. Mater. Continua **70**(2), 2171–2186 (2022). https://www.techscience.com/cmc/v70n2/44626/html
7. Kumar, R., Singh, R.: Iot-based monitoring system for agriculture. Bull. Electr. Eng. Inform. **9**(4), 1631–1642 (2020). https://www.beei.org/index.php/EEI/article/view/3096
8. Lee, K.H., Kim, H.J., Lee, J.S., Kim, S.B.: Data-driven approach for predicting the remaining useful life of bearings using complex system modeling and machine learning algorithms. Appl. Sci. **12**(2), 593 (2022). https://www.mdpi.com/2076-3417/12/2/593
9. Li, X., Wang, Q.: Artificial intelligence in agriculture: opportunities and challenges. Comput. Mater. Continua **70**(1), 195–209 (2021). https://cdn.techscience.cn/ueditor/files/cmc/TSP_CMC_70-1/TSP_CMC_18562/TSP_CMC_18562.pdf
10. Li, Y., Zhang, J., Chen, H.: Complexity analysis of networked systems: a review of recent advances and applications. Complexity **2021**, 9736179 (2021). https://www.hindawi.com/journals/complexity/2021/9736179/
11. Martin, C., Liu, J.: Drones in agriculture: current and future trends. Comput. Electron. Agric. **170**, 105251 (2020). https://www.sciencedirect.com/science/article/abs/pii/S0168169920300442
12. University of Minnesota Extension: Anthracnose of cucurbits (2020). https://es.extension.umn.edu/disease-management/anthracnose-cucurbits#:~:text=La%20antracnosis%20puede%20infectar%20todas,como%20resultado%20una%20apariencia%20irregular. Accessed 30 May 2024

13. Muñoz, C.A., et al.: El virus del mosaico del pepino (cmv) en la región andina: situación actual y perspectivas. Universitas Scientiarum **26**(1), 92–106 (2021). http://www.scielo.org.co/scielo.php?script=sci_arttext& pid=S1692-35612021000100092#:~:text=El%20virus%20del%20mosaico%20del, de%20manejo%20adecuadas%20y%20oportunas
14. Rodriguez, L., Garcia, M.: Machine vision in precision agriculture: a review. Comput. Electron. Agric. **142**, 103–123 (2017). https://www.sciencedirect.com/ science/article/abs/pii/S0168169917300820
15. Science, B.C.: Manejo de downy mildew en pepinos (2022). https://www. vegetables.bayer.com/mx/es-mx/recursos/agronomic-spotlights/manejo-de- downy-mildew-en-pepinos.html#:~:text=%C2%BB%20El%20mildiu%20velloso %20o%20downy,cultivo%2C%20resistencia%20y%20tratamientos%20qu%C3 %ADmicos. Accessed 19 May 2024
16. Singh, S., Kumar, P.: Machine learning for network security: a comprehensive review and future directions. Math. Probl. Eng. **2022**, 8909121 (2022). https:// www.hindawi.com/journals/misy/2022/8909121/
17. Smith, J., Doe, A.: Smart agriculture using IoT technology: a case study of pre- cision farming and predictive analytics. Int. J. Adv. Comput. Sci. Appl. **12**(2) (2021). https://journals.sagepub.com/doi/full/10.1177/15501477211007407
18. Systems, K.B.: Pepino: Enfermedades bajo cultivo protegido. https://www. koppert.es/cultivos/hortalizas-bajo-cultivo-protegido/pepino/#enfermedades
19. de Tecnología Agropecuaria (INTA), I.N.: Enfermedades de la hoja de yuca. https://www.inta.gob.ar/documentos/enfermedades-de-la-hoja-de-yuca
20. Thanaporn, P., Nuchprayoon, S.: Sustainable development in asean countries: a comparative analysis. GMSARN Int. J. **16**(3), 10–20 (2021). http://gmsarnjournal. com/home/wp-content/uploads/2021/10/vol16no3-10.pdf
21. Zamora: CORYNESPORA CASSIICOLA CUCUMBER (ENFPL-CP-002) (2015). https://dagus.unison.mx/Zamora/CORYNESPORA%20CASSIICOLA %20CUCUMBER%20(ENFPL-CP-002).pdf

Automatic Disease Detection in Physalis Peruviana Based on Image, a Review Systematic

Marco Yandún-Velasteguí[1,2]([envelope]) [ORCID], Luis Rivera[3] [ORCID], and José Herrera[1] [ORCID]

[1] Programa de Doctorado en Sistemas e Informática, Universidad Nacional Mayor de San Marcos, Decana de América, La Universidad del Perú, Av. Carlos Germán Amezaga #375, Lima, Peru
{marco.yandun, jherreraqu}@unmsm.edu.pe
[2] Universidad Politécnica Estatal del Carchi, Tulcán, Ecuador
[3] Center of Sciences and Technology (CCT), Laboratory of Mathematical Sciences (LCMAT), State University of North Fluminense, Rio de Janeiro, Brazil
rivera@uenf.br

Abstract. In this study, we present a bibliographic compilation on the detection of plant diseases using computer vision, focusing on Physalis peruviana, known as Uvilla, Uchuva, Golden Berry, and Aguaymanto. A problem has been identified in the manual detection of diseases and fruit damage due to the unique characteristics of the plant, which has a protective covering that creates uncertainty about the condition of the fruit. The research gathered articles from journals indexed in Scopus, IEEE Xplorer, and Web of Science, selecting 57 documents, of which 49 provide detailed information on computer vision in disease detection and Physalis peruviana, representing 13.2% of the total. Various techniques and methods, as well as image processing algorithms that improve quality before analysis, were classified. The types of images used, along with the detection of pests and diseases affecting the plant, were also described. Additionally, the visible effects on leaves, fruits, and stems according to the caused disease were shown. An initial experiment in image processing was conducted to pave the way for a subsequent application of disease detection algorithms using computer vision.

Keywords: Physalis peruviana · computer vision · image processing · Uchuva · Aguaymanto

1 Introduction

Technology, image processing, and machine learning algorithms, including the application of mathematical morphology for image analysis, are applicable in the agricultural sector, gaining significance in the so-called Agriculture 4.0 to incorporate technologies into the processes of automatic decision-making in Physalis peruviana plants, commonly known as Golden Berry, Uchuva, Uvilla, and Aguaymanto.

The detection of agricultural diseases in a preliminary phase reduces economic losses and improves crop quality. Manual identification of agricultural pests is often evident in plants; however, it takes more time and is a costly technique. [1] mention that timely

N. D. Duque-Méndez et al. (Eds.): CCC 2024, CCIS 2208, pp. 168–183, 2024.
https://doi.org/10.1007/978-3-031-75233-9_13

monitoring of disease characteristics during the plant growth process is crucial for implementing timely control measures. The limitation is that plant images are easily altered by environmental factors such as dust, low light, clouds, or leaf shadows. In response to this, [2] provided a modified rotation kernel transform (MRKT), based on a targeted feature extraction approach, to address difficulties caused by color, structure, or other deceptive aspects during crop disease identification. All of this is complemented with machine learning (ML) techniques in the supervised learning paradigm for image classification. This enables early detection of disease in plants. [3] applied this method for late blight detection in potatoes, using a convolutional neural network and a support vector machine (SVM).

On the other hand, we have the evolution of the Internet of Things (IoT) and deep learning models (DLM) that are suitable for the implementation of smart agriculture, allowing continuous monitoring of plant development [4–6], particularly for the efficient determination of the exact location of the affected part of the crop leaf [7]. For this purpose, among others, the application of image processing and pattern recognition methods, such as Convolutional Neural Networks (CNN), is required, in this case, due to the complex morphological characteristics of Golden Berry plant leaves for disease detection. There are several features that need to be automatically analyzed by the technology, including, for example, Sigatoka spots and Fusarium wilting [8]. This disease leads to a reduction in production and harvest quality. There is also a demand for efficient methods in fruit state detection, which, due to the fragile nature of the fruit's outer covering, makes traditional computer vision analysis complex.

The low performance of image processing technology in Golden Berry plantations hinders timely damage prediction in the plants, consequently leading to economic losses. These cases create a negative perception among farmers regarding the use of technology to improve their products, and they may continue with manual inspections, which require processing time and action that, in some cases, can be too time-consuming. This issue is highlighted by [9], who indicate that even an experienced person in disease recognition can fail in the detection process.

The limited emerging technologies in these processes result in false positives that are often observed during harvesting. This is because certain scenarios can occur, such as ripe fruit with an immature coloration and vice versa, immature coloration with ripe fruit. Therefore, a method is needed to carry out this process without compromising the integrity of the fruit.

The objective of this study is to analyze the use of computer vision with the algorithms involved in plant disease detection. To achieve this, the methodology of the review is outlined in Sect. 2. Section 3 reviews the relevant literature to address the formulated questions. Section 4 discusses the findings, and the conclusion is presented in Sect. 5.

2   Materials and Methods

There are multiple scientific articles that mention the use of computer vision and image processing for plant disease detection. One notable example of such research, taking into consideration that plant diseases have an impact on agricultural production and quality, is the study on Mango plants conducted by [4].

Additionally, [5] proposed a method for detecting Apple defects based on the fuzzy c-means algorithm and nonlinear programming genetic algorithm (FCM-NPGA), combined with multivariate image analysis. Similarly, the study by [6] presented a method for classifying defects on the surface of citrus fruits based on computer vision.

And other articles that can be mentioned where different image processing techniques are applied, along with the use of various algorithms. All of this leads to answering different research questions such as:

Firstly, these research questions are proposed:

Q1: Computer vision techniques or methods used for detecting diseases in plants.

Q2: Image processing algorithms applicable to enhance image quality in processing.

Q3: Types of images that can be used for processing and feature extraction.

Q4: Pests and diseases that can affect the leaves and fruits of Physalis peruviana.

Q5: Contribution to the improvement of image processing for disease detection in Physalis peruviana.

2.1 Application of Systematic Review - Application of Inclusion Criteria

Article Age: 5 years ago.

Publication Language: Articles in English and Spanish were used.

Article Database: Articles published in journals indexed in Scopus, Web of Science, IEEE, ACM.

Access Type: Filtered to select those freely accessible.

Relevance to the Topic: Articles related to Physalis Peruviana, plant diseases, and computer vision were selected.

2.2 Application of PRISMA Methodology

For the application of the systematic literature review, the following search string is used, adaptable to different databases: (ALL(physalis AND peruviana) AND ALL(computer AND vision) OR ALL(Process)) AND (LIMIT-TO(OA,"all")) AND (LIMIT-TO(DOCTYPE,"ar")) AND (LIMIT-TO(PUBYEAR,2024) OR LIMIT-TO(PUBYEAR,2023) OR LIMIT-TO(PUBYEAR,2023) OR LIMIT-TO(PUBYEAR,2021) OR LIMIT-TO(PUBYEAR,2020)) (see Fig. 1).

Bibliographic research has been conducted using the Bibliometrics technique, where multiple relevant articles dating from 2018 have been analyzed. These articles were obtained from journals indexed in Scopus and other scientific sources, including Google Scholar, using various search strings with the following keywords: "disease detection" and "plants," "Physalis peruviana," "image processing," "computer vision," "CNN." In Fig. 2, you can observe the distribution of areas where articles on image processing are published, with agriculture and engineering both having a 13.2% share of applications.

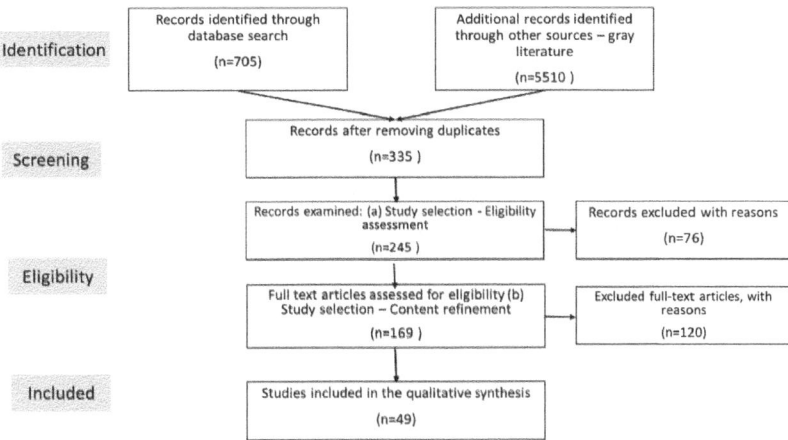

Fig. 1. Application of the PRISMA Methodology

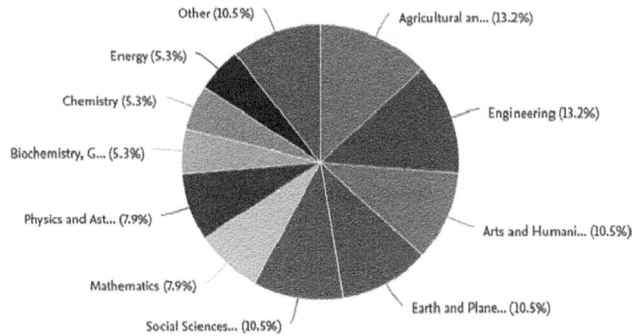

Fig. 2. Agriculture and engineering are the leading fields where image processing is applied.

3 Theoretical Foundations Compiled in the Systematic Literature Review

3.1 Artificial Neural Networks

[10] mention that artificial neural networks are a mathematical computational model based on a large set of simple neural units, where artificial intelligence and statistical psychology are used. Based on the above, it can be said that the ANN (Artificial Neural Network) is nothing more than a pattern of connections that generates an artificial model inspired by the biological behavior of the human brain. Therefore, artificial neural networks (ANNs) are of great help when comparing an input image with a database of diseased leaves to obtain more accurate results.

The classification of Artificial Neural Networks (ANN) can be performed. Neural networks are artificial intelligence programs based on the functioning of biological neural networks in the human brain. These computational models are applied to classification problems and time series, identifying connections that other techniques may not, meaning they utilize both linear and nonlinear relationships without requiring extensive experience in the subject matter's execution and distribution [11]. These networks are applied for feature extraction, which serves as the foundation for image retrieval, including textual and visual characteristics such as color, texture, shape, and faces [12].

Training a Neural Network can be done. According to [13] as cited by [11], it is mentioned that to train an artificial neural network, parameters such as loss need to be considered and minimized before training. It is necessary to calculate how good the network is initially so that it can be improved. The Mean Squared Error (MSE) is commonly used to measure loss, and its Eq. (1) is shown below[1]:

$$MSE = \frac{1}{n} + \sum_{i=1}^{n} (ytrue - ypred)^2 \qquad (1)$$

n: number of samples
y: prediction variable
y_true: true value of the variable ("correct response")
y_pred: value predicted by the variable. Result from the network
$(y\_true\text{-}y\_pred)^2$: squared error. The loss function takes the average of all other squared errors. The loss will be lower when we have better predictions.
Better prediction = less loss
Training a neural network = trying to minimize the lost

Then, pattern detection can be applied, similar to what [14] describe. They suggest that facial feature recognition is not limited to face recognition; it can also be applied to pattern detection for recognizing abnormalities in plant leaves.

3.2 Machine Learning

One of the processes is Machine Learning, which is known as a set of techniques applied in artificial intelligence for the autonomous recognition or learning of a specific aspect. In this case, the computer or application has programmed precise instructions, as in the classic example of being programmed to identify a circular object when it sees a specific image of a circular object. This algorithm is feasible when an image contains a single circular figure, but it can become confused when there are multiple circular objects or patterns [15].

Similarly, Deep Learning, known as deep learning, is a process within Machine Learning where an artificial neural network is used. In this process, a series of iterations are carried out, where the treatment of the first layer or level generates the result for the subsequent levels, combining information and creating increasingly complex representations. Deep Learning is useful for recognizing data within images, identifying product brands or logos, and recognizing individuals, objects, or other elements [16].

Also, the use of learning model repositories like TensorFlow (TF-CNN), also known as a series of libraries applied in convolutional neural networks, allows for image classification and training of associated models. This is highly useful for processing multiple images, whether in RGB or grayscale. The results of these models are mutually exclusive, with no overlap between them [17].

Another model that uses artificial intelligence is DeepLabV3 with an encoder-decoder architecture. The encoder consists of a convolutional neural network (CNN) trained to encode feature maps of the input image. The decoder is used for up-sampling and reconstructing the output using important information extracted by the encoder [7]. Additionally, the model CED-Net is mentioned. CED-Net consists of four small encoder-decoder networks divided into two levels. The encoder-decoder networks in each level are independently trained for leaf segmentation, as shown in Fig. 3. In this context, "Encoder" refers to encoding, and "Decoder" refers to decoding.

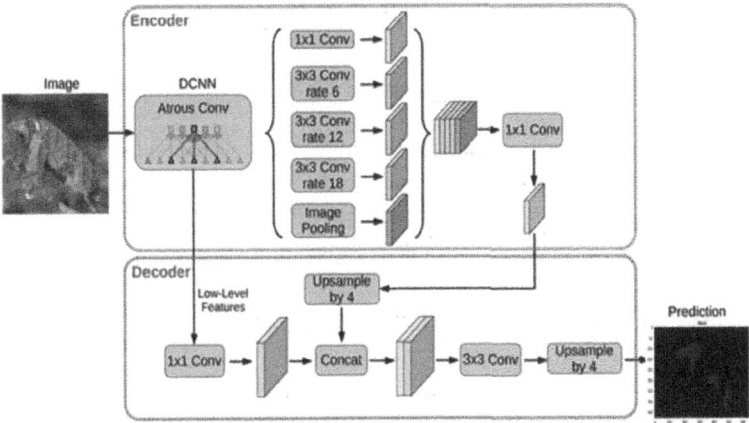

Fig. 3. Architecture of the DeepLabV3 + Model - The architecture of the DeepLabV3 + model, where the image is input, and the Encoder and Decoder algorithms are applied [7]

3.3 Tools Related to Computer Vision that Are Applied for Image Processing

[1] State that MF-RAnet is a framework consisting of two frameworks, one being the main framework and the other for displaying detailed main features. It employs the cross-stacking structure of ResNet50 and RAM to extract the main features in the image dataset. RAM is used to extract attention weights in the feature layer, enabling it to assign higher weight to disease-related features.

[2] Introduced the Modified Rotated Kernel Transformation (MRKT), based on a targeted feature extraction approach, to address challenges posed by color, structure, or other deceptive aspects during crop disease identification.

Image analysis using multi-SVM and Gray-Level Co-Occurrence Matrix (GLCM), as explained by [18], was used to identify anthracnose, borers, and sooty mold with a processing reliability of 85%.

Tools like the You Only Look Once (YOLO) method allow for the detection of various shapes within a single image, enabling quick prediction of the type of disease in leaves. While it does not guarantee faultless results as it may not determine the specific plant disease in a single image capture, YOLO provides two actions: the first recognizes leaves, fruits, or other plant parts, and the second predicts the type of disease based on the model's highest probability [12].

The Robust Mean Shift (ROMS) method, as explained by [19], is a proposed method that can be applied directly to restore damaged images, even in the presence of a strong combination of Gaussian and impulsive noise. It can be used in various practical image generation tasks.

The Pulse-Coupled Neural Network (PCNN) Coupled Neural Network (PCCN) by [20] uses the image segmentation method of the Pulse-Coupled Neural Network (PCNN) based on minimum cross-entropy to segment apple images. The disease recognition accuracy for apples after segmentation is 93%.

The Entropy-ELM algorithm with Whale Optimization Algorithm (WOA) is used for feature extraction, where the training and testing processes were carried out. After collecting the optimal feature subset, it employs an SVM classifier for pest classification [4].

There is also the HSV (Hue, Saturation, Value) image processing and segmentation algorithm, among the tested image processing and segmentation algorithms. The method that yielded the best results for background removal in the image was the HSV color range segmentation algorithm [3].

3.4   Methods that Can Be Used for the Recognition of Diseases in Physalis Plants

A model known as EM-ERNet, a disease recognition method, where [21] proposed a banana disease recognition model based on EM-ERNet, which improved the model's adaptability to banana disease image samples. However, different types of diseases also exhibit similar characteristics, and these characteristics are more challenging to distinguish due to the influence of image quality.

Another method worth mentioning is the Fast Disease Detection Based on FCM-KM Fusion, as proposed by [13]. They introduced a fast rice disease detection method based on the fusion of FCM-KM and Fast R-CNN to address issues such as noise, blurry edges, significant background interference, and low detection accuracy in rice disease images. A two-dimensional filter mask combined with a multi-stage weighted median filter (2DFM-AMMF) is used to reduce noise, and a faster two-dimensional Otsu threshold segmentation algorithm is employed to reduce complex background interference in the detection of target leaves in the image.

4   Analysis of Factors and Discussion of Research Questions

4.1   Computer Vision Methods Used in Plant Disease Detection

[1] proposed the MF-RANet method, which was evaluated on the citrus disease image dataset from the University of Science and Technology of China. It achieved a classification accuracy of 97.8%, surpassing the performance of other citrus disease detection methods, based on the Multiscale Retinex algorithm.

Computer vision methods used in the detection of various plant aspects can be listed as follows: the CED-Net model. CED-Net consists of four small encoder-decoder networks divided into two levels. The encoder-decoder networks in each level are independently trained for leaf segmentation [7].

Hyperspectral images, which combine imagery and spectral information to detect external or internal quality attributes, are a promising detection method [22]. Spectral signatures can be used to uniquely characterize, identify, and quantify the chemical composition of agricultural products, offering advantages in detecting early defects that are invisible to the naked eye [23]. The works of other authors with their various algorithms used are detailed in Table 1.

Table 1. Computer vision methods applicable in plant disease detection.

Author	Vision method/algorithm
Yang et al. (2022)	The AMSR algorithm (Multiscale Retinex Algorithm) MF-RANet multi-factorial attention neural network for citrus disease detection
Zhang et al. (2020)	FCM: Fuzzy C-Means (clustering algorithm) NPGA: Nonlinear Programming Genetic Algorithm FCM-NPGA: Nonlinear Programming Genetic Algorithm for Fuzzy C-Means
Zhang et al. (2021)	State Transition Algorithm (STA)
Mohanty et al. (2016)	CNN, for image classification, detection, and segmentation
Xu L et al. (2017)	Pulse-Coupled Neural Network (PCNN) for image segmentation, detection, and classification
Zhou G, et al. (2019)	Fusion of FCM-KM and R-CNN (Region Convolutional Neural Network)
Wang, J.-X. et al. (2017)	Near-infrared (NIR), near-infrared image processing algorithm
R. Jayaraj y S. Lokesh (2023)	Artificial vision
Muthaiah y Chitra (2023)	Entropy-ELM with Whale Optimization Algorithm (WOA) It allows quantifying the relationship between the gray levels of adjacent pixels in an image

4.2 Applicable Image Processing Algorithms to Improve Image Quality in Processing

A technique called ESFO-EALD (Apple Leaf Disease Detector) has been developed, which is a powerful and versatile machine learning algorithm. It is efficient and accurate in solving classification and regression problems. The algorithm is easy to implement and flexible. It is robust against data noise. However, the algorithm can be sensitive to initial parameters and may be more complex than other machine learning algorithms used to enhance image quality for determining affected regions through imaging [24].

Another algorithm used is micro magnetic resonance (MR) imaging, a non-destructive and non-invasive spectroscopic technique that provides detailed morphostructural images, information on the spatial distribution of proton density, relaxation parameters (spin-lattice relaxation time (T1) and spin-spin relaxation time (T2)), and self-diffusion coefficient within the sample. Magnetic resonance emerges as the technique of choice for investigating the internal structure and water state in biological and food tissues [25].

Among the tested image processing and segmentation algorithms, the method that yielded the best results in the background removal process was the HSV color range segmentation algorithm. Regarding classification, the CNN model achieved better results when trained with the augmented dataset. On the other hand, SVM models performed better when the algorithm was trained with color features, demonstrating that color features play an important role in the development of SVM classification models. [3]. Additionally, Table 2 shows other authors indicating algorithms used to improve and preprocess images for feature extraction.

Table 2. Methods or algorithms applicable in image preprocessing that enhance the image prior to feature extraction

Author	Vision method/algorithm
Dae-Hyun Jung et al. (2022)	CycleGAN is a CNN algorithm used for image quality improvement and visual content generation
Durgabai et al. (2019)	Modified Rotation Kernel Transformation (MRKT)
Nermeen et al. (2022)	Fully Convolutional Networks for Semantic Segmentation (FCN), FCN32s, FCN16s, and FCN8s are machine learning models used to generate image label maps
Suárez et al. (2022)	Support Vector Machine (SVM) algorithm used in image recognition Hue Saturation Value (HSV) technique to identify and segment objects in an image and create a brighter tone
Nermeen et al. (2022)	Fully Convolutional Networks for Semantic Segmentation FCN-8 Convolutional Neural Network for Defect Detection and Segmentation in Manufacturing Products CED-Net, used for feature extraction from images SegNet, a type of FCN network used in image labeling DeepLabv3 U-Net
Lateef, F., Ruichek, Y. (2019)	UNet, and UNets are CNN-like networks used for image segmentation, feature extraction, and reconstruction
Sathit Prasomphan	You Only Look Once (YOLO) algorithm, used for real-time image detection
Damian Kusnik y Bogdan Smolka (2022)	Robust Mean Shift (ROMS) can group images before their conversion to grayscale or numerical formats
R. Jayaraj y S. Lokesh (2023)	Adaptive Convolutional Deep Learning Model (ACDLM) for automatic image annotation Honey Badger Algorithm (HBA) CNN (Convolutional Neural Network)
Zhenhua Cui et al. (2021)	nanoVoxel-3502E system (Sanying Precision Instruments Co., Ltd., Tianjin)
Arora et al. (2008)	Multilevel thresholding methods
Afsana Mimi et al. (2023)	The CNN-SVM training algorithm
Afsana Mimi et al. (2023)	MobileNetV2, a CNN-based network used for image classification
Zhou Y. et al. (2023) [33]	The fusion of visible and infrared images (IVIF)
Xiao et al. (2023) [34]	Adaptive Instance Normalization Modulator (INAM) in adaptable 3D LUTs for image enhancement

(continued)

Table 2. (*continued*)

Author	Vision method/algorithm
Y. Jiang et al. (2023) [35]	Multiscale Retinex (MSR) enhances SSR through a multiscale Gaussian filter and color restoration DEANet CNN (Convolutional Neural Network) uses the high-frequency component of an image to remove noise and the low-frequency component to adjust the brightness of details in the image
Wang, T. et al. (2023)	Terahertz (THz) image enhancement based on the noise2noise strategy (Learning Image Restoration without Clean Data)
Lu et al. (2023)	Retinex to obtain an enhanced image with vivid color

4.3   Types of Images that Can Be Used for Processing and Feature Extraction

Different types of images can be used in computer vision applications. These include traditional images (TI), laser backscattering images (LBI), ultrasound images (UI), fluorescence images (FI), Raman images (RI), microwave images (MI), odor images (OI), and others as per the criteria specified by the authors in Table 3.

Table 3. Types of images applicable in image processing

Author	Article	Image type
Hyeyeon et al. (2022) Yuping Huang (2020)	Nondestructive classification of soft rot disease in napa cabbage using hyperspectral imaging analysis Measurement of Early Disease Blueberries Based on Vis/NIR Hyperspectral Imaging System	Hyperspectral images (HSI),
Taglienti Anna et al. (2020)	Study on ultra-structural effects caused by Onion yellow dwarf virus infection in 'Rossa di Tropea' onion bulb by means of magnetic resonance imaging	Magnetic resonance images (MRI),
Zhenhua Cui et al. (2021)	High-Resolution Microstructure Analysis of Cork Spot Disordered Pear Fruit "Akizuki" (Pyrus pyrifolia Nakai) Using X-Ray CT	X-ray images (XRI),
Yuping Huang (2020)	Measurement of Early Disease Blueberries Based on Vis/NIR Hyperspectral Imaging System	Thermal images (TI),

4.4   Pests and Diseases that Can Affect the Leaves and Fruits of Physalis Peruviana

One of the pests that causes substantial damage to plants during their growth phase is the Agrotis epsilon larva, which can appear from October to December and hatch in February and March. Its main food source is the newly emerging plant stems [26].

Other pests and diseases that can affect plants, specifically Physalis peruviana, include the following: Agrotis epsilon larva, Thrips, mites, fleas, worms, Cutting, Bird's Eye, Gray spot, Alternaria, Fusarium [27].

Studies of specialized metabolism in Physalis species can be applied based on RNA transcriptome analysis. [28], to determine resistance to different pests and diseases.

Exploring the virome in Physalis peruviana to assess the health status of planting material and crops. [29] would help establish new pest resistance methods. And those mentioned by other authors as indicated in Table 4.

Table 4. Pests that can be found in Physalis peruviana

Author	Article	Pest/disease/aspect
Sarkar et al. 2021	Morphological Alterations in Haemocytes of Adult American Cockroach, Periplaneta Americana (Linnaeus)(Insecta: Blattodea: Blattidae)	Larva Agrotis
Torres et al. 2019	Diagnóstico de la problemática actual de enfermedades y plagas observadas en el cultivo de la Uchuva (Physalis peruviana L.)	larva Agrotis épsilon, Trips, ácaros, pulguillas, gusanos, Cortado, Ojo de pollo, Man-cha gris, Alternaría, Fusarium
Fukushima A, et al. (2016)	Comparative Characterization of the Leaf Tissue of Physalis alkekengi and Physalis peruviana Using RNA-seq and Metabolite Profiling	Trasncriptoma analizado de ARN

In the Physalis peruviana plant, the diseases described in Table 5 can be identified, where the visible effects on the stem, fruit, and leaf are also shown.

Table 5. Visible effects on the plant according to the identified disease

STEM	FRUIT	LEAF	DISEASE
N/A	Circular spots, brown or grayish-black	Circular spots, brown or grayish-black	GRAY SPOT
Cankers	Small plants	N/A	Yellowish wilting
Rot	Mantle rot	Brown spots	SCLEROTINIA
Decomposition	Necrosis	Necrosis	WHITE MOLD
N/A	Dry fruits	Falling mantle	N/A
Oval black spots	Oval black spots	Fruit deformation	BLACK SPOT
Gray mold	Fruit rot	Gray mold	GRAY MOLD
Oiliness	Oiliness	GREASY SPOT	
Wilt	Deformed fruit	Deformed leaves	BACTERIAL WILT
N/A	Holes in the mantle	Leaves eaten by blight	LEAF BLIGHT
N/A	Small fruits	Wrinkled leaves	LEAF SMUT

4.5   Contribution to the Improvement in Image Processing for Disease Detection in Physalis Peruviana

The MF-RANet algorithm consists of a main feature framework and a detailed feature framework. The main feature framework uses the cross-stacking structure of ResNet50 and RAM to extract the main features in the plant image dataset, as in the study by [1].

CycleGAN is a type of convolutional neural network (CNN) algorithm trained to perform image-to-image translation by learning from two unpaired images, cycling between the two. It involves generators and discriminators, with a representative application example being the transformation of a zebra image into that of a regular horse. This technology can alter patterns in two images, such as a depth-informed image and a general RGB image. CycleGAN is a learning process that generates images through self-learning and requires a relatively small amount of labeled image data, as mentioned by [30]. Additionally, the contribution to image processing applicable to Physalis peruviana is presented in Table 6.

Table 6. Contribution to the improvement of image processing in disease detection in Physalis peruviana

Author	Article	Contribution
Revathy et al. (2018)	Clasificación de datos de plagas de cultivos agrícolas mediante el uso de C5 basado en hadoop mapreduce. 0	Gray-Level Co-Occurrence Matrix (GLCM) image classification, edge detection, and image segmentation
Licodiedoff et al. (2013)	Use of Image Analysis for Monitoring the Dilution of Physalis peruviana Pulp	Color gradient in R, G, and B image
Antúnez-Ocampo et al. (2020)	Growth dynamics of morphological and reproductive traits of Physalis peruviana L. M1 plants obtained from seeds irradiated with Gamma rays	Mutation induction is used in plant breeding
De-la-Torre et al. (2019)	Multivariate Analysis and Machine Learning for Ripeness Classification of Cape Gooseberry Fruits	Seven maturity levels, visual scale Fischer et al. (2005)
Zhang et al. (2014)	developments and applications of computer vision for external quality inspection of fruits and vegetables	Visual inspection of color, size, and shape parameters

5   Experimentation

Application of PCA in the image compression process prior to processing Loading of images using the Python language in the Colab environment, and its results are shown in Fig. 4.

```
import numpy as np
import matplotlib.pyplot as plt
import seaborn as sns
sns.set()

!wget
https://repovirtual.upec.edu.ec/index.php/s/LFP485nTiwDtLGJ/download/20221203_154226.jpg

from matplotlib.image import imread
plt.rcParams['axes.grid'] = False
image = imread('20221203_154226.jpg')
print(image.shape)
plt.figure(figsize=(12,8))
plt.imshow(image)


pca = PCA()
pca.fit(image_bw)
print(pca.n_components_)
print(pca.explained_variance_ratio_[:20])

var_cum = np.cumsum(pca.explained_variance_ratio_) * 100
print(var_cum[:61])
#¿cuantos PCs explican el 95% de los pixeles?
k = np.argmax(var_cum >= 95) + 1
print('La cantidad de PCs que explican el 95% de pixeles es',k)

x = k * 100 / 391
print(x)

def plot_k(k):
  pca_tmp = PCA(n_components=k)
  trans = pca_tmp.fit_transform(image_bw)
  image_reconst = pca_tmp.inverse_transform(trans)
  plt.imshow(image_reconst, cmap=plt.cm.gray)
  return trans.nbytes

k = 61 #el valor de PCs para obtener 95%
plt.figure(figsize=(12,8))
plot_k(k)

ks = [10,25,50,100,150,250,300,391]
plt.figure(figsize=(15,9))
for i in range(8):
  plt.subplot(2,4,i+1)
  n_bytes = plot_k(ks[i])
  plt.title('#PCs ' + str(ks[i]) + ' - ' + str(n_bytes/1024) + 'KB')
plt.subplots_adjust(wspace=.2,hspace=.0)
plt.show()
```

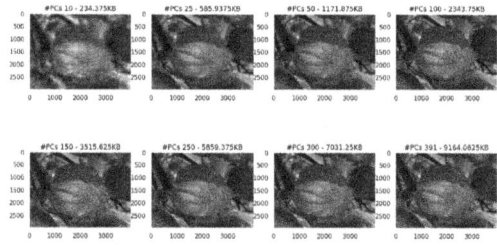

Fig. 4. Application of the PCA algorithm that allows resizing of the image to proceed with the feature extraction process. It is applied with an example image of Physalis Peruviana. The exercise can be viewed at: https://colab.research.google.com/drive/1xQ8_uW1KDsB4hApzF_aKQ 6sFC2KjZ6c2?usp=sharing

6 Discussions and Conclusions

Pest control by farmers is not always carried out effectively and often relies on manual methods such as visual inspection. It is necessary to implement control measures in plantations, both physically and chemically, especially when quality parameters must be met for export.

Applying technological methods such as X-ray Computed Tomography is a three-dimensional visualization technique that has long been used to non-destructively acquire 3D images of the internal structure of agricultural products [31].

The microstructure of fruits determines the mechanical and transport properties of tissue in products like pears. Intercellular spaces can occupy up to 5.1% of the total fruit volume, which is the most important pathway for the diffusion of gases such as O2, CO2 (carbon dioxide), and H2O (water) in plants [32].

[3] conducted their study in the state of Boyacá, the second-largest in Colombia, where the economy is based on agriculture, with potato cultivation being the most influential. In the 2019/2020 period, the department accounted for 27.30% of the total planted area and 26.84% of the national production, making it a major player in this crop.

[4] explored the use of efficient plant disease recognition systems based on the Internet of Things (IoT). They employed semantic segmentation methods such as SegNet, CED-Net, U-Net, DeepLabv3, and FCN 8s, along with a post-processing enhancement method known as Conditional Random Field (CRF) to assign disease portions in leaf crops. In this study, they introduced a post-processing method with theoretical simplicity within a unified framework to improve the results.

Notes

1. The Mean Squared Error.

References

1. Yang, R., Liao, T., Zhao, P., Zhou, W., He, M., Li, L.: Identification of citrus diseases based on AMSR and MF-RANet. Plant Methods, **18**(1) (2022). https://doi.org/10.1186/s13007-022-00945-4
2. Durgabai, P., Bhargavi Y S. Jyothi, "Clasificación de plagas de cultivos de algodón mediante análisis de big data", en Proc. En t. Conf. On Computational and Bio Engineering, Coimbatore, India, págs. 37–45 (2019)
3. Suárez, M., Gómez, A., Espíndola, J.: Supervised learning-based image classification for the detection of late blight in potato crop. MDPI Open Access J. **12**(18) (2022). https://doi.org/10.3390/app12189371
4. Muthaiah, U., Chitra, S.: Mango pest detection using entropy-ELM with whale optimization algorithm. Intell. Autom. Soft Comput. **35**(3), 3447–3458 (2023). https://doi.org/10.32604/iasc.2023.028869
5. Zhang, W., Hu, J., Zhou, G., He, M.: Detección de defectos en manzanas basada en FCMNPGA y un análisis de imágenes multivariante. Acceso IEEE, **8**(2), 38833–45 (2020). https://doi.org/10.1109/ACCESO.2020.2974262

6. Zhang, W., Tan, A., Zhou, G., Chen, A., Hu, Y.: Un método para clasificar Defectos de la superficie de los cítricos basados en visión artificial. J Food Meas Caract. **15**(3), 2877–2888 (2021)
7. Rezk, N.G., Attia, A.F., El-Rashidy, M.A., El-Sayed, A., Hemdan, E.E.D.: An efficient plant disease recognition system using hybrid convolutional neural networks (CNNs) and conditional random fields (CRFs) for smart IoT applications in agriculture. Int. J. Comput. Intell. Syst. **15**(1), 65 (2022). https://doi.org/10.1007/s44196-022-00129-x
8. Anasta, N., Setyawan, F.X.A., Fitriawan, H.: Disease detection in banana trees using an image processing-based thermal camera. In: IOP Conference Series: Earth and Environmental Science, vol. 739, no. 1, p. 012088). IOP Publishing (2021)
9. Song, H., Yoon, S.R., Dang, Y.M., et al.: Nondestructive classification of soft rot disease in napa cabbage using hyperspectral imaging analysis. Sci. Rep. **12**, 14707 (2022). https://doi.org/10.1038/s41598-022-19169-6
10. Priya, S., Abinaya, M.: Feature selection using random forest technique for the prediction of pest attack in cotton crops. Int. J. Pure Appl. Math. **118**(18), 2899–2903 (2018)
11. Gonzales, L., Abraham, A.: «Predicción de renuncia de socios de una Cooperativa utilizando técnicas supervisadas de aprendizaje automático». Universidad Católica de Santa María (2019)
12. Prasomphan, S.: Rice bacterial infection detection using ensemble technique on unmanned aerial vehicles images. Comput. Syst. Sci. Eng. **44**(2), 991–1007 (2023)
13. Zhou, G., Zhang, W., Chen, A., et al.: Detección rápida de la enfermedad del arroz basada en FCM-KM y una fusión R-CNN más rápida. Acceso IEEE **7**(9), 143190–143206 (2019)
14. Zhang, H., Qu, Z., Yuan, L., Li, G.: A face recognition method based on LBP feature for CNN. In: 2017 IEEE 2nd Advanced Information Technology, Electronic and Automation Control Conference (IAEAC), pp. 544–547. IEEE (2017)
15. García Santillán Iván Danilo, Visión artificial y Procesamiento Digital de imágenes usando Matlab, 2008, Seritex- PUCE-SI, Tulcán Ecuador
16. Gil, R.: Desarrollo de un sistema de reconocimiento de emociones faciales en tiempo real (2017)
17. Taqi, A.M., Awad, A., Al-Azzo, F., Milanova, M.: The impact of multi-optimizers and data augmentation on TensorFlow convolutional neural network performance. In: 2018 IEEE Conference on Multimedia Information Processing and Retrieval (MIPR), pp. 140–145. IEEE (2018)
18. Revathy, R., Balamurali, S., Lawrance, R.: Clasificación de datos de plagas de cultivos agrícolas mediante el uso de C5 basado en hadoop mapreduce 0. J. Cyber Secur. Mob. **8**(3), 393–408 2019
19. Kusnik, D., Smolka, B.: Robust mean shift filter for mixed Gaussian and impulsive noise reduction in color digital images. Sci. Rep. **12**, 14951 (2022). https://doi.org/10.1038/s41598-022-19161-0
20. Xu, L., Lv, J.: (2017), Método de reconocimiento de manzanas basado en SUSAN y PCNN. Aplicación de herramientas multimedia. **77**, 7205–7219 (2017)
21. Lin, H., Lin, H., Zhou, G.: EM-ERNet for image-based banana disease recognition. J Food Meas. Charact. **15**, 4696–4710 (2021). https://doi.org/10.1007/s11694-021-01043-0
22. Jiang, H., Jiang, X., Ru, Y., Wang, J., Zhou, H.: Application of hyperspectral imaging for detecting and visualizing leaf lard adulteration in minced pork. Infrared Phys. Technol. **110**, 103467 (2020)
23. Huang, Y., Lu, R., Chen, K.: Detection of internal defect of apples by a multichannel Vis/NIR spectroscopic system. Postharvest Biol. Technol. **161**, 111065 (2020)
24. Alqahtani, M., et al.: Sailfish optimizer with efficientnet model for apple leaf disease detection. Comput. Mater. Continua, **74**(1), 217–233 (2023). https://doi.org/10.32604/cmc.2023.025280

25. Taglienti, A., et al.: Study on ultra-structural effects caused by onion yellow dwarf virus infection in 'Rossa di Tropea' onion bulb by means of magnetic resonance imaging. Sci. Hortic. **271**(109486), 109486 (2020). https://doi.org/10.1016/j.scienta.2020.109486

26. Sarkar, S., Ghosh, J.: Morphological alterations in haemocytes of adult american cockroach, periplaneta Americana (Linnaeus) (Insecta: Blattodea: Blattidae) in Response to Thermal Stress and Induced Infection. Int. J. Sci. Res. Biol. Sci. **8**(5) (2021)

27. Torres Panqueva, W.F., Cuéllar Meneses, J.G.: Diagnóstico de la problemática actual de enfermedades y plagas observadas en el cultivo de la Uchuva (Physalis peruviana L.) en dos unidades productivas del municipio de San José de Isnos del departamento del Huila

28. Fukushima, A., et al.: Comparative characterization of the leaf tissue of physalis alkekengi and physalis peruviana using RNA-seq and metabolite profiling. Front. Plant Sci. **7**, 1883 (2016). https://doi.org/10.3389/fpls.2016.01883

29. Corrales-Cabra, E., Higuita, M., Hoyos, R., Gallo, Y., Marín, M., Gutiérrez, P.: Prevalence of RNA viruses in seeds, plantlets, and adult plants of cape gooseberry (Physalis peruviana) in Antioquia (Colombia). Physiol. Mol. Plant Pathol. **116**, 101715 (2021)

30. Jung, D., Kim, C., Lee, T., et al.: Depth image conversion model based on CycleGAN for growing tomato truss identification. Plant Methods **18**, 83 (2022). https://doi.org/10.1186/s13007-022-00911-0

31. Schoeman, L., Williams, P., Plessis, A.D., Manley, M.: X-ray micro-computed tomography (μCT) for non-destructive characterisation of food. Trends Food Sci. Technol. **47**, 10–24 (2016). https://doi.org/10.1016/j.tifs.2015.10.016

32. Kader, A.: Postharvest Technology of Horticultural Crops, 3rd edn. University of California, Davis, CA (2002)

33. Zhou, Y., Xie, L., He, K., Xu, D., Tao, D., Lin, X.: Low-light image enhancement for infrared and visible image fusion. IET Image Process. **00**, 1–19 (2023). https://doi.org/10.1049/ipr2.12857

34. Xiao, X., Gao, X., Hui, Y., Jin, Z., Zhao, H.: INAM-based image-adaptive 3D LUTs for underwater image enhancement. Sensors. **23**(4), 2169 (2023). https://doi.org/10.3390/s23042169

35. Jiang, Y., Li, L., Zhu, J., Xue, Y., Ma, H.: DEANet: decomposition enhancement and adjustment network for low-light image enhancement. Tsinghua Sci. Technol. **28**(4), 743–753 (2023). https://doi.org/10.26599/TST.2022.9010047

Computational Statistics and Formal Methods

On Missing Values and the Imputation in Learning Neural Networks

Jefferson A. Peña-Torres[1]($\boxtimes$) (iD) and Cristian E. Garcia[2] (iD)

[1] Universidad del Valle, Cali, Colombia
jefferson.amado.pena@correounivalle.edu.co
[2] Institución Universitaria Antonio José Camacho, Cali, Colombia
cegarcia@admon.uniajc.edu.co

Abstract. Effective handling of missing values and imputation is critical in data-driven learning. Various approaches are available, including imputation, deletion, and direct training with incomplete data. Each method has its own set of advantages and disadvantages, impacting learning performance differently. This paper aims to evaluate the performance of a neural network model using two complete data sets, where different percentages of missing values are later introduced and imputed with zero and mean methods. We compare model performance with complete, imputed data and present two adapted versions to handle missing values. The experiments involve introducing missing values into the data sets and applying zero and mean imputation methods. Additionally, we introduce two adapted models, namely Masked and MissNN, specifically designed to handle missing values. Our experiments demonstrate that the adapted models, Masked and MissNN, exhibit greater consistency and minor spread in performance, achieving results comparable to training with complete data.

Keywords: Multilayer perceptron · Imputation · Missing data · Regression model

1 Introduction

The presence of missing values (MVs) emerges as a notable issue that has gained interest in research areas such as data mining, machine learning and statistics [8]. In practice, MV means that there is no access to all data and although reasons for missing values are diverse, their presence can be random (MAR), completely at random (MCAR), or not at random (MNAR) [10,31].

The presence of MVs is common in real-life data and can be challenging because it alters the data distribution, adds uncertainty, and consequently leads to overfitting, biased estimates, poor predictive performance, and undesirable training times, among other issues [43]. Therefore, numerous imputation methods have been developed to complete and restore dataset integrity [6]. Although the imputation is challenging, statistical and machine learning-based techniques

© The Author(s), under exclusive license to Springer Nature Switzerland AG 2024
N. D. Duque-Méndez et al. (Eds.): CCC 2024, CCIS 2208, pp. 187–200, 2024.
https://doi.org/10.1007/978-3-031-75233-9_14

consistently estimate or infer values. However, the presence of missing values in learning predictive models raises other performance issues, whether or not the training and testing data have been imputed [7,8,16,29,44]. Furthermore, in several cases the treatment, imputation or impact on performance is rarely reported [23].

Artificial Neural Networks (ANN) are a fundamental model in modern Machine Learning. In particular, for their usage in Deep learning methods that are entirely based on artificial neural networks and that have achieved notable results in various areas, such as Natural Language Processing [1,13] and Computer Vision [5,32]. A relevant ANN architecture is the multilayer perceptron (MLP), a feedforward ANN with multiple neurons fully-connected, a supervised learning algorithm that learns from the dataset and allows a non-linear approximator model for either classification or regression tasks [14,21,39]. The MLP follows the supervised approach and for learning completed data are needed. However, a challenge arises when these methods find missing values [4,18,19,41].

To build supervised-models using data with MVs presence a way is i) fill data with statistical or machine learning-based imputation method, and the explanations and results can vary and ii) train, test and validate using incomplete data if the technique allows it. While imputation is an essential step towards more reliable data analysis, an approach to address the situation could be the design, implementation and adaptation of the techniques that allow the learning with MVs presence during training. This approach provides insight into the subspace correlated with the observed data and allows for a comparison of the performance of different machine learning (ML) models. In this paper, we build a NN-based model and assess the performance with and without imputation. We also assess two adapted versions of the model to handle the presence of missing values. We compare the performance on two datasets with presence of 5%, 10%, 20% and 30% of missing values.

The rest of this paper is organized as follows. Section 2, we present a comprehensive overview of relevant research and related work. Section 3 presents datasets, simulated scenarios with missing values, explains our baseline model and their adaptations to deal with the presence of missing values, and the metrics used for comparison. Section 4 shows the results and discussion of our experiments. Section 5 the Conclusion and Future Research.

2   Related Works

The MVs within the dataset poses a relevant challenge for classification tasks. Several research studies have explored learning classifiers from incomplete data, resulting in the identification of three possible approaches.

Imputation before learning, involves the usage of some imputation method, statistical such as mode, mean, or median of the available value, regression and more-sophisticated machine learning-based methods. Following this approach, mixed-models and maximum likelihood density estimation for imputation before learning were the basis to explore with missing patterns, expectation-maximization principle, likelihood, and other bayesian-based imputation methods has been key [2, 30]. Estimation of MVs using distribution based analysis [11], subspace analysis [40] and regression based analysis [12, 35] allows the imputation task. Although, it is a separate task to the learning and the impact on performance and imputation outcomes are rarely reported [23].

Consequently, following the imputation during learning approach, machine learning-based methods are increasingly being employed to handle imputation tasks. By integrating imputation within the learning process, machine learning models can learn to simultaneously impute missing values and perform classification tasks. Kernel and Support vector machines have shown relevant results [17, 25, 45] which admit vectors with missing values and compute similarities and learn on available data. K-Nearest neighbors that using a quantitative attribute or qualitative attribute of all the non-missing values find the missing ones [9, 22, 38]. Random Forest an ensemble method which learns on available data and performs the classification [26, 37]. Recently deep learning (DL) methods have been widely used for imputation but conventional methods outperform results [36]. The reasons are associated with limitation of available data, representative subspace among other statistical features that DL methods do not take into account. Combining the math, statistical and deep learning knowledge have demonstrated the robustness and feasibility of DL-based methods for imputation tasks [20, 43].

The third approach is to work with raw data, without imputation. This approach creates uncertainty in the prediction results due to the presence of missing values despite the robustness of certain algorithms [28]. The kernel and SVM based has been extensively used under this approach, because they maximized the margin of the subspace related to available data and allows optimization of the nonlinear classifier to seek the best data subspace [33, 42]. Following the same approach, several researchers have chosen to delete, remove or discard the missing values after the learning [6]. But it has been demonstrated that this option introduces bias especially when MVs are not randomly distributed [3, 15, 34].

3   Neural Network Architectures

Neural network architectures, including Multi-Layer Perceptrons (MLPs), are fundamental components of modern machine learning systems. These networks are characterized by their ability to learn complex nonlinear relationships within data by iteratively adjusting the weights and biases of connections between neurons during training. MLPs similar to other machine learning (ML) methods are sensitive to presence of missing values.

To handle missing values, it is necessary to modify the NN architecture to accept inputs with MVs. Fortunately, NN have a high degree of customizability, making them adaptable to various tasks and datasets and their architecture is flexible in terms of the number of layers, the number of neurons in each layer, and the choice of activation functions. Here, we describe the training of an NN model after the mean imputation and compare it with an adapted NN-based model to handle missing values.

3.1   Datasets

In order to compare results, we select a publicly available dataset and a own scrape from the web. Both used in other researches studies, red wine quality [3] and coffee quality [24]. The selection of these two datasets is based on the popularity, dimensionality and multiple target categories. Moreover, they have been used widely for classification, regression and missigness analysis.

3.2   Missingness Scenarios

To simulate the presence of missing values complete data is required. We systematically introduce 5%, 10%, 20% and 30% missing values according to the MCAR and MNAR mechanism and build multiple datasets with missing data. Additionally, we train our model using incomplete data and compare its performance with that of other similar models trained on imputed data.

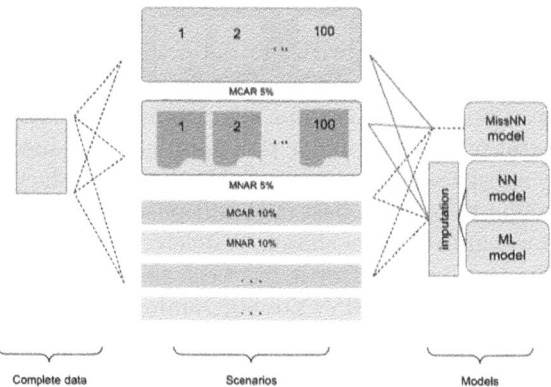

Fig. 1. Overview of the missing scenarios and learning models with and without imputation

Figure 1 depict each scenario, we simulate from complete datasets percentages of 5%, 10%, 20% and 30% missing values and we generate ten (10) incomplete datasets by each percentage using a missing completely at random (MCAR) or missing non-random (MNAR) mechanism.

3.3 Multilayer NN Model

A multilayer perceptron (MLP) is a type of ANN that consists of multiple layers of neurons, including an input layer, one or more hidden layers, and an output layer. Figure 2 shows a schematic representation of our MLP design. The proposed architecture comprises Xd input neurons, as input layer, two hidden linear layers each with ReLU activation functions, and a dropout layer, then, these layers are followed by a single linear output layer.

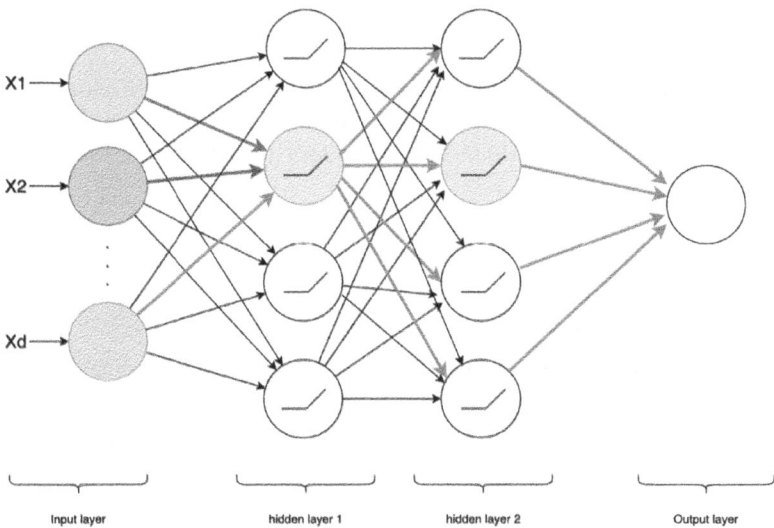

Fig. 2. Neural network architecture. Source:Authors

A simple neural network architecture which from the input layer propagates dataset representation through subsequent layers and eventually, while training and testing, allows the classification. Each layer feeds into the layer above it, until we generate outputs and the final layer is the linear predictor. Our implementation uses the Pytorch library [27], and all linear layers are initialized with He/Kaiming method. We used this NN-architecture as the baseline for our experiments.

3.4 Adapted Models

MVs are challenging and for our experiments we modify the baseline architecture for missing handling. First, we build MissNN, a variant with a masking layer which acts as filters selectively altering MVs before they reach subsequent layers in the network. For each missing value, the masked layer generates a random number from a normal distribution calculated for the column. The masked layer offers several advantages. It allows the NN architecture to handle missing values, ensuring that multiple imputation methods can be used as part of the network. Moreover, combining with imputation based on the normal distribution provides a consistent way to learn data patterns.

Our second modification, MissNN which includes a function that during forward propagation handles missing values and prevents them from being used during backpropagation. This function is responsible for computing the output of each neuron and propagating it forward through networks. In the baseline model, neurons correspond to features of the input and if there are presence of missing values, they are imputed according to distributional context.

3.5 Performance Metrics

The performance evaluation of machine learning models can be done using different criteria. We consider the most used. Mean Absolute Error (MAE), Mean Squared Error (MSE), Root Mean Squared Error (RMSE) and Area Under the Curve (AUC).

$$\frac{1}{n} \sum_{i=1}^{n} |y_i - \hat{y}_i| \tag{1}$$

Mean Absolute Error (MAE). Measure the average difference between the predicted and actual values in a dataset. Mathematically, Eq. 1 shown how MAE is calculated. Where, n is the number of samples in the dataset, yi represents the target value in the i-th sample, yi represents the predicted value in the i-th sample and closes with (|) symbol denotes the absolute value.

$$\frac{1}{n} \sum_{i=1}^{n} (y_i - \hat{y}_i)^2 \tag{2}$$

Mean Squared Error (MSE): Measure the average of the squared differences between the predicted and actual values in a dataset. Mathematically, Eq. 2 shown how MSE is calculated. Where n is the number of samples in the dataset, yi represents the target value in the i-th sample, yi represents the predicted value in the i-th sample.

$$\sqrt{\frac{1}{n} \sum_{i=1}^{n} (y_i - \hat{y}_i)^2} \tag{3}$$

Root Mean Square Error (RMSE): Provides a measure of the average magnitude of the errors made by the model, expressed in the same units as the target variable. Mathematically, Eq. 3 shown how RMSE is calculated. Where n is the number of samples in the dataset, yi represents the target value in the i-th sample, yi represents the predicted value in the i-th sample.

R-Square (R^2): is calculated. TRP is the True positive rate, also known as Recall and calculated as TP/ (TP + FN), FPR is the false positive rate, calculated as FP/(FP + TN), where TP is the number of true positives, FN is the number of false negatives, FP is the number of False positives and TN is the number of true negatives.

4   Results and Discussion

Table 1 shows the performance across both datasets. Wine quality and coffee quality previously used for classification/regression and missing values studies respectively. Both datasets were converted to incomplete versions and divided into training and test subsets.

Table 1. R^2 measurement with 5% of MVs with MCAR mechanisms using wine dataset

Data	Model	Missing value ratio			
		5	10	20	30
Wine [3]	NN + Zero imputation	0.493 ± 0.262	0.552 ± 0.215	0.481 ± 0.240	0.421 ± 0.257
	NN + Mean Imputation	0.520 ± 0.261	0.590 ± 0.209	0.517 ± 0.238	0.465 ± 0.247
	MissNN Masked	0.508 ± 0.213	0.360 ± 0.153	0.310 ± 0.172	0.273 ± 0.180
	MissFF	0.504 ± 0.236	0.481 ± 0.176	0.416 ± 0.205	0.362 ± 0.225
Coffee [24]	NN + Zero imputation	0.749 ± 0.072	0.728 ± 0.079	0.690 ± 0.103	0.663 ± 0.107
	NN + Mean Imputation	0.773 ± 0.080	0.719 ± 0.092	0.714 ± 0.073	0.704 ± 0.077
	MissNN Masked	0.608 ± 0.094	0.529 ± 0.102	0.500 ± 0.090	0.362 ± 0.225
	MissFF	0.655 ± 0.143	0.586 ± 0.041	0.576 ± 0.072	0.575 ± 0.075

We analyzed the distribution of performance and error metrics, such as the mean ± standard deviation obtained through the cross-validation process. We consider the R^2 as the central metric to compare the performance of models. Our baseline model (NN) yields a score of 0.325 ± 0.014 and 0.598 ± 0.273 using complete data from wine and coffee datasets respectively.

Using MCAR (5%) mechanism and imputing with zero and mean before the training stage, it achieves 0.586 ± 0.252, 0.597 ± 0.245 for wine dataset and 0.749 ± 0.072, 0.773 ± 0.080 for coffee dataset while with MNAR(5%) 0.526 ± 0.270, 0.541 ± 0.266 for wine and 0.546 ± 0.275, 0.550 ± 0.264 for coffee. The proposed models showcasing a coherent performance under the same conditions. With MCAR (5%), the Masked model yields 0.508 ± 0.214 and 0.608 ± 0.094 for the wine and coffee dataset. The forward (FF) function yields 0.504 ± 0.236 and 0.655 ± 0.143 for wine and coffee respectively.

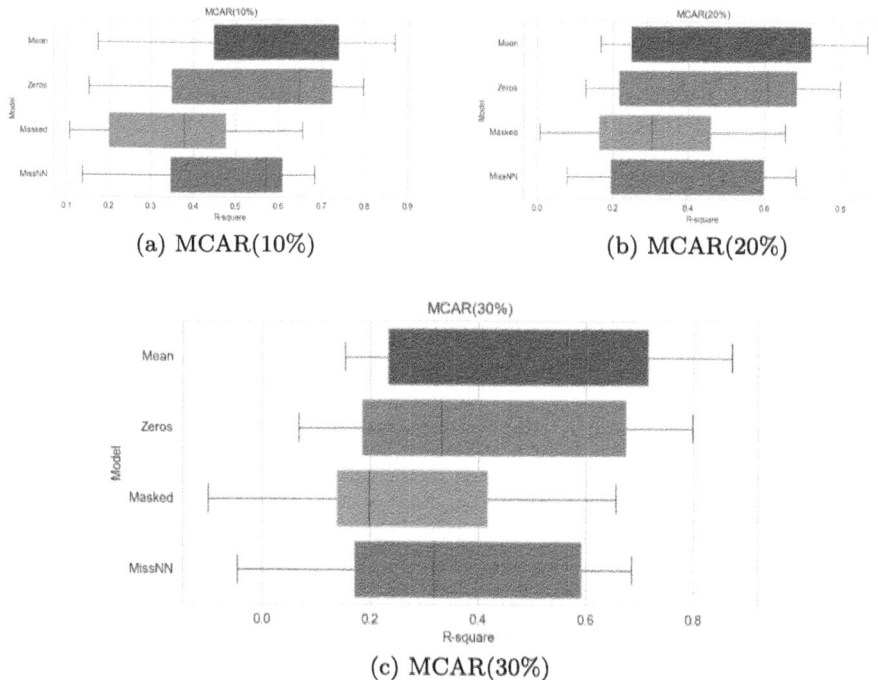

(a) MCAR(10%)   (b) MCAR(20%)

(c) MCAR(30%)

Fig. 3. R^2 with MCAR (10%), MCAR (20%) and MCAR (30%) for wine quality dataset. Source: authors

In regression analysis, a potential consequence of mean imputation is that estimated values are more similar to available data. While that with zero imputation variability is reduced. As a result, the model trained on imputed data learns more discernible patterns, which contribute to an enhanced R^2 metric. Boxplots in Fig. 3 illustrate the measurements obtained for each model under varying missing percentages, showcasing R^2 values superior to those of the baseline model. Mean imputation (Mean) shows a narrow spread of the metric, the presence of more similar values biases the evaluation and presents this better way when there are missing values. Imputation with mean (Mean) and zero value

(Zeros) exhibits a wider spread and suggests a major variability in the performance. While Masked and NissNN models have a consistent spread, maintaining a significant variability and slightly lower median values.

The interquartile range for Masked and NissNN models are slightly narrower compared to the mean imputation. This suggests that, on average, the R^2 values for the "Masked" model have less dispersion than the R^2 values for the "Mean" based model. A relevant finding because models are consistent and with lower variability although it does not achieve the major R^2 measurement.

In order to assess consistency, we set the coffee dataset with different percentages of missing data and train and test our models. In Fig. 4 we also found that NissNN models are slightly narrower compared to the mean imputation. The Masked model for this dataset displays the widest spread.

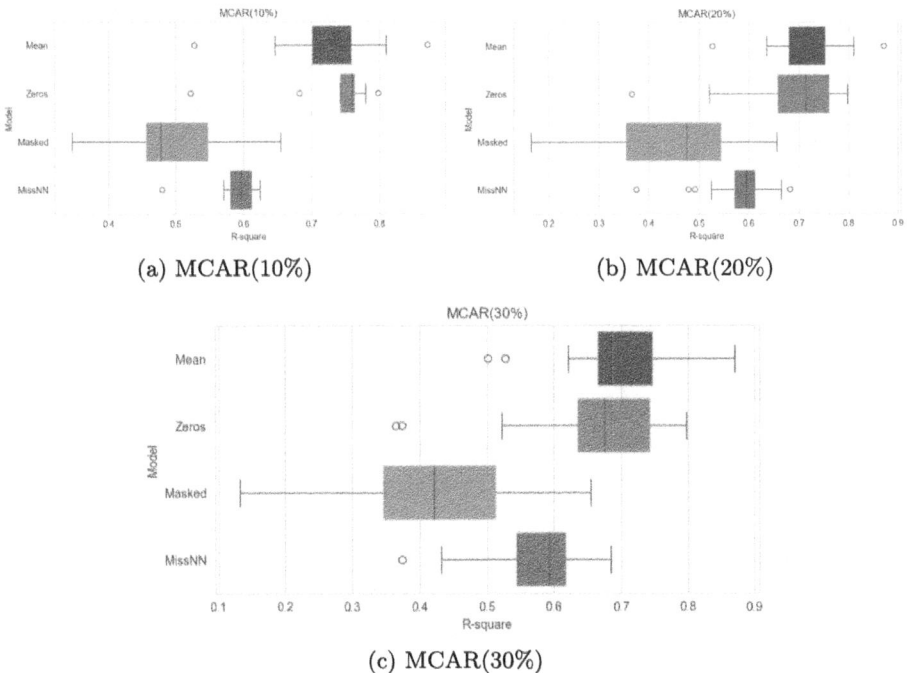

(a) MCAR(10%)   (b) MCAR(20%)

(c) MCAR(30%)

Fig. 4. R^2 with MCAR (10%), MCAR (20%) and MCAR (30%) for coffee quality dataset. Source: authors

Notably, the proposed models demonstrate consistent performance. Specially, in MCAR datasets with short interquartile ranges and comprehensible variability. Our MissNN model in MNAR incomplete datasets yield similar behavior and the Masked model varies across the presence of missing values. This suggests that performance and the behavior can vary according to the type, size, and distribution of values into the dataset.

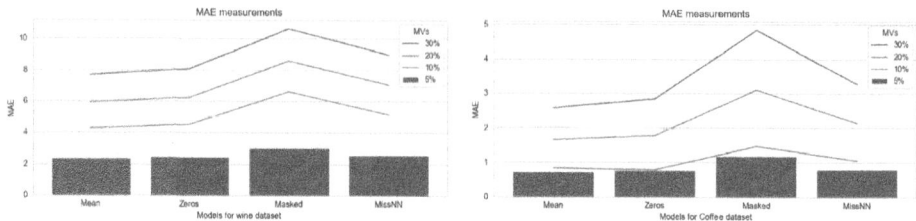

Fig. 5. MAE in presence of missing values. Source: Authors

In the MissNN model the function to handle missing values and avoid propagation leads to the stability of the model and the alteration of input with a previous layer leads to a major spread. We also assess the error for each model. The plot of MAE, MSE, RMSE metrics for different percentages of the missing values in wine and coffee dataset are shown in Fig. 5, Fig. 6 and Fig. 7. The errors are coherents to the presence of missing values. When the percentage of MVs increases the amount of available data decreases and as a result poor model performance.

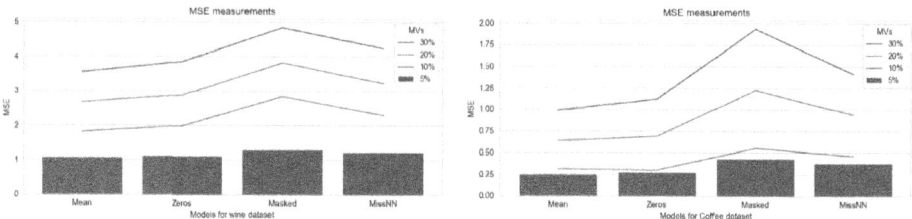

Fig. 6. MSE in presence of missing values. Source: Authors

Our models are sensitive to missing values, mainly the Masked. MissNN maintains a similar error rate to the Mean model. The distribution of Missing values impacts the model performance according to the MCAR or MNAR mechanism used. Figure 8 shows that MNAR datasets lead to a minor difference among error ratios when the percentage increases.

The numerical findings suggest that Neural networks can be adapted, modified to handle the presence of missing values. For regression analysis the mean and zero imputations can lead to bias and results associated with the similarity of values or by patterns added in the data. The experiments with NN models are promising, yet the generalizability remains challenging because test, validation and other adaptations and comparisons are possible.

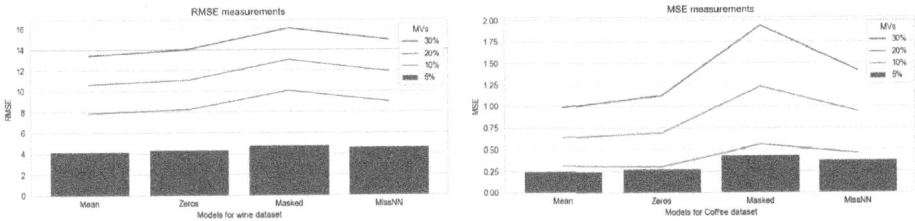

Fig. 7. RMSE in presence of missing values. Source: Authors

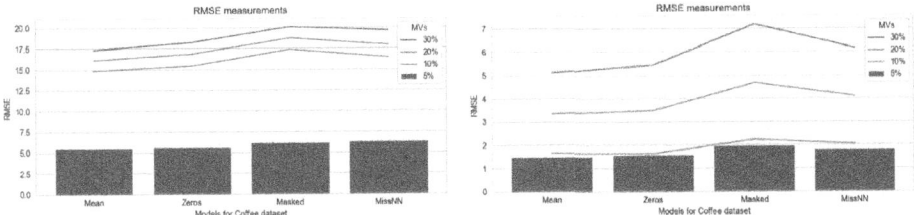

Fig. 8. RMSE with the MNAR(left) and MCAR(right) mechanism. Source: Authors

5   Conclusions

In this paper, we implemented a NN-based model to learn in the presence of missing values. We used two datasets and introduced different missing percentages. We evaluate the performance of the baseline model after imputation with zero and mean methods. The present research contributes to advancing direct training and learning with Incomplete data approach by providing a comprehensive comparison among various Neural network-adaptations and other ways to impute and remove missing values. Although the issues persist with the presence of the missing values, we consider that this approach offers a viable path toward performance analysis and exploration of other methods more trustworthy to handle missing values during training.

In the future, advancements in neural network (NN) technology are expected to continue, leading to various modifications and enhancements. These developments may involve the integration of other missing imputation methods, such as random-sample and distribution-based techniques. Neural network adaptations and modifications which include theoretical and statistical considerations, such as Neuman series approximations, regularization, dropout, capture missing values as features, arbitrary value replacement and so forth.

Notes:

- Code for reproducing paper experiments is available at: https://github.com/japeto/neuraln-missing-values
- Wine quality dataset from UCI: https://archive.ics.uci.edu/dataset/186/wine+quality

Acknowledgement. The authors gratefully acknowledge the support provided by Universidad del Valle and Universidad Valparaíso for their assistance and resources in facilitating this research. We extend our sincere appreciation to GUIA Research group, whose expertise and collaboration were invaluable to the successful completion of this project.

References

1. Baroni, M.: Linguistic generalization and compositionality in modern artificial neural networks. Philos. Trans. Roy. Soc. B: Biol. Sci. **375**(1791), 20190307 (2019)
2. Batista, G.E.A.P.A., Monard, M.C.: An analysis of four missing data treatment methods for supervised learning. Appl. Artif. Intell. **17**(5-6), 519–533 (2003). https://doi.org/10.1080/713827181
3. Berrevoets, J., Imrie, F., Kyono, T., Jordon, J., van der Schaar, M.: To impute or not to impute? Missing data in treatment effect estimation. In: Proceedings of The 26th International Conference on Artificial Intelligence and Statistics, pp. 3568–3590. PMLR (2023). ISSN 2640-3498
4. Bhattacharya, A., Bhose, S., Choudhury, S.J.: Classification of incomplete data using augmented MLP. In: 2023 International Conference for Advancement in Technology (ICONAT), pp. 1–5 (2023)
5. Celard, P., Iglesias, E.L., Sorribes-Fdez, J.M., Romero, R., Seara Vieira, A., Borrajo, L.: A survey on deep learning applied to medical images: from simple artificial neural networks to generative models. Neural Comput. Appl. **35**(3), 2291–2323 (2023)
6. Emmanuel, T., Maupong, T., Mpoeleng, D., Semong, T., Mphago, B., Tabona, O.: A survey on missing data in machine learning. J. Big Data **8**(1), 140 (2021)
7. Feng, R., Calmon, F., Wang, H.: Adapting fairness interventions to missing values. In: Oh, A., Neumann, T., Globerson, A., Saenko, K., Hardt, M., Levine, S. (eds.) Advances in Neural Information Processing Systems, vol. 36, pp. 59388–59409. Curran Associates, Inc. (2023)
8. Fernando, M.-P., Cèsar, F., David, N., José, H-O.: Missing the missing values: the ugly duckling of fairness in machine learning. Int. J. Intell. Syst. **36**(7), 3217–3258 (2021). https://onlinelibrary.wiley.com/doi/pdf/10.1002/int.22415
9. García-Laencina, P.J., Sancho-Gómez, J.-L., Figueiras-Vidal, A.R., Verleysen, M.: K nearest neighbours with mutual information for simultaneous classification and missing data imputation. Neurocomputing **72**(7), 1483–1493 (2009)
10. Gomer, B.: MCAR, MAR, and MNAR values in the same dataset: a realistic evaluation of methods for handling missing data. Multivariate Behav. Res. **54**(1), 153–153 (2019). https://doi.org/10.1080/00273171.2018.1557033
11. Jafrasteh, B., Hernández-Lobato, D., Lubián-López, S.P., Benavente-Fernández, I.: Gaussian processes for missing value imputation. Knowl.-Based Syst. **273**, 110603 (2023)
12. Khan, S.I., Hoque, A.S.M.L.: SICE: an improved missing data imputation technique. J. Big Data **7**(1), 37 (2020)
13. Khemani, B., Patil, S., Kotecha, K., Tanwar, S.: A review of graph neural networks: concepts, architectures, techniques, challenges, datasets, applications, and future directions. J. Big Data **11**(1), 18 (2024)
14. Krogh, A.: What are artificial neural networks? Nat. Biotechnol. **26**(2), 195–197 (2008)

15. Kwak, S.K., Kim, J.H.: Statistical data preparation: management of missing values and outliers. Korean J. Anesthesiol. **70**(4), 407–411 (2017)
16. Morvan, M.L., Josse, J., Scornet, E., Varoquaux, G.: What's a good imputation to predict with missing values? In: Ranzato, M., Beygelzimer, A., Dauphin, Y., Liang, P.S., Wortman Vaughan, J. (eds.) Advances in Neural Information Processing Systems, vol. 34, pp. 11530–11540. Curran Associates, Inc. (2021)
17. Liu, X., et al.: Multiple kernel kk-means with incomplete kernels. IEEE Trans. Pattern Anal. Mach. Intell. **42**(5), 1191–1204 (2020)
18. Lu, H., Zhang, L.: Incomplete data classification based on the tracking-removed autoencoder. In: 2023 42nd Chinese Control Conference (CCC), pp. 8394–8400 (2023). ISSN 1934-1768
19. Markey, M.K., Tourassi, G.D., Margolis, M., DeLong, D.M.: Impact of missing data in evaluating artificial neural networks trained on complete data. Comput. Biol. Med. **36**(5), 516–525 (2006)
20. Morvan, M.L., Josse, J., Moreau, T., Scornet, E., Varoquaux, G.: NeuMiss networks: differentiable programming for supervised learning with missing values. arXiv:2007.01627 [cs, stat] (2020)
21. Murotzhonovich, T.S.: Introduction to artificial neural networks. Web Synergy: Int. Interdisc. Res. J. (2023)
22. Murti, D.M.P., Pujianto, U., Wibawa, A.P., Akbar, M.I.: K-nearest neighbor (k-NN) based missing data imputation. In: 2019 5th International Conference on Science in Information Technology (ICSITech), pp. 83–88 (2019)
23. Nijman, S.W.J., et al.: Missing data is poorly handled and reported in prediction model studies using machine learning: a literature review. J. Clin. Epidemiol. **142**, 218–229 (2022)
24. Ochoa-Muñoz, A.F., Peña-Torres, J.A., García-Bermúdez, C.E., Mosquera-Muñoz, K.F., Mesa-Diez, J.: On characterization of sensory data in presence of missing values: the case of sensory coffee quality assessment. Ingeniare. Rev. chilena ingeniería **30**(3), 564–573 (2022)
25. Palanivinayagam, A., Damaševičius, R.: Effective handling of missing values in datasets for classification using machine learning methods. Information **14**(2), 92 (2023)
26. Pantanowitz, A., Marwala, T.: Missing data imputation through the use of the random forest algorithm. In: Yu, W., Sanchez, E.N. (eds.) Advances in Computational Intelligence, pp. 53–62. Springer, Heidelberg (2009)
27. Paszke, A., et al.: PyTorch: an imperative style, high-performance deep learning library. In: Advances in Neural Information Processing Systems, vol. 32. Curran Associates, Inc. (2019)
28. Pelckmans, K., De Brabanter, J., Suykens, J.A.K., De Moor, B.: Handling missing values in support vector machine classifiers. Neural Netw. **18**(5), 684–692 (2005)
29. Poulos, J., Valle, R.: Missing data imputation for supervised learning. Appl. Artif. Intell. **32**(2), 186–196 (2018). https://doi.org/10.1080/08839514.2018.1448143
30. Ramoni, M., Sebastiani, P.: Robust learning with missing data. Mach. Learn. **45**(2), 147–170 (2001)
31. Rubin, D.B.: Inference and missing data. Biometrika **63**(3), 581–592 (1976)
32. Schiatti, L., et al.: Modeling visual impairments with artificial neural networks: a review. In: Proceedings of the IEEE/CVF International Conference on Computer Vision (ICCV) Workshops, pp. 1987–1999 (2023)
33. Sheikholesalmi, F., Mardani, M., Giannakis, G.B.: Classification of streaming big data with misses. In: 2014 48th Asilomar Conference on Signals, Systems and Computers, pp. 1051–1055 (2014). ISSN 1058-6393

34. Soley-Bori, M.: Dealing with missing data: key assumptions and methods for applied analysis. Boston University (2013)
35. Song, Q., Shepperd, M.: Missing data imputation techniques. Int. J. Bus. Intell. Data Min. **2**(3), 261–291 (2007)
36. Sun, Y., Li, J., Yifan, X., Zhang, T., Wang, X.: Deep learning versus conventional methods for missing data imputation: a review and comparative study. Expert Syst. Appl. **227**, 120201 (2023)
37. Tang, F., Ishwaran, H.: Random forest missing data algorithms. Stat. Anal. Data Min.: ASA Data Sci. J. **10**, 363–377 (2017)
38. Tran, C.T., Nguyen, B.P.: Random subspace ensemble for directly classifying high-dimensional incomplete data. Evol. Intell. (2024)
39. Walczak, S.: Artificial neural networks. In: Advanced Methodologies and Technologies in Artificial Intelligence, Computer Simulation, and Human-Computer Interaction, pp. 40–53. IGI Global (2019)
40. Wang, H., Kim, J.K.: Statistical inference using regularized M-estimation in the reproducing kernel Hilbert space for handling missing data. Ann. Inst. Stat. Math. **75**(6), 911–929 (2023)
41. Williams, D., Liao, X., Xue, Y., Carin, L., Krishnapuram, B.: On classification with incomplete data. IEEE Trans. Pattern Anal. Mach. Intell. **29**(3), 427–436 (2007)
42. Xu, Z., Liu, Y., Li, C.: Distributed semi-supervised learning with missing data. IEEE Trans. Cybern. **51**(12), 6165–6178 (2021)
43. Yang, J., Wang, Y., Yang, Y., Ding, K., Na, C., Yang, Y.: Effects of single and multiple imputation strategies on addressing over-fitting issues caused by imbalanced data from various scenarios. Appl. Intell. **54**(3), 2812–2830 (2024)
44. Zaffran, M., Dieuleveut, A., Josse, J., Romano, Y.: Conformal prediction with missing values. In: Krause, A., Brunskill, E., Cho, K., Engelhardt, B., Sabato, S., Scarlett, J. (eds.) Proceedings of the 40th International Conference on Machine Learning. Proceedings of Machine Learning Research, , vol. 202, pp. 40578–40604. PMLR (2023)
45. Zhang, S., Jin, Z., Zhu, X.: Missing data imputation by utilizing information within incomplete instances. J. Syst. Softw. **84**(3), 452–459 (2011)

Formalisation of Hall's Theorem for Countable Infinite Graphs

Fabián Fernando Serrano Suárez[1] , Mauricio Ayala-Rincón[2][(✉)] ,
and Thaynara Arielly de Lima[3]

[1] Universidad Nacional de Colombia - Sede Manizales, Manizales, Colombia
ffserranos@unal.edu.co
[2] Universidade de Brasília, Brasília, D.F., Brazil
ayala@unb.br
[3] Universidade Federal de Goiás, Goiânia, Brazil
thaynaradelima@ufg.br

Abstract. This work presents two formalisations in Isabelle/HOL of the extension of Hall's marriage theorem for finite graphs to countable infinite graphs. The proofs use a formalisation of the authors' countable set-theoretical version of Hall's theorem, which was proved using a formalisation in Isabelle/HOL of the compactness theorem for propositional logic by dealing with finite families of sets through the well-known marriage-condition characterisation. The first formalisation focuses on maintaining specifications and proofs as closely as possible to textbook proofs. The second one states the theorem directly in terms of the existence of perfect matchings over finite and infinite graphs, profiting from the conciseness of Isabelle/HOL locales' technology. The development contributes to mechanising countable infinite versions of properties equivalent to Hall's marriage theorem in contexts other than set theory.

Keywords: Interactive Theorem Proving · Graph Theory · Set Theory · Combinatorics · Automated deduction

1 Introduction

Hall's marriage theorem is a landmark result established primarily by Philip Hall [17], and it is equivalent to several other significant theorems in combinatorics and graph theory (cf. [8,9,28]), namely: Menger's theorem (1929), König's

Research funded by grants: CNPq Universal 409003/21-2, and Productivity 313290/2021-0, FAPDF DE 00193-00000229/2021-21, FAPEG 202310267000223, and CAPES PrInt under financial code 001. The second and third authors are grateful to the Hausdorff Institute für Mathematik (Universität Bonn) for hosting them as part of the trimester "Prospects of formal Mathematics", funded by the Deutsche Forschungsgemeinschaft (DFG) under Germany Excellence Strategy - EXC-2047/1 - 390685813. A preliminary version of this work was presented at the 24th Int. Conf. on Logic for Programming Artificial Intelligence and Reasoning (available as preprint [33]).

minimax theorem (1931), König-Egerváry theorem (1931), Dilworth's theorem (1950), Max Flow-Min Cut theorem (related to the well-known Ford-Fulkerson algorithm), among others. Consequently, any mechanisation of Hall's theorem allows one to prove any of those equivalent results formally.

Two well-known versions of Hall's theorem exist, one for *finite* families of finite sets and another for *finite* graphs. The proofs of any previously cited equivalences can be more adapted to a specific version of Hall's Theorem, either the set-theoretical or the graph-theoretical version. For example, König-Egerváry theorem states that the minimum cover in a finite bipartite graph has the same cardinality as a maximum matching. Thus, if we assume Hall's theorem for finite graphs, one possible way to infer König-Egerváry theorem will consist of building a reduction from the latter to the former. Considering the nature of König-Egerváry theorem, it is clear that the graph-theoretical version of Hall's theorem is more appropriate than the set-theoretical version to establish the equivalence between these theorems.

Although we referred to the *finite* versions of the mentioned results in the previous paragraphs, we point out that extensions to *infinite* sets and graphs are of primary interest [2].

Mechanisations such as those presented in this work aim to pave the way to develop formalisations of *infinite* versions of some theorems in combinatorics related to Hall's Theorem. For example, the authors formalised the set-theoretical version of Hall's Theorem for a countable (infinite) collection of finite subsets $\{S_i\}_{i \in I}$ of a set S [32]. Such a development applied a formalisation of the compactness theorem for propositional logic, developed by Serrano in [31], and Jiang and Nipkow's formalisation for the finite case of the set-theoretical version of Hall's theorem [21].

As main results, this work discusses how applying authors' development in [32], the *infinite* graph-theoretical version of Hall's theorem is mechanised in Isabelle/HOL. The result applies to a general class of *infinite* bipartite graphs with finite neighbourhoods regarding one of the sets of vertices of the vertex bipartition. Additionally, a second succinct formalisation of the same result that uses Isabelle locales is also discussed. The formalisations are of practical interest since they can be used to establish the mechanisation of other combinatorial results, as the previous ones discussed, over *infinite* sets and graphs.

Interestingly, other combinatorial well-known results equivalent to Hall's theorem in the finite case are not straightforwardly equivalent in the infinite case; for instance, the infinite version of König-Egerváry theorem that as reported in [2] cannot be inferred from the compactness theorem.

Organisation. Section 2 discusses Hall's marriage theorem for finite and infinite countable sets and graphs and explains the equivalence between the versions for graphs and sets. Then, Sect. 3 presents the two formalisations in Isabelle/HOL of the graph-theoretical version of Hall's theorem for countable graphs. Section 4 discusses related work before concluding in Sect. 5. The paper includes links to the formalisation highlighted by the symbol 🔗.

2 Hall's Theorem for Sets and Graphs

2.1 Finite and Infinite Versions of Hall's Theorem

Hall's theorem for sets establishes that a finite family $\{S_i\}_{i \in I}$ of finite sets not necessarily disjoint, of elements in a set S, has a system of distinct representatives (SDR) if and only if the so-called *marriage condition* holds. The marriage condition states that:

$$\text{For any } J \subseteq I, |J| \leq \left| \bigcup_{j \in J} S_j \right|$$

Above, an SDR for the family $\{S_i\}_{i \in I}$ is understood as a subset of elements of S that contains exactly an element for each set in the family. This can be formalised as an injective function $f : I \to S$, such that $f(i) \in S_i$, for $i \in I$.

Definition 1 (SDR). *Let S be an arbitrary set and $\{S_i\}_{i \in I}$ a collection of not necessarily distinct subsets of S with indices in the set I. An injective function $f : I \to \bigcup_{i \in I} S_i$ is an SDR for $\{S_i\}_{i \in I}$ if for all $i \in I$, $f(i) \in S_i$.*

Using the compactness theorem, a proof of a countable infinite version of this theorem was formalised in Isabelle/HOL [32]. The infinite version states that a countable family of finite sets, indexed by a set I, has a set of distinct representatives if and only if Hall's *marriage condition* below holds:

$$\text{For any } J \subseteq I, J \text{ finite}, |J| \leq \left| \bigcup_{j \in J} S_j \right|$$

Hall's theorem for finite graphs states that in a bipartite graph $G = \langle X, Y, E \rangle$, (where $E \subseteq X \times Y$,) there is a perfect matching covering X if and only if $|J| \leq |N(J)|$ for all $J \subseteq X$. Here, for $x \in X \cup Y$, the neighbourhood of x is the set of vertices $N(x) = \{y \mid (x, y) \in E, \text{ or } (y, x) \in E\}$. N is extended straightforwardly to sets.

Definition 2 (Directed bipartite digraph and perfect matching). *Let X and Y be nonempty sets. The triple $G = \langle X, Y, E \rangle$ is a directed bipartite digraph if and only if the following conditions hold.*
1. $X \cap Y = \emptyset$, and   2. $E \subseteq (X \times Y)$.
A subset of arcs $E' \subseteq E$ is a perfect matching of $G = \langle X, Y, E \rangle$ if and only if
1. $X = \{x \mid (x, y) \in E'\}$, and   2. E' is an injective relation.

The infinite version of Hall's theorem for graphs states that in a countable bipartite graph $G = \langle X, Y, E \rangle$, where for all $x \in X$, $N(x)$ is finite, there is a perfect matching covering X if and only if $|J| \leq |N(J)|$ for all J finite, $J \subseteq X$. It may be directly graph-theoretically stated as "there exists a perfect matching if and only if for any finite subgraph there is a perfect matching." Indeed, the graph and each of their subgraphs translate into a family of indexed sets, $\{N(j)\}_{j \in J}$, and the system of distinct representatives allows the construction of a perfect matching consisting of the set of edges $\{(j, y) \mid j \in J$, where y is the representative of $j\}$.

Notice that for the infinite version of this theorem, the finiteness of $N(x)$ cannot be relaxed; in fact, the graph $G = \langle \mathbb{N}, \mathbb{N}^+, \{(0, i) \mid i \in \mathbb{N}^+\} \bigcup \{(i, i) \mid i \in \mathbb{N}^+\}\rangle$ is an easy counterexample. In G, the sets of vertices $\mathbb{N}$ and $\mathbb{N}^+$ are seen as different copies of natural numbers.

The formalisation of Hall's Theorem for countable families in [32] uses Nipkow's formalisation of Hall's theorem for finite families of sets [20] and Serrano's formalisation of the compactness theorem for propositional logic [31].

2.2   Countable Versions of Hall's Theorem for Sets and Graphs

The equivalence between countable versions of this theorem for sets and graphs is clear intuitively.

On the one side, a countable bipartite graph $G = \langle X, Y, E \rangle$ gives a countable family of neighbourhoods $\{N(x)\}_{x \in X}$, which are finite sets under the constraint that neighbourhoods of vertices in X are finite. If M is a perfect matching of G, thus one builds an SDR by considering the injective function $f : X \rightarrow Y$ such that, for each $x \in X$, $f(x) = y$, where $(x, y) \in M$.

On the other side, if one has a countable family of finite sets $\{S_i\}_{i \in I}$ satisfying the marriage condition, then there exists a distinct set of representatives for $\{S_i\}_{i \in I}$, given by f. We consider the countable bipartite graph built as $G = \langle I, \bigcup_{i \in I} S_i, E \rangle$, where $E = \{(i, y) \mid i \in I, y \in S_i\}$. Since the sets in the countable family of sets $\{S_i\}_{i \in I}$ are finite the set of neighbourhoods in G, for each $i \in I$, $N(i)$, is finite; indeed, $|S_i| = |N(i)|$. Since f is injective, the perfect matching covering I is given by the set of arcs $M = \{(i, f(i)) \mid i \in I\}$.

3   Formalisation of Hall's Theorem for Graphs

Initially, we discuss how infinite families of sets and infinite bipartite graphs are specified. Afterwards, we explain how the proof of correction of the specialised construction of an SDR from a perfect matching over an infinite directed bipartite graph is used to conclude the infinite graph-theoretical version of Hall's theorem. Finally, a formalisation using Isabelle locales is presented.

3.1   Formalising Relations Between Sets and Graphs

The formalisation is constructive, and its kernel is the transformations of indexed infinite families of sets to and from directed bipartite digraphs. One of the vital features of our formalisation is how we build a *system of distinct representatives* (SDR) for a family of sets from a perfect matching over arbitrary directed bipartite graphs. Such transformations are more general than those discussed in the previous section since neither the family of sets need to be countable nor the sets in the family must be restricted to finite sets. Thus, the bipartite graph may also be non-countable, and the neighbourhoods of the vertices do not need to be finite. Theorems 1 and 2 present the reductions from a problem to another one and state that from the existence of a perfect matching, the resulting transformation is an indexed family of sets that has an SDR, and vice-versa.

Theorem 1 (SDR associated to a directed bipartite digraph). *Let $G = \langle X, Y, E \rangle$ be a directed bipartite digraph.*

The collection of sets associated to G is built as $\{V_i\}_{i \in I}$, where $I = X$, and for all $i \in I$, $V_i = \{y \mid (i, y) \in E\}$.

Therefore, if E' is a perfect matching of G, the function $R : I \to \bigcup_{i \in I} V_i$, defined as $R(i) = y$, where y is the unique element in V_i such that $(i, y) \in E'$, is an SDR of $\{V_i\}_{i \in I}$.

Theorem 2 (Perfect matching associated to a collection of sets). *Let $\{S_i\}_{i \in I}$ be a collection of non-necessarily distinct subsets of an arbitrary set S.*

The directed bipartite digraph associated to $\{S_i\}_{i \in I}$ is built as the graph $G = \langle X, Y, E \rangle$ where $X = I$, $Y = \bigcup_{i \in I} S_i$ and $E = \{(i, x) \mid i \in I \text{ and } x \in S_i\}$.

Therefore, if R is an SDR of $\{S_i\}_{i \in I}$, then the subset of arcs $E' = \{(i, x) \mid i \in I \text{ and } x = R(i)\}$ is a perfect matching of G.

Preliminaries and Definitions. The Isabelle Archive of Formal Proofs contains a collection of theories regarding Graph Theory [25]. In particular, Noschinski and Neumann specified, in the theory *Digraph.thy*, the primary data structure *pre_digraph* as the basis to develop complex formalisations such as Kuratowski theorem and the existence of a Eulerian path on directed finite graphs. We also apply such a *record* to establish our formalisation.

```
record ('a, 'b) pre_digraph =
  verts :: "'a set"        arcs :: "'b set"
  tail :: "'b ⇒ 'a"       head :: "'b ⇒ 'a"
```

Such a record from the theory mentioned above is used since the formalisation established in [25] contains specialised concepts intrinsic to the specific results formalised in it. For example, in the Isabelle AFP theory, *Kuratowski.thy* and *complete bipartite digraphs* are defined. However, there is no general specification of complete bipartite digraphs. Consequently, a small variety of basic concepts for graphs were specified. For instance, specifications of the *neighbourhood* of a vertex and the notion of *bipartite_digraph*, among others, are necessary to our development. In the following, some preliminary definitions are presented that were specified to establish the equivalence between the infinite versions of Hall's Theorem.

Arcs of a graph G have tails and heads in the set of vertices of the graph. The binary predicate neighbour🔗 on pairs of vertices u, v, holds if there exist and arc (u, v) or (v, u) in the graph. A bipartite_digraph🔗 is a *pre_digraph* G with two disjoint sets of vertices X and Y, whose union is the set of vertices of the graph, and such that all arcs in the graph have tails in X and heads in Y or vice versa.

definition tails:: "('a,'b) pre_digraph ⇒ 'a set" **where**
"tails G ≡ { tail G e |e. e ∈ arcs G }"

definition tails_set :: "('a,'b) pre_digraph ⇒ 'b set ⇒ 'a set" **where**
"tails_set G E ≡ { tail G e |e. e ∈ E ∧ E ⊆ arcs G }"

definition heads:: "('a,'b) pre_digraph ⇒ 'a set" **where**
"heads G ≡ { head G e |e. e ∈ arcs G }"

definition heads_set:: "('a,'b) pre_digraph ⇒ 'b set ⇒ 'a set" **where**
"heads_set G E ≡ { head G e |e. e ∈ E ∧ E ⊆ arcs G }"

definition neighbour:: "('a,'b) pre_digraph ⇒ 'a ⇒ 'a ⇒ bool" **where**
"neighbour G v u ≡
∃e. e∈ (arcs G) ∧ ((head G e = v ∧ tail G e = u) ∨
(head G e = u ∧ tail G e = v))"

definition neighbourhood:: "('a,'b) pre_digraph ⇒ 'a ⇒ 'a set" **where**
"neighbourhood G v ≡ {u |u. neighbour G u v}"

definition bipartite_digraph:: "('a,'b) pre_digraph ⇒ 'a set ⇒ 'a set
⇒ bool" **where**   "bipartite_digraph G X Y ≡
(X ∪ Y = (verts G)) ∧ X ∩ Y = {} ∧
(∀e ∈ (arcs G).(tail G e) ∈ X ⟷ (head G e) ∈ Y)"

The specialised notion of directed bipartite digraphs used is specified in definition dir_bipartite_digraph[↗] Such a graph is a bipartite digraph, consisting of a bi-partition of vertices X and Y in which all *arcs* have *tails* in the set X and *heads* in the set Y. Arcs with the same tail and head are equal.

definition dir_bipartite_digraph:: "('a,'b) pre_digraph ⇒ 'a set ⇒
'a set ⇒ bool"   **where**   "dir_bipartite_digraph G X Y ≡
(bipartite_digraph G X Y) ∧ ((tails G = X) ∧
(∀e1 ∈ arcs G. ∀e2 ∈ arcs G. e1 = e2 ⟷
head G e1 = head G e2 ∧ tail G e1 = tail G e2))"

Definition dirBD_matching[↗] specifies a matching in a directed bipartite digraph G is specified as a subset E of the arcs of the graph, such that any pair of distinct arcs in E have neither the same head nor the same tail. A perfect matching, specified in definition dirBD_perfect_matching[↗], is a matching in the digraph G that covers the set of vertices X.

definition dirBD_matching:: "('a,'b) pre_digraph ⇒ 'a set ⇒ 'a set ⇒
'b set ⇒ bool"   **where**   "dirBD_matching G X Y E ≡
dir_bipartite_digraph G X Y ∧ (E ⊆ (arcs G)) ∧
(∀e1∈E. (∀e2∈ E. e1 ≠ e2 ⟶
((head G e1) ≠ (head G e2)) ∧
((tail G e1) ≠ (tail G e2))))"

```
definition dirBD_perfect_matching::
  "('a,'b) pre_digraph ⇒ 'a set ⇒ 'a set ⇒ 'b set ⇒ bool"
  where    "dirBD_perfect_matching G X Y E ≡
                  dirBD_matching G X Y E ∧ (tails_set G E = X)"
```

The theory background_on_graphs🔗 includes all definitions in this subsection. It specialised graphs according to the target formalisation requirements.

Building SDRs from Perfect Matchings. Theorem 1, is specified as theorem dir_BD_to_Hall🔗 below. It uses the definition E_head🔗 that for any set of arcs E in a digraph and any vertex x, tail of some arc in E, selects the head, y, of an arc in E with tail x. The theorem states that for any directed bipartite digraph, $G = \langle X, Y, E \rangle$ with a perfect matching $E' \subseteq E$, the arcs of G, the family of sets given by the neighbourhoods of vertices $x \in X$ in G, $\{N(x)\}_{x \in X}$, the set of indices given by the set of vertices in X, and the representatives given by E_head using the perfect matching E', is an SDR. Since E' is a perfect matching, a unique arc with tail x in E' exists.

The required properties on the operator E_head on directed bipartite digraphs is that it gives an injective function over matchings also covering X over perfect matchings, which is stated as the crucial lemma dirBD_matching_inj_on🔗. The proof requires proving a chain of auxiliary lemmas, including one stating the unicity of the operator E_head over matchings and then constructing an injective function that univocally maps tails into heads on the set of arcs E'.

Then, after unfolding definitions, one concludes that $(E\_head \ G \ E)$, as an injective function on X, gives an SDR for the family of neighbourhoods of vertices in X, $\{N(X)\}_{x \in X}$, built from the graph G and the perfect matching E'.

```
definition E_head🔗   :: "('a,'b) pre_digraph ⇒ 'b set ⇒ ('a ⇒ 'a)"
  where   "E_head G E =
              (λx. (THE y. ∃ e. e ∈ E ∧ tail G e = x ∧ head G e = y))"
```

```
theorem dir_BD_to_Hall🔗:
  "dirBD_perfect_matching G X Y E ⟶
  system_representatives (neighbourhood G) X (E_head G E)"
```

3.2   Formalising the Graph-Theoretical Version of Hall's Theorem

Here, we explain how the graph-theoretical version of Hall's theorem is obtained from its set-theoretical version formalised in [32]. The graph-theoretical version is stated as Theorem 3.

Theorem 3 (Hall - marriage-conditioned graph-theoretical version).
Let $G = \langle X, Y, E \rangle$ be a directed bipartite digraph. G contains a perfect matching covering the set of vertices X if and only if

$$|J| \leq |N(J)| \quad \text{for all} \quad J \subseteq X$$

As mentioned in the introduction, the theorem may be stated without paraphrasing the marriage condition to the context of graph theory.

Theorem 4 (Hall - graph-theoretical version). *Let $G = \langle X, Y, E \rangle$ be a directed bipartite digraph. G contains a perfect matching covering the set of vertices X if and only if for all finite $X_s \subset X$ the induced bipartite digraph has a perfect matching.*

These theorems are usually stated for finite graphs only. Also, in contrast to proofs presented in classical textbooks on (finite) graph theory (e.g., [10,38]), their formalisations, given as the theorems *Hall_digraph*, at the end of this section, and *Hall_Graph*, in Subsect. 3.3, apply the combinatorial set-theoretical version of this theorem, obtained through application of the compactness theorem for propositional logic, extended for countable sets and published in [32].

The formalisation of Theorem 3 uses the Theorem 1 proved in Isabelle/HOL as described in Subsect. 3.1 as theorem *dir_BD_to_Hall* and that states the correctness of the reduction of a directed bipartite digraph $G = \langle X, Y, E \rangle$ with a perfect matching E, to the family of neighbourhoods of vertices X, concluding that the operator *E_head* indeed builds an SDR from the perfect matching E.

The formalisation is based on applying two auxiliary lemmas relating the marriage condition for directed bipartite digraphs to perfect matchings.

The first auxiliary lemma, marriage_necessary_graph🔗, states that if a directed bipartite graph has a perfect matching, then the marriage condition holds. Indeed, this lemma holds for arbitrary infinite graphs. Furthermore, relaxing the restriction on countable families to infinite families is possible since the lemma is proved as a consequence of the mechanisation of the fact that the existence of an SDR for arbitrarily infinite indexed families of finite sets implies the marriage condition. The last result was formalised through the theorem marriage_necessity🔗, part of the mechanisation reported in [32].

lemma `marriage_necessary_graph`🔗:
assumes `"(dirBD_perfect_matching G X Y E)"` **and**
      `"∀ i ∈X. finite (neighbourhood G i)"`
shows `"∀ J⊆X. finite J ⟶ (card J) ≤ card (⋃ (neighbourhood G ' J))"`

Applying the transformation *(system_representatives (neighbourhood G) X (E_head G E))* through theorem *dir_BD_to_Hall* is the tricky part of this lemma. So, from the SDR, one obtains an injective function R from any subset J to their representatives in the union of neighbourhoods of elements $j \in J$ such that: $card\ J \leq card\ (\bigcup_{j \in J} N(j))$. The injectivity of R, guaranteed by theorem *dir_BD_to_Hall*, implies the desired inequation.

The second auxiliary lemma, marriage_sufficiency_graph🔗 below, states that if the marriage condition holds for a countable directed bipartite graph, then there exists a perfect matching.

lemma `marriage_sufficiency_graph`🔗:
fixes `G :: "('a, 'b) pre_digraph"` **and** `X:: "'a set"`

assumes *"dir_bipartite_digraph G X Y"* **and**
        "$\forall\, i \in X$. finite (neighbourhood G i)"
and *"$\exists\, g$. enumeration (g:: nat $\Rightarrow$ 'a)"*
and *"$\exists\, h$. enumeration (h:: nat $\Rightarrow$ 'b)"*
shows
  "($\forall\, J \subseteq X$. finite J $\longrightarrow$ (card J) $\leq$ card ($\bigcup$ (neighbourhood G ' J))) $\longrightarrow$
  ($\exists\, E$. dirBD_perfect_matching G X Y E)"

This lemma applies the formalisation of the countable set-theoretical version of Hall's theorem ([32]) to infer the existence of an SDR R for the countable indexed family of sets $\{N(i)\}_{i \in X}$. Applying the lemma is possible since the marriage condition for this family of sets is the premise of the target implication. From the system of representatives, it is possible to build the perfect matching as the set of arcs $\{(i,\ R(i))\}_{i \in X}$. Through two additional auxiliary lemmas, it is proved that this set covers the set of vertices X (lemma perfect🔗) and is indeed a matching (lemma dirBD_matching🔗). Therefore, one concludes that $(dirBD\_perfect\_matching\ G\ X\ Y\ \{(i, R(i))\}_{i \in X})$.

Finally, the countable graph-theoretical version of Hall's theorem (Theorem 3), specified as theorem Hall_digraph🔗, is formalised as below. The use of necessity and sufficiency auxiliary lemmas is highlighted in the mechanisation.

theorem *Hall_digraph*🔗:
fixes *G :: "('a, 'b) pre_digraph"* **and** *X:: "'a set"*
assumes *"dir_bipartite_digraph G X Y"*
and *"$\forall\, i \in X$. finite (neighbourhood G i)"*
and *"$\exists\, g$. enumeration (g:: nat $\Rightarrow$ 'a)"*
and *"$\exists\, h$. enumeration (h:: nat $\Rightarrow$ 'b)"*
shows *"($\exists\, E$. dirBD_perfect_matching G X Y E) $\longleftrightarrow$*
       ($\forall\, J \subseteq X$. finite J $\longrightarrow$ card J $\leq$ card ($\bigcup$ (neighbourhood G ' J)))"
proof
 assume *hip1: " $\exists\, E$. dirBD_perfect_matching G X Y E"*
 show *"$\forall\, J \subseteq X$. finite J $\longrightarrow$ card J $\leq$ card $\bigcup$ (neighbourhood G ' J)"*
   using *hip1 assms(1-2) marriage_necessary_graph[of G X Y]* **by** *auto*
next
 assume *hip2 :"$\forall\, J \subseteq X$. finite J $\longrightarrow$ card J $\leq$card$\bigcup$ (neighbourhood G ' J)"*
 show *"$\exists\, E$. dirBD_perfect_matching G X Y E"*
   using *assms marriage_sufficiency_graph[of G X Y] hip2*
  proof-
   have *"$\forall\, J \subseteq$ X. finite J$\longrightarrow$ card J $\leq$ card $\bigcup$ (neighbourhood G ' J)*
      $\longrightarrow$ ($\exists\, E$. dirBD_perfect_matching G X Y E)"
    using *assms marriage_sufficiency_graph[of G X Y]* **by** *auto*
  thus *?thesis* **using** *hip2* **by** *auto*
 qed
qed

3.3   Alternative Formalisation Using Isabelle *locales*

Locales are an extension of Isabelle (Isar) that provide support for modular reasoning allowing dependent typing in a straight forward manner. Locales were initially developed by Kammüller [22] to support reasoning in abstract algebra, but are applied in a variety of domains [6,7]. This section discusses the formalisation of the difficult direction of the graph version of Hall's theorem directly over graph notions, as stated by Theorem 4, using locales.

Initially, locales are used to specify an indexed family of sets from which the notion of SDR is specified providing an injective function *repr* from the set of indices I to distinct elements in each set of the family.

```
locale set_family🗗  =
   fixes I :: "'a set" and X :: "'a ⇒ 'b set"
```

```
locale sdr🗗  = set_family +
   fixes repr :: "'a ⇒ 'b"
   assumes  inj_repr: "inj_on repr I" and
            repr_X: "x ∈ I ⟹ repr x ∈ X x"
```

Then, the notions of bipartite digraph and countable bipartite digraph $G = \langle X, Y, E \rangle$ with finite sets of neighbourhoods for each vertex $x \in X$ are defined as below.

```
locale bipartite_digraph🗗  =
   fixes X :: "'a set" and Y :: "'b set" and E :: "('a × 'b) set"
   assumes E_subset: "E ⊆ X × Y"
```

```
locale Count_Nbhdfin_bipartite_digraph🗗  =
   fixes X :: "'a:: countable set" and Y :: "'b:: countable set"
            and E :: "('a × 'b) set"
   assumes E_subset: "E ⊆ X × Y"
   assumes Nbhd_Tail_finite: "∀x ∈ X. finite {y. (x, y) ∈ E}"
```

In the sequel, matching over bipartite digraphs and perfect matching are specified using locales. The succinctness of locales is observed clearly in the definition of perfect matching. For this, it is only required to add to the notion of matching the assumption that the matching covers the set of vertices X.

```
locale matching = bipartite_digraph🗗  +
   fixes M :: "('a × 'b) set"
   assumes M_subset: "M ⊆ E"
   assumes M_right_unique: "(x, y) ∈ M ⟹ (x, y') ∈ M ⟹ y = y'"
   assumes M_left_unique: "(x, y) ∈ M ⟹ (x', y) ∈ M ⟹ x = x'"
```

```
locale perfect_matching🗗  = matching +
   assumes M_perfect: "fst ' M = X"
```

Then, using the locales for systems of distinct representatives, *sdr*, and for perfect matchings, respectively, two lemmas can be easily established, proving how a perfect matching can be built from an SDR, and how a perfect matching gives rise to an SDR. The former lemma uses the injective function *repr* in the locale for *sdr*, building the perfect matching as the set of edges $\{(x, repr\ x)\mid x \in I\}$. The latter lemma uses the set $M \subseteq E$ in the locale for perfect matching to build the SDR using as a set of indices the vertices X, as the family of indexed sets the function mapping indices into their finite neighbourhoods, $x \in X$, $\lambda x.\{y \mid (x,y) \in E\}$, and as injective function $\lambda x.\{y \mid (x,y) \in M\}$.

lemma (**in** `sdr`) `perfect_matching`⧉ :
  `"perfect_matching I (⋃i∈I. X i) (Sigma I X) {(x, repr x)|x. x ∈ I}"`

lemma (**in** `perfect_matching`) `sdr`⧉ :
  `"sdr X (λx. {y. (x,y) ∈ E}) (λx. the_elem {y. (x,y) ∈ M})"`

Finally, the difficult direction of Hall's theorem, as stated by Theorem 4, for countable infinite graphs is specified using the locale for countable bipartite digraphs with finite neighbourhoods for all vertices $x \in X$. The theorem below formalises that if the subgraph induced by any finite subset X_s of X has a perfect matching, then the whole graph has a perfect matching.

theorem (**in** `Count_Nbhdfin_bipartite_digraph`) `Hall_Graph`⧉ :
  shows `"(∀ Xs ⊆ X. (finite Xs) ⟶`
          `(∃ Ms.   perfect_matching Xs`
                              `{y. x ∈ Xs ∧ (x,y) ∈ E}`
                              `{(x,y). x ∈ Xs ∧ (x,y) ∈ E}`
                              `Ms))`
  `⟶ (∃ M.  perfect_matching X Y E M)"`

The proof uses the hypotheses of the existence of a perfect matching, M_S, for each bipartite digraph induced by any finite $X_S \subset X$. Using the previous lemma *(in perfect_matching) sdr*, it is possible to construct an SDR for the associated family of sets of neighbourhoods of vertices incident to vertices in X_S. Then, the existence of different images of the injective function to the distinct representatives, *repr* in the locales for *sdr*, permits inferring that $|X_S| \leq \cup_{x \in X_s}\{y \mid (x,y) \in E\}$. Notice that this condition corresponds to the set-theoretical marriage condition. Thus, applying the set-theoretical version of Hall's theorem⧉, one concludes that the whole digraph has an SDR. Finally, the existence of a perfect matching for the whole digraph is concluded by applying the previous lemma *(in sdr) perfect_matching*.

Notice that the other direction of Theorem 4 is easy; indeed, the restriction of the perfect matching of the whole graph to the subgraph induced by any subset $X_S \subset X$ is a perfect matching of the induced subgraph.

4   Related Work

4.1   Automation Versus Interactive Comprehensive Proofs

As mentioned in the abstract, our primary interest in developing such a detailed formalisation is to provide insight to Mathematicians and Computer Scientists about the usefulness of proof assistants. So, the high granularity used in presenting definitions and proof steps is essential. Using the Isabelle Sledgehammer [26,37] the user may infer proofs without having a clear idea of how these proofs are obtained, which is not our objective. To summarise the steps inferred by the Sledgehammer, it is recommended to restrict it to *isar* proofs. Such an alternative approach, oriented towards automation, is presented at the end of the formalisation using locales [6,7].

In synthesis, our educational goal prioritises the application of proof assistants as *interactive theorem provers* and not as *automated theorem provers*. This is the spirit we have followed teaching for years computer science and Math students in our institutions as reported in [4] (on the adequate application of interactive theorem provers to motivate mathematicians), [3] (on the application of the proof assistant PVS to teach computer science, mathematicians, and engineering students to verify algorithms), and in [5] (on teaching computational logic to computer science, engineering and mathematics students, illustrating the application of the Gentzen's sequent-style calculus implemented in the proof assistant PVS).

4.2   On Hall's Theorem and Other Combinatorial Theorems

Extensions to the infinite case from theorems equivalent to Hall's marriage theorem in the finite case are generally not straightforward. In addition to the infinite version of Hall's marriage theorem, our development includes formalisations of infinite versions of De Bruijn-Erdös graph colouring theorem ([11]) and König lemma ([23]), obtained from the compactness theorem for predicate logic (theorems available through the links k_coloring🔗 and Koenig_Lemma🔗, respectively). Moreover, even such extensible theorems would not necessarily be proven by the compactness theorem and elementary techniques. An example is König's duality theorem, proved by Aharoni [1], and subsequently studied in detail by Aharoni et al. [2]. This theorem states that in every bipartite graph $G = \langle X, Y, E \rangle$, *there exists* a matching $M \subseteq E$ such that selecting one vertex from each arc in M one has a cover of the graph. König duality theorem is a strong form of the finite, well-known König-Egerváry theorem that states that in a finite bipartite graph, the size of a maximal matching is equal to the size of a minimal cover [24]. The vital difference of the duality theorem is that such a cover of the graph cannot be extracted from an arbitrary matching. Indeed, from a matching, it is possible to build a cover of the same cardinality as the cardinality of the matching, but not that it covers the graph. So, the notion of *König cover* came to arise, which is defined as a cover of the graph that consists of a selection of one vertex from each arc of a matching.

Lifting results from the finite to the infinite through the application of compactness (of König's lemma) corresponds to a recursive construction of a procedure that produces the target solution in the degree of unsolvability of the halting problem [2]. Such a recursive construction is possible for Dilworth's theorem (restricting the maximal anti-chains in infinite partial ordered sets to be finite - [12], see also Sec. 2.5 in [19]) but not for König's duality theorem. Indeed,

Aharoni et al. [2] proved that the complexity of constructing covers exceeds the complexity of the halting problem; it is even a problem of higher complexity than answering all first-order questions about arithmetic. Also, they proved that the compactness theorem and König's lemma do not suffice to prove the duality theorem and other related results in matching theory.

The first formalisation of the finite version of Hall's Theorem was developed in Mizar by Romanowicz and Grabowski [29]. Also, there are formalisations in Isabelle/HOL by Jiang and Nipkow [21]. These formalisations follow Rado's proof [27], but the last one also includes a mechanisation based on Halmos and Vaughan's proof [18]. In addition, Coq has a formalisation that uses formalisations of Dilworth's decomposition theorem and bi-partitions in graphs [34]. An earlier formalisation of Dilworth's theorem in Mizar is presented in [30]. Recently, Gusakov, Mehta and Miller [16] presented three different proofs of the finite version of Hall's theorem in Lean in terms of indexed families of finite subsets, of the existence of injections that saturate binary relations over finite sets and of matchings in bipartite graphs. Related combinatorial results are reported in recent works by Doczkal et al. in their graph theory Coq library (e.g., [13,15], and [14]). Additionally, Singh and Natarajan formalised in Coq other combinatorial results as the perfect graph theorem and a weak version of this theorem (e.g., [35,36]).

Known mechanisations of the enumerable version of the set-theoretical version of Hall's theorem appear in the formalisation used in the authors' work, previously discussed, [32], and in Gusakov, Mehta, and Miller's work [16]. The former work uses the compactness theorem for predicate logic. In the latter work, the authors apply an *inverse limit* version of the König's lemma. This lemma states that if $\{X_i\}_{i \in \mathbb{N}}$ is an indexed family of nonempty finite sets with functions $f_i : X_{i+1} \to X_i$, for each $i \in \mathbb{N}$, then there exists a family of elements $x \in \prod_i X_i$ such that $x_i = f_i(x_{i+1})$, for all $i \in \mathbb{N}$. König's lemma follows from this infinite limit version by choosing as set X_i the paths of length i from the root vertex v_0 in a tree. So, the function f_i maps paths in X_{i+1} into the paths without their last arc that belong to X_i. The inverse limit consists of the infinite chain of functions $f_1, f_2, \ldots$. König's lemma is applied to prove the enumerable version of Hall's theorem by taking M_n as the set of all matchings on the first n indices of I (i.e., the set of all possible SDRs for the sets $S_1, \ldots, S_n$), and $f_n : M_{n+1} \to M_n$ as the restriction of a matching to a smaller set of indices. Since the marriage condition holds for the finite indexed families, each M_n is nonempty, and by König's lemma, an element of the inverse limit gives a matching on I.

5   Conclusions and Future Work

This paper presented two formalisations in Isabelle/HOL of the graph-theoretical version of Hall's theorem for countable (infinite) graphs. The prominent feature of the first formalisation is following a presentation close to pen-and-paper proofs but dissecting all minimal required steps in the assisted proof. Exhibiting minimal details, usually omitted in practice, is relevant to highlight to Math and CS students and professionals the relevance of mechanised proofs. On the other hand, the second one is more succinct and uses Locales, which are powerful mechanisms to deal with parametric theories in Isabelle/HOL.

These developments will enable other mechanisations of infinite combinatorial, set-theoretical, and graph-theoretical results related to the compactness theorem for predicate logic and its derivations, such as König lemma, Hall's marriage theorem, and de Bruijn-Erdös k-colouring theorem, as well as generalisations of Dilwort's theorem.

An exciting challenge for future research consists in developing the required formal background in proof assistants to enable the formalisation of other theorems which do not extend straightforwardly from the results mentioned above, such as the König duality theorem, among others.

Acknowledgements. We want to thank Cezary Kaliszyk, René Thiemann, Fabian Huch, and Yutaka Nagashima, who kindly shared their expertise on Isabelle/HOL.

References

1. Aharoni, R.: König's duality theorem for infinite bipartite graphs. J. London Math. Soc. **s2-29**(1) (1984). https://doi.org/10.1112/jlms/s2-29.1.1
2. Aharoni, R., Magidor, M., Shore, R.A.: On the strength of König's duality theorem for infinite bipartite graphs. J. Combin. Theory Ser. B **54**(2), 257–290 (1992). https://doi.org/10.1016/0095-8956(92)90057-5
3. Almeida, A.A., Rocha-Oliveira, A.C., Ramos, T.M.F., de Moura, F.L.C., Ayala-Rincón, M.: The computational relevance of formal logic through formal proofs. In: Dongol, B., Petre, L., Smith, G. (eds.) FMTea 2019. LNCS, vol. 11758, pp. 81–96. Springer, Cham (2019). https://doi.org/10.1007/978-3-030-32441-4_6
4. Ayala-Rincón, M., de Lima, T.A.: Teaching interactive proofs to mathematicians. In: Proceedings of the 9th International Workshop on Theorem Proving Components for Educational Software, ThEdu. EPTCS, vol. 328, pp. 1–17 (2020). https://doi.org/10.4204/EPTCS.328.1
5. Ayala-Rincón, M., de Moura, F.L.C.: Applied Logic for Computer Scientists - Computational Deduction and Formal Proofs. Undergraduate Topics in Computer Science. Springer (2017). https://doi.org/10.1007/978-3-319-51653-0
6. Ballarin, C.: Locales and locale expressions in Isabelle/Isar. In: Berardi, S., Coppo, M., Damiani, F. (eds.) TYPES 2003. LNCS, vol. 3085, pp. 34–50. Springer, Heidelberg (2004). https://doi.org/10.1007/978-3-540-24849-1_3
7. Ballarin, C.: Locales: a module system for mathematical theories. J. Autom. Reason. **52**(2), 123–153 (2014). https://doi.org/10.1007/s10817-013-9284-7
8. Borgersen, R.D.: Equivalence of seven major theorems in combinatorics. Department of Mathematics, University of Manitoba, Canada (2004). https://home.cc.umanitoba.ca/~borgerse/Presentations/GS-05R-1.pdf
9. Cameron, P.J.: Combinatorics: Topics, Techniques, Algorithms. Cambridge University Press, Cambridge (1994)
10. Chartrand, G., Lesniak, L., Zhang, P.: Graphs and Digraphs, 5th edn. Chapman & Hall/CRC (2010)
11. Bruijn, N.G.D., Erdös, P.: A colour problem for infinite graphs and a problem in the theory of relations. Indagationes Math. (Proc.) **54**, 371–373 (1951)

12. Dilworth, R.P.: A decomposition theorem for partially ordered sets. Ann. Math. **51**(1), 161–166 (1950). https://doi.org/10.2307/1969503

13. Doczkal, C., Combette, G., Pous, D.: A formal proof of the minor-exclusion property for treewidth-two graphs. In: Avigad, J., Mahboubi, A. (eds.) ITP 2018. LNCS, vol. 10895, pp. 178–195. Springer, Cham (2018). https://doi.org/10.1007/978-3-319-94821-8_11

14. Doczkal, C., Pous, D.: Completeness of an axiomatization of graph isomorphism via graph rewriting in Coq. In: Proceedings of the 9th ACM SIGPLAN International Conference on Certified Programs and Proofs - CPP, pp. 325–337. ACM (2020)

15. Doczkal, C., Pous, D.: Graph theory in Coq: minors, treewidth, and isomorphisms. J. Autom. Reason. **64**(5), 795–825 (2020). https://doi.org/10.1007/s10817-020-09543-2

16. Gusakov, A., Mehta, B., Miller, K.A.: Formalizing hall's marriage theorem in lean. arXiv abs/2101.00127 [math.CO] (2021). https://doi.org/10.48550/arxiv.2101.00127

17. Hall, P.: On representatives of subsets. London Math. Soc. **10**, 26–30 (1935). https://doi.org/10.1112/jlms/s1-10.37.26

18. Halmos, P.R., Vaughan, H.E.: The marriage problem. Am. J. Math. **72**(1), 214–215 (1950)

19. Harzheim, E.: General relations between posets and their chains and antichains. In: Ordered Sets. Advances in Mathematics, vol. 7. Springer (2005)

20. Jiang, D., Nipkow, T.: Hall's marriage theorem. Archive of Formal Proofs (2010)

21. Jiang, D., Nipkow, T.: Proof pearl: the marriage theorem. In: Jouannaud, J.-P., Shao, Z. (eds.) CPP 2011. LNCS, vol. 7086, pp. 394–399. Springer, Heidelberg (2011). https://doi.org/10.1007/978-3-642-25379-9_28

22. Kammüller, F.: Modular Reasoning in Isabelle. In: Proceedings Automated Deduction - CADE-17, 17th International Conference on Automated Deduction. Lecture Notes in Computer Science, vol. 1831, pp. 99–114. Springer (2000)

23. König, D.: Über eine schlussweise aus dem endlichen ins unendliche. Acta Sci. Math. (Szeged) **3**(2–3), 121–130 (1927)

24. König, D.: Theorie Der Endlichen Und Unendlichen Graphen: Kombinatorische Topologie Der Streckenkomplexe. Mathematik und ihre Anwendungen in Monographien und Lehrbüchern, vol. 16. Chelsea (1936)

25. Noschinski, L.: Graph Theory. Archive of Formal Proofs (2013). http://isa-afp.org/entries/Graph_Theory.html. Formal proof development

26. Paulson, L.C., Blanchette, J.C.: Three years of experience with sledgehammer, a practical link between automatic and interactive theorem provers. In: The 8th International Workshop on the Implementation of Logics, IWIL 2010, Yogyakarta, Indonesia, 9 October 2011. EPiC Series in Computing, vol. 2, pp. 1–11. EasyChair (2010). https://doi.org/10.29007/36dt

27. Rado, R.: Note on the transfinite case of Hall's theorem on representatives. London Math. Soc. **S1−42**(1), 321–324 (1967). https://doi.org/10.1112/jlms/s1-42.1.321

28. Reichmeider, P.F.: The Equivalence of Some Combinatorial Matching Theorems. Polygonal Publishing House (1985)

29. Romanowicz, E., Grabowski, A.: The hall marriage theorem. Formalized Math. **12**(3), 315–320 (2004). https://fm.mizar.org/2004-12/pdf12-3/hallmar1.pdf

30. Rudnicki, P.: Dilworth's decomposition theorem for posets. Formalized Math. **17**(4), 223–232 (2009). https://doi.org/10.2478/v10037-009-0028-4

31. Suárez, F.F.S.: Formalización en Isar de la Meta-Lógica de Primer Orden. Ph.D. thesis, Departamento de Ciencias de la Computación e Inteligencia Artificial, Uni-

versidad de Sevilla, Spain (2012). https://idus.us.es/handle/11441/57780. In Spanish

32. Serrano Suárez, F.F., Ayala-Rincón, M., de Lima, T.A.: Hall's theorem for enumerable families of finite sets. In: Buzzard, K., Kutsia, T. (eds.) CICM 2022. LNCS, vol. 13467, pp. 107–121. Springer, Cham (2022). https://doi.org/10.1007/978-3-031-16681-5_7

33. Suárez, F.F.S., Ayala-Rincón, M., de Lima, T.A.: Mechanising hall's theorem for countable graphs. EasyChair Preprint no. 10365. Easy Chair Preprint of talk presented at LPAR 2023 (2023). https://easychair.org/publications/preprint/g3F7

34. Singh, A.K.: Formalization of some central theorems in combinatorics of finite sets. arXiv abs/1703.10977[cs.Lo] (2017). Short presentation at the 21st International Conference on Logic for Programming, Artificial Intelligence and Reasoning - LPAR. https://doi.org/10.48550/arxiv.1703.10977

35. Singh, A.K., Natarajan, R.: Towards a constructive formalization of perfect graph theorems. In: Khan, M.A., Manuel, A. (eds.) ICLA 2019. LNCS, vol. 11600, pp. 183–194. Springer, Heidelberg (2019). https://doi.org/10.1007/978-3-662-58771-3_17

36. Singh, A.K., Natarajan, R.: A constructive formalization of the weak perfect graph theorem. In: Proceedings of the 9th ACM SIGPLAN International Conference on Certified Programs and Proofs - CPP, pp. 313–324. ACM (2020). https://doi.org/10.1145/3372885.3373819

37. Sultana, N., Blanchette, J.C., Paulson, L.C.: LEO-II and satallax on the sledgehammer test bench. J. Appl. Log. 11(1), 91–102 (2013). https://doi.org/10.1016/j.jal.2012.12.002

38. West, D.B.: Introduction to Graph Theory. Pearson Modern Classics for Advanced Mathematics. Pearson Education, Inc. (2001)

Efficient Strategies for Finding the Minimum Information Partition in Integrated Information Theory 3.0

Luz Enith Guerrero Mendiesta[1,2]([✉]) [iD], Jeferson Arango-López[1] [iD],
Luis Fernando Castillo Ossa[1,2] [iD], and Jorge Alberto Jaramillo-Garzón[1] [iD]

[1] Universidad de Caldas, Calle 65 # 26-10, Manizales, Colombia
{luzenith_g,jeferson.arango,luis.castillo,
Jorge.jaramillo}@ucaldas.edu.co, {leguerrerom,
lfcastilloos}@unal.edu.co
[2] Universidad Nacional de Colombia – Sede Manizales, Cra 27 # 64-60, Manizales, Colombia

Abstract. The problem of finding the Minimum Information Partition (MIP) in the context of Integrated Information Theory (IIT) 3.0 presents significant computational challenges due to the complexity of calculating integrated information in large, interconnected systems. Addressing the need for efficient solutions, this work explores bounded approximations to the formalism proposed by IIT 3.0, aiming to establish the theory's specified properties in more complex systems and to scale with system size and architecture.

In this study, we present two strategies aimed at efficiently solving the problem of finding the MIP. The first strategy is based on classical searches with a top-down approach supported by memoization. Beyond classical searches, the second technique introduces a metaheuristic inspired by biological evolution and genetics to reduce processing times and computational complexity. Several test cases for systems of different sizes were conducted, and their results were compared with those presented by the PyPhi application.

The proposed optimizations were based on probability properties, approximations using a top-down approach supported by memoization, and metaheuristic techniques aimed at solving combinatorial optimization problems.

Keywords: IIT · MIP · algorithms · optimization

1 Introduction

The mind-body relationship has captivated and been studied throughout human history, yet defining consciousness precisely remains challenging. Various definitions have been suggested, such as the ability to assign feelings to mental states or the capacity for self-reflection and environmental awareness [1]. At the close of the 20th century, driven by neuroscientists and physicists, the scientific study of consciousness began, despite its subjective nature seemingly conflicting with scientific objectivity [2]. Within this context, IIT, proposed by [3], offers a rigorous theoretical framework to define, quantify,

© The Author(s), under exclusive license to Springer Nature Switzerland AG 2024
N. D. Duque-Méndez et al. (Eds.): CCC 2024, CCIS 2208, pp. 217–233, 2024.
https://doi.org/10.1007/978-3-031-75233-9_16

and identify consciousness in various systems. IIT is based on two core postulates: subjective experience contains information, and this information is integrated irreducibly. This suggests that a system's degree of consciousness is determined by its interconnectivity. The integration measure, known as Phi (Φ), is essential for assessing a system's irreducibility by identifying the Minimum Information Partition (MIP), a numerically intractable problem due to the combinatorial explosion of possible partitions [4].

This challenge is crucial not only for neuroscience but also for cognitive sciences and artificial intelligence. Solving the MIP problem could aid in identifying conscious systems, with practical applications in medicine, advanced technology, and decision-making systems. Therefore, given the problem's significance across various fields, this study aims to explore alternatives to address the MIP problem.

The remainder of this document is organized as follows: Sect. 2 presents a description of the MIP problem in IIT. Section 3 provides a contextualization of the research topics. Section 4 describes the methods and strategies employed in this proposal. Section 5 shows the analysis of the results and a discussion of the obtained data. Finally, a review of future work and the challenges posed are found in Sect. 6.

2 Motivation

Integrated Information Theory (IIT) proposes that to measure information integration within a system, various partitions must be evaluated to identify the Minimum Information Partition (MIP). The MIP is defined as the partition where the loss of information due to disconnecting subsystems is minimized [4]. Metrics such as mutual information (ΦMI), stochastic interaction (ΦSI), and integrated geometric information (ΦG) quantify integrated information by comparing the current probability distribution with a disconnected one [5].

The MIP is the partition that divides a system into the least interdependent subsystems so that the loss of information caused by removing the interactions between the subsystems is minimized. The MIP, π_{MIP} is defined as a partition where integrated information is minimized, although there could be more than one partition that complies with the request.

$$\pi_{\text{MIP}} := \arg\min_{\pi \in \mathcal{P}} \Phi(\pi) \tag{1}$$

In general, for a set of partitions, $\mathcal{P}$ is the universal set of partitions, which includes the different k-partitions. Although, the simplest ways to partition a system are bipartitions since they are only determined by specifying a subset S (since by partitioning a set C into two, one obtains S and S' = C\S), the integrated information can be considered as a function of a set S, $\Phi(S)$, so finding the MIP is the same as finding the subset S_{MIP}, which achieves the minimum integrated information.

$$S_{\text{MIP}} := \arg\min_{S \subset \Omega, S \neq \emptyset} \Phi(S) \tag{2}$$

Finding the MIP involves optimizing a set function where the number of partitions for a system with N elements is 2^N-1 - 1, posing computational challenges due to exponential growth in partition combinations [6]. This complexity limits practical applications, especially in neuroscience where estimating Φ for real systems requires significant computational resources [7]. To enhance practicality and scalability, bounded approximations to IIT's formalism are necessary, enabling broader application in complex networks beyond the brain.

Recently, the exploration of how to find the MIP has been undertaken by Hidaka and [5], who investigated the submodularity of integrated information. Following the guidelines of previous work, [6] employed Queyranne's algorithm for stochastic interaction (ΦSI) and geometric integrated information (ΦG). On the other hand, [8] proposed a method based on spectral clustering with correlation of neural time series data.

3 Background

This section will briefly present descriptions of the relevant areas of this research, such as artificial intelligence, and integrated information theory, to identify alternative solutions to the MIP problem.

3.1 Artificial Intelligence (AI)

For [9], the primary goal of general artificial intelligence (GAI) is to enable a machine to perform general intelligence tasks like humans. This could theoretically be achieved through approaches such as machine learning and cross-domain optimization. The concept of consciousness is intriguing for achieving both approaches because it not only encodes but also processes varied information with perfect integration. If a machine is to develop its own rules, the concept of a learning machine will be required, and it has been said that for this, a machine needs consciousness. Additionally, the development in this field requires more efficient computational applications for information processing to learn from the environment, which in turn drives the development of AI [10].

Metaheuristics – Genetic Algorithms. These algorithms are pertinent for problems where the solution state is important, rather than the cost of reaching it. Included in these algorithms are methods inspired by evolutionary biology. These algorithms perform a stochastic hill-climbing search, maintaining a large population of states. New states are generated through mutation and crossover, which combine pairs of states from the population.

3.2 Computational Neuroscience

Computational neuroscience is the field where issues overlap both computational sciences and neuroscience [11]. Its main goal is to digitally represent neural networks in the brain and their interactions to understand how functions such as perception, processing, and reaction to stimuli arise from the electrochemical communication between individual neurons. This involves creating computer models to simulate neural activity using mathematical models based on statistical estimates, thereby enabling various visualizations of brain activity. Data obtained from experiments using neuroimaging and other techniques are utilized to build models and simulations, allowing predictions about the functions and networks involved [12]. Computational neuroscience also draws parallels between the learning processes of living organisms and forms of computerized learning or machine learning [13].

Conversely, numerous neuroscience models employ probabilistic calculations, and the stochastic behavior of neurons at their peak enables the network to solve problems heuristically. Therefore, a network state can represent a potential solution to a problem. This implies a possible relationship between the fluctuating internal states of brain networks and responses in terms of perception and behavior [14].

Advances in image recognition have been achieved, recently, through deep convolutional networks, and progress has been made in natural language processing with the help of recurrent networks [15].

3.3 Integrated Information Theory (IIT)

IIT attempts to characterize consciousness mathematically in both quantity and quality [16]. More specifically, it provides a mathematical framework to fully characterize the cause-effect structure of a physical system. It was proposed by [3] and has undergone several revisions: $IIT_{2.0}$ [16], then $IIT_{3.0}$[4], and its most recent version, $IIT_{4.0}$ [17].

IIT is derived from thought experiments leading to phenomenological axioms and ontological postulates. IIT aims to explain what consciousness consists of and how it can be associated with certain physical systems. The theory determines whether such systems possess consciousness, the degree of consciousness, and the particular experience they are having [3]. According to IIT, a system comprises a set of elements that can assume discrete states, and depending on the interaction between the elements, there will be a specification of the transition rules between these states. Although IIT aims to describe sets of elements that influence each other's states through direct physical causal interactions, it can be applied to any set of elements that take discrete states [18, 19]. The consciousness of a system is determined by its causal properties; therefore, consciousness is an intrinsic and fundamental property of any physical system [19].

4   Methods

Given the need to find efficient solutions for calculating integrated information and to achieve practical relevance of the theory beyond the study of the human brain, it would be very useful to deduce bounded approximations to the formalism proposed by $IIT_{3.0}$. These approximations should help establish the properties specified by the theory in more complex networks and scale according to the system size.

In this work, several strategies were employed to reduce processing times and computational complexity. To this end, a review of different test cases was conducted and compared with the results obtained using the PyPhi[1] application. The proposed optimizations were based on metaheuristics, specifically a genetic algorithm, and approximations using a top-down approach supported by memoization, which will be explained in this section.

---

[1] PyPhi is a Python software package that implements this framework for causal analysis and unfolds the full cause-effect structure of discrete dynamical systems of binary elements.

To evaluate the impact of this proposal, various systems with different structures, as indicated by [4], were tested to assess the effectiveness of the optimizations. The process begins with a discrete Markovian dynamic system S, as described by [20], composed of n elements. These elements have an input-output function that determines their state at time $t + 1$, based on the state of their parents at time t. This function establishes a transition probability matrix (TPM). The TPM is obtained from a graphical representation by perturbing the system in each possible state and observing the resulting state at the subsequent time point. Therefore, the TPM serves as the starting point, defining a conditional probability distribution over the next timestep $t + 1$(S_{t+1}), given the current timestep t (s_t), as shown in Eq. 1:

$$P(S_{t+1}|S_t = s_t), \forall \text{st} \in \Omega S, \tag{3}$$

where Ω_S represents the set of all possible states.

Elements within the dynamic system can only influence each other from one timestep to the next, that is, there is no instantaneous causation. The system must satisfy the conditional independence property (Markov condition): given the state S_t, the state of each element at $t + 1$ must be independent of the states of the other elements, as shown in Eq. 2:

$$P(S_{t+1}|S_t = s_t) = \prod_{N \in S} P(N_{t+1}S_t = st), \forall s_t \in S \tag{4}$$

As an illustration, consider a system with mechanisms A, B, and C. The conditional independence of these mechanisms can be depicted in the following manner:

$$P\left(ABC^{t+1}|ABC^t\right) = P\left(A^{t+1}|ABC^t\right) \times P\left(B^{t+1}|ABC^t\right) \times P\left(C^{t+1}|ABC^t\right) \tag{5}$$

This implies that, given the state of the system at time t, the probabilities of A, B, and C can be determined separately. After identifying a candidate set (distinguishing which elements are included and which are excluded), the elements outside this set are treated as external conditions with fixed values. The integrated information of a mechanism in each state over a set of interest variables is assessed by comparing its cause-effect relationship in the whole system with that of a partitioned system. For example, for the effect of the variable $A = 1$ (mechanism) on the variables A, B, and C (repertoire) at the next moment, the behavior of $P(ABC^{t+1}|A^t = 1)$ is studied. This is calculated by fixing the current state of A at 1, while the remaining elements B and C are independently perturbed into all their possible states with equal probability. So, the effect repertoire, according to the conditional independence, can be calculated as follows:

$$P\left(ABC^{t+1}|A^t = 1\right) = P\left(A^{t+1}|A^t = 1\right) \times P\left(B^{t+1}|A^t = 1\right) \times P(C^{t+1}|A^t = 1 \tag{6}$$

That is, the effect repertoire of a single element in the future, for example, $P(A^{t+1}|A^t = 1)$, is given by the Eq. 5:

$$P\left(A^{t+1}|A^t = 1\right) = \sum_{b,c} P\left(A^{t+1}|A^t = 1, BC^t = bc\right) P^{per}(BC^t = bc) \tag{7}$$

Initially, we have a system in some state at a given moment in time, and we want to evaluate a subset of its elements (candidate system). For this, the TPM of these elements is used, conditioned on the state of the external background elements and marginalization. To calculate the effect repertoire ABCt + 1|ABCt the product of all individual effect repertoires of the purview elements is taken. In other words, if ABC are elements belonging to the system, we can select a subset of these elements (mechanism) and evaluate the causal properties over the states of the system at t − 1(cause) and t + 1(effect).

This essentially describes how the mechanism in its current state t causally constrains the other elements in the past and/or future. Our focus will be on the effect repertoire, but the cause repertoire can be derived similarly. Generally, knowing the current state of a mechanism allows us to determine how the next state of a subset of elements or the entire system (referred to as the purview) is constrained. In this context, the effect repertoire of the mechanism AC in the current state 10 over the next state of the purview ABC is denoted as $P\left(ABC^{t+1}|AC^t = 10\right)$. Since IIT is concerned with the system's intrinsic perspective, the goal is to determine whether the effects of a mechanism can be reduced to the effects of its parts. If this is possible, no additional information is gained by grouping its parts in the first place, or in other words, there is no information loss due to the system's division. The aim is to find the partition of the effect repertoire that results in the least information loss.

Earth Mover's Distance (EMD). It is the amount of "earth" (in this case a portion of the probability) that needs to be moved multiplied by the "distance" (in this case, the Hamming distance, which is the distance between binary states, counting the number of positions by which two strings differ)[19]. Also known as the Wasserstein metric. Its operation is based on finding the minimum Hamming distance between multiple values of the vectors. The algorithm focuses on sorting and subtracting the deltas in both vectors until reaching zero.

4.1 Strategy 1

Taking the system from the perspective of the repertoire effect, the different ways of factoring the repertoire are taken as a starting point. Practically speaking, the cuts made on the system (purview-mechanism) generate two disjoint partitions. Each cut is a division of the system and has its own probability distribution (repertoire), which is then grouped by means of the tensor product whose result is the probability distribution of the "union" of these two parts. Finally, this result is compared with the original probability distribution (of the system without splitting) by means of a distance measure which in our case is the EMD and which allows to determine the distance between two probability distributions. For this, we proceed as follows:

1. The system's bipartitions are generated, each having two sub-partitions resulting from a cut. Future states are masked to identify the variables for calculating the tensor of that sub-partition. For a future state like (1111), a tree is generated to select each variable using a mask, marginalize it with respect to the present state, and then perform the tensor of the distributions from the marginalization process. The tree for a future state

would look like this:

1111

1100 0011

1000 100 0010 0001

2. The top-down approach with memoization checks if a result has been previously calculated. If so, it returns the saved result. If the future state is zero, another function checks if the solution is already in memory, marginalizes each state, saves the result, and returns the repertoire effect.
3. If no stored solution exists, the mask algorithm subdivides the purview using indices k and j, incrementing until only one masked element remains or the mechanism is marginalized.
4. When the mask is returned, if the left part is zero, it validates the right part. If both sides are non-zero, the tensor product of the two distributions is performed and saved for later use.

In this strategy, by storing the results, each decomposition that is generated is calculated only once and can be used many times.

Strategy 2

This strategy is based on a genetic algorithm, the main components of which are shown in Fig. 1. Any programming algorithm can be approached under the following reasoning of ideas to lead to an optimum:

```
BEGIN
      Generate an initial population.
      Compute evaluation function for every individual.
      WHILE fails to comply with the termination condition DO
            Select individuals to cross them.
            Cross selected individuals.
            Mutate some individuals (+infectious).
            Test the current generation.
            Replace the current population with the new ge neration.
      END WHILE
END
```

Genotype: It is the coded representation of a solution in the search space, using binary strings (0s and 1s) to represent solutions.

Phenotype: It is the external manifestation or decoded solution resulting from the genotype, interpreting the genotype in the context of the problem being solved.

Genes: These are the basic units of information in a chromosome, representing specific characteristics of the solution. Boolean genes will be used due to their high performance in terms of computational spatial complexity and the high number of logical operators in this system.

Chromosome: It is a structure containing a sequence of genes and represents a potential solution to the problem. It is denoted with masks over Markov chains in conjunction with the constraints given to the coupling system to a partitioned subsystem.

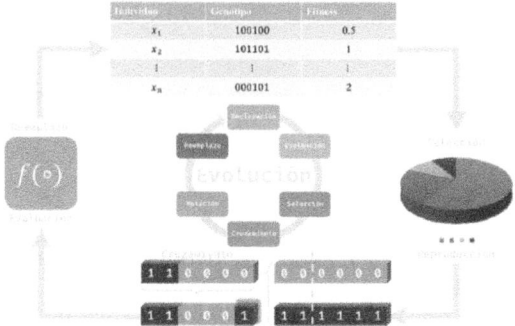

Fig. 1. Components of a genetic algorithm

The definition of variables associated to the purview with m elements, where $F_i \in F$ *in the range of* $i \in \{0 \to m-1\}$. Let n be the variables associated with the mechanism, where each $C_j \in C$ *such that* $j \in \{m \to m + n-1\}:(F_0 F_1..F_{m-1}|C_m C_{m+1}..C_{m+n-1} = b_0 b_1..b_{n-1})$.

The solution representation will be denoted as $(B_0, B_1, \cdots, B_{m+n-1})$, The solution representation will be denoted as where m is the number of purview elements existing in the worst-case scenario (each time $(t + 1)$ has a time (t)) y $B \in \mathbb{B}$ where $\mathbb{B}$ is the set of Boolean states of size 01. It is defined **n** for the number of mechanism elements present in the system.

Locus. It is the specific position of a gene on a chromosome. Each locus may contain different values or alleles. The partitioning state is defined as $\mathbb{P} = [0, 1]$, the associated locus will be a boolean linked to each variable of the original system, which indicates in which partitioning state the variable belongs (*False* $\equiv 0$, *True* $\equiv 1$). In this sense, it is understood how the variables from $B_0 \to B_{m-1}$ will refer to the elements of the purview and, since $B_m \to B_{m+n-1}$ to the elements of the mechanism.

Diversity. Diversity in the initial population is critical to avoid premature convergence and to ensure that the algorithm explores a wide range of possible solutions. A diverse population allows the genetic algorithm to cover more areas of the search space and increase the probability of finding an optimal solution. This problem has a search space of km + n-1−1 with k = 2, and n, m defined as the number of mechanism and purview variables representing the system.

Population Size. A scenario is represented with 3 variables for the mechanism and 3 variables for the purview of which the following analysis is made (see Fig. 2).

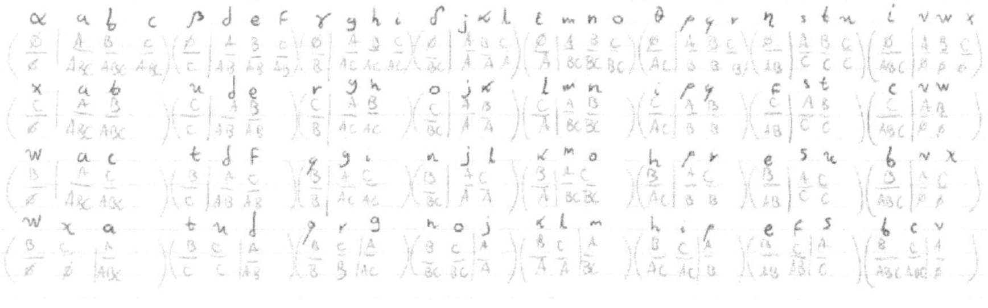

Fig. 2. Model for calculating probability distribution.

As can be seen in Fig. 2, it is not necessary to generate 8 rows but 4, because each primary partition has its dual adjacent, so that we will have 2^{n-1} combinations of the first row with respect to the search space, $2^{m+n-1}-1$, of which are identified as the basis for obtaining the other combinations according to the variation they offer, since it can be seen how the other partitions are nothing more than repetitions of the different combinations of the first row (this can be seen in how we denote each bipartition with the variables from a $\rightarrow$ x as recombination's of the variables of the first row, except for the void scenarios in $\alpha \rightarrow \iota$).

Regarding genetic operators, particularly within the different algorithms applicable for crossover, uniform crossover is chosen. In this type of crossover, chromosomes are not divided into segments; rather, each gene is treated separately. For each gene, a coin is flipped to decide from which parent it will be inherited in the offspring. This allows for a finer mixing of the parents' genetic information. For example: mask: 011010, ind0: 100110, ind1: 011000. As illustrated, a mask is generated that indicates which individual's bit will be selected; this bit is then stored in a primary offspring, while the bit not taken (from the other parent) is placed in a secondary offspring. For the termination operator, a combination of two criteria will be used: the maximum number of generations and several generations without improvement. Given that this problem is NP-hard and has a large search space, it is likely that many generations will be required to find an acceptable solution.

5 Results

Our proposals are based on strategies to address the combinatorial explosion problem involved in finding the MIP, based on the characteristics of a physical system that defines a cause-effect structure.

5.1 Strategy 1

Strategy 1 is based on a top-down approach with memoization. Se basa en particionar cada combinación de elementos. This strategy takes a dictionary that contains System_schema y Margination (see Table 1).

Table 1. Data entry for Strategy 1

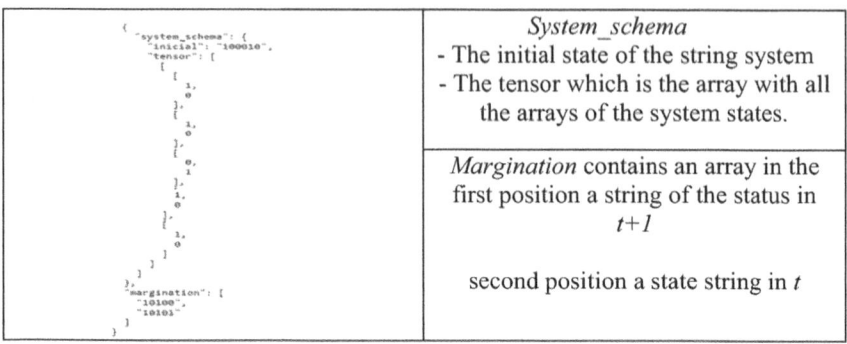

	System_schema - The initial state of the string system - The tensor which is the array with all the arrays of the system states.
	Margination contains an array in the first position a string of the status in *t+1* second position a state string in *t*

A summary of results are presented in Table 2:

Table 2. Results model according to strategy 1

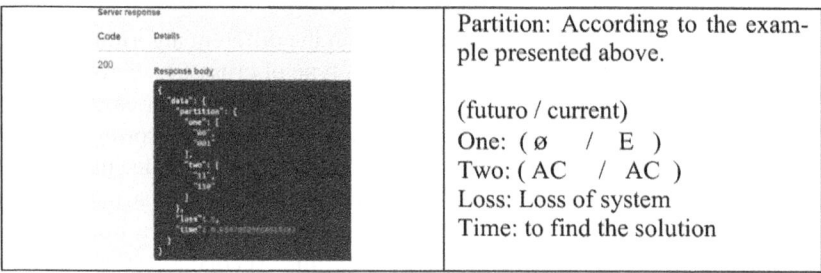

	Partition: According to the example presented above. (futuro / current) One: (ø / E) Two: (AC / AC) Loss: Loss of system Time: to find the solution

5.2 Strategy 2

It has been proposed to find the partition with the smallest loss using a genetic algorithm. It receives as input data a dictionary containing System_schema and Margination.

System_schema defines the initial system state as a string corresponding to the number of arrays in the tensor. The tensor consists of arrays representing probability distributions for system elements. Margination includes strings for future (t + 1) and current (t) states. Controls encompass parameters like crossRate, initPopSize, maxGensStr, mutateRate, and streakOfNoImprovement, governing crossover rate, initial population size, maximum iterations, mutation rate, and termination criterion based on generations without efficiency function improvement. The Fig. 3 gives a summary model of results:

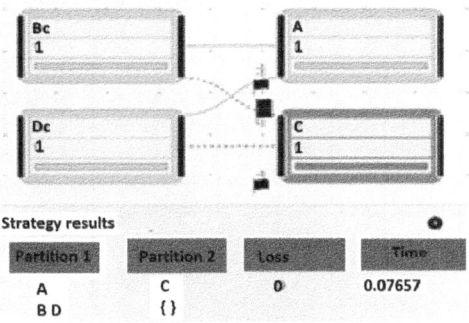

Fig. 3. Results Genetic Algorithm

5.3 Test Cases

A series of tests have been performed with 4-, 6-, 8-, and 10-element systems. Each of these systems is represented by a transition probability matrix, TPM, which has the probabilities in a system of transition from a state at time t to a state at time t + 1. In principle, the matrices are in the node-state form, as given in PyPhi. The data for the tests are presented below:

— 4-element system initial state =1000

1)ABCDt+1|ABCDt   2)ABCt+1|ABCDt   3)ABCDt+1|ACt   4)ACt+1|ABCt   5)ABCt+1|ABCt

— 6-element system initial state =100000

1)(ABCt+1|ACt)   2)ABCt+1|ABCt   3)ABCt+1|ABCDEt   4)ABCt+1|ABCEt   5)ABCt+1|ACDEt

— 8-element system initial state =10000000

1)(ABCDEFGHt+1| 2)(ABCGHt+1| 3)ABCDEFGHt+1| 4)ABCDEFGHt+1| 5)ABCDt+1|
ABCDEFGHt)   ABCDEFGHt)   BCDEFGHt   DEFGHt   DEFGHt

— 10-element system initial state =1000000000

1)(ABCDEFGHIJt+1| 2)(ABCDEFJt+1| 3)(ABCDEFJt+1|ADE-
ABCDEFGHIJt)   ABCDEFGHIJt)   FGHIJt)   4)ABCDEFGJt+1| 5)ABCDt+1|
                                                          ABDEFGHIJt   DEFGHt

The results are presented, illustrating graphically the resulting partition, using the representation of causal graphical models of the system with elements at t and at t + 1 (see Table 3, 4).

Results- Strategy 1.

Table 3. Results for some cases in 4-element system

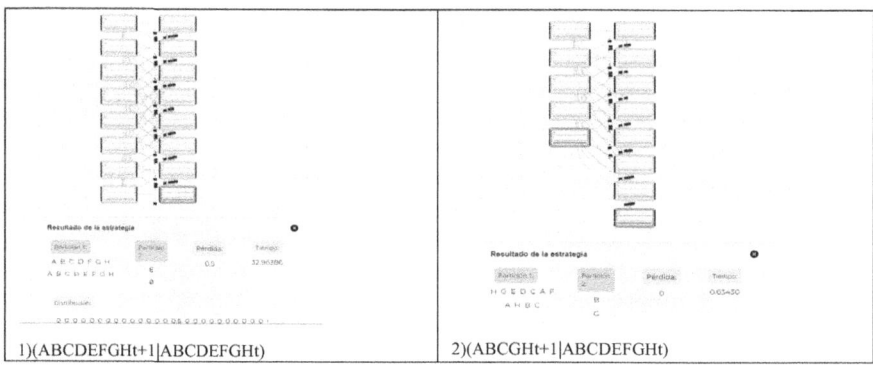

Table 4. Results for some cases in 8-element system

Results- Strategy 2. The following parameters are used for the genetic algorithm tests. Table 5 presents some results for the 4-element system:

Crossover rate: 0.6 Mutation rate: 0.1 Maximum generations: 100 Streak without improvement:10.

Table 5. Results for some cases in 4-element system with GA

1)ABCDt+1|ABCDt   2)ABCt+1|ABCDt   3)ABCDt+1|ACt

5.4   Analysis of Results

The two strategies were tested on different systems and compared with Pyphi using EMD as the distance measure. Comparative graphs show the performance based on the loss function, which measures the distance between two probability distributions. For the 4-node system, Strategy 1 (Top-Down with Memoization) was effective in all test cases, while the GA performed worse in smaller problems. For the 6-element network, Strategy 1 succeeded in 5 out of 7 cases, whereas the GA only succeeded in 2 cases but found the partition with zero loss (Figs. 4, 5, 6 and 7).

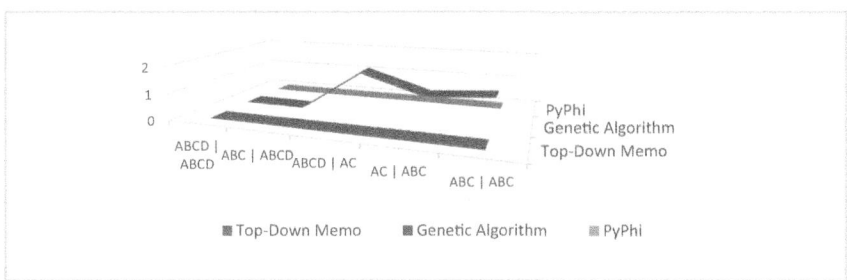

Fig. 4. Comparison of strategies for a 4-element system with different repertoires

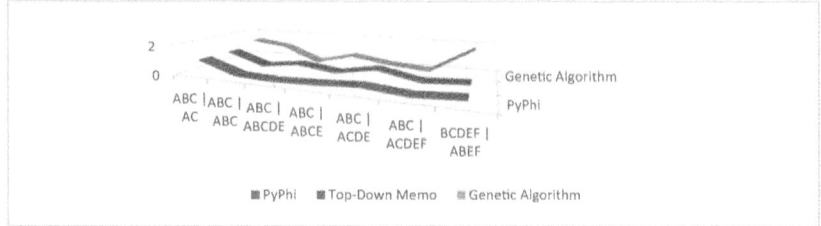

Fig. 5. Comparison of strategies for a 6-element system with different repertoires

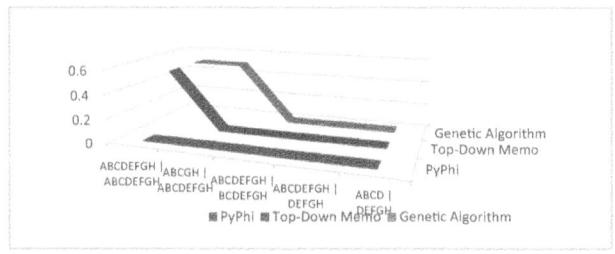

Fig. 6. Comparison of strategies for a 8-element system with different repertoires

In the case of the 8-element system, the responses of both algorithms were in good agreement with the PyPhi results (see Fig. 8). Only one case was unsuccessful according to strategy 1, and 2 according to strategy 2.

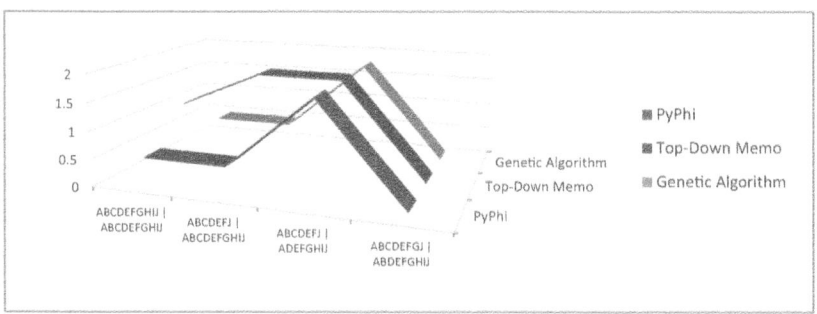

Fig. 7. Comparison of strategies for a 10-element system with different repertoires

Finally, for the 10-element system, the genetic algorithm was effective for the 4 cases analyzed while the TDM was successful in two of the cases (see Fig. 9). Table 6 presents an error analysis for each of the systems studied.

Table 6. Error analysis for the different systems

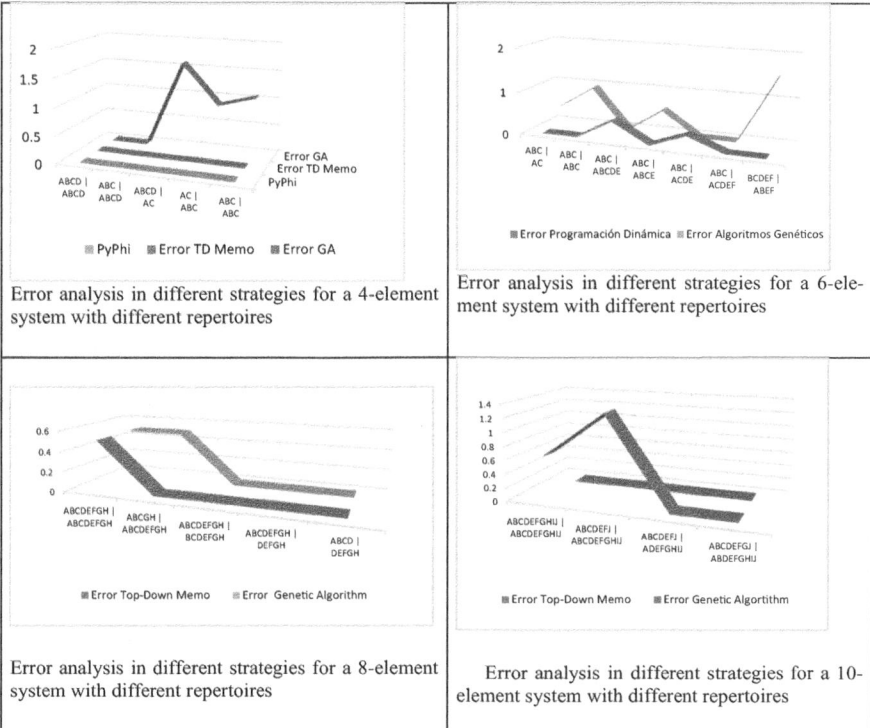

Error analysis in different strategies for a 4-element system with different repertoires	Error analysis in different strategies for a 6-element system with different repertoires
Error analysis in different strategies for a 8-element system with different repertoires	Error analysis in different strategies for a 10-element system with different repertoires

Given that our interest is to find the partition with the lowest loss and that in this problem several partitions with the same loss value may result, in this study we focused on presenting the analysis of the values, rather than on the partitions themselves, since we were interested in knowing whether the strategies were on track.

6 Future Work

In our proposal, we have examined strategies for problems beyond the classic cases in discrete, deterministic, and observable environments. Although many comparisons of genetic algorithms with other approaches have found that genetic algorithms are slower to converge, recent attempts, such as a Bayesian learning approach, could help bridge that gap. The proposed method marks a significant improvement in finding the MIP more efficiently by examining, among other things, specific aspects of the data representation to achieve faster operations on probability distributions. By leveraging conditional independence, we create a model supported by a top-down strategy utilizing memorization. We are currently developing features that aim to set better bounds based on the distances between the probability distributions of the original and partitioned systems. This enhancement will allow us to discard partitions early if superior options are already available. Looking ahead, we are exploring additional heuristic techniques to refine our method, focusing on the system's information loss properties to enable MIP generation in polynomial time.

Acknowledgments. We would like to thank Colombian Center for Bioinformatics and Computational Biology, BIOS, for their contribution and support with the computational resources necessary to execute the algorithms for this research.

References

1. Patnaik, L.M., Kallimani, J.S.: Promises and limitations of conscious machines. En: self, culture and consciousness: interdisciplinary convergences on knowing and being. In: Menon S., Nagaraj N., Binoy V.V. (Eds.). Springer Singapore: Singapore, pp. 79–92 (2017).
2. Chalmers, D.J.: Facing up to the problem of consciousness. J. Conscious. Stud. **2**, 19 (1995)
3. Tononi, G.: An information integration theory of consciousness. BMC Neurosci. **5**(1), 42 (2004)
4. Oizumi, M.., Albantakis, L., Tononi, G.: From the phenomenology to the mechanisms of consciousness: integrated information theory 3.0. PLOS Comput. Biol. **10**(5) (2014)
5. Oizumi, M., Tsuchiya, N., Amari, S.I.: Unified framework for information integration based on information geometry. Proc. Nat. Acad. Sci. **113**(51), 14817–14822 (2016)
6. Kitazono, J., Kanai, R., Oizumi, M.: Efficient algorithms for searching the minimum information partition in integrated information theory. Entropy **20**, 173 (2018)
7. Kim, H., et al.:.Estimating the integrated information measure Phi from high-density electroencephalography during states of consciousness in humans. Front. Hum. Neurosci. **12**(42) (2018)
8. Toker, D., Sommer, F.T.: Information integration in large brain networks. PLoS Comput. Biol. **15**(2), e1006807 (2019)
9. Nazri, A., Abdul Ghani, A., Hafez, I., Ng, K.-Y.: A new theoretical framework for testing consciousness in a machine, pp. 330–339 (2018)
10. Kriegeskorte, N., Douglas, P.K.: Cognitive computational neuroscience. Nat. Neurosci. **21**(9), 1148–1160 (2018)
11. Schwartz E.L.: Computational neuroscience. System Development Foundation Benchmark Series Mit Press (1990)
12. Marte, H.: Neurociencia computacional: el futuro de la investigación.Disponible en (2020). https://neuro-class.com/neurociencia-computacional-el-futuro-de-la-investigacion/. [Visitada en de 2020]
13. Makin, J.G.: Statistical learning theory in computational neuroscience (2019)
14. Wendin, G.: Can biological quantum networks solve NP-Hard problems? Adv. Quantum Technol. 1800081 (2019)
15. Liao, Y., Yan, M., Tang, S.: The robot consciousness based on empirical knowledge. J. Phys. Conf. Ser. **1861**(1), 012103 (2021)
16. Tononi, G.: Consciousness as integrated information: a provisional manifesto. Biol. Bull. **215**(3), 216–242 (2008)
17. Albantakis, L., et al.: Integrated information theory (IIT) 4.0: Formulating the properties of phenomenal existence in physical terms. PLoS Comput. Biol. **19**(10), e1011465 (2023)
18. Arrabales, R., Ledezma, A., Sanchis, A.: ConsScale: a pragmatic scale for measuring the level of consciousness in artificial agents (2010)

19. Fekete, T., van Leeuwen, C., Edelman, S.: System, subsystem, hive: boundary problems in computational theories of consciousness. Front. Psychol. **7**(1041) (2016)

20. Mayner, W.G.P., Marshall, W., Albantakis, L., Findlay, G., Marchman, R., Tononi, G.: PyPhi: a toolbox for integrated information theory. PLoS Comput. Biol. **14**(7), e1006343 (2018)

Decomposition Models for Agricultural Commodity Price Time Series: A Comparative Research

Adelaida Ojeda-Beltran[1]([⊠]) (iD), Emiro De-La-Hoz-Franco[2],
and José Escorcia-Gutierrez[2] (iD)

[1] Faculty of Economic Sciences, Universidad del Atlántico, Barranquilla 080020,
Colombia
`adelaidaojeda@mail.uniatlantico.edu.co`
[2] Department of Computational Science and Electronic, Universidad de la Costa,
CUC, Barranquilla 080020, Colombia
`jescorci56@cuc.edu.co`

Abstract. This paper explores applying decomposition models to deconstruct agricultural commodity price time series. It compares data decomposition using empirical wavelet transform (EW), empirical modal decomposition (EMD), singular spectral analysis (SSA), and variational mode decomposition (VMD), which could also decompose time series into trends and detailed components. The analysis is based on daily data from the Chicago Board of Trade (CBOT) corn closing prices from January 1980 to August 2021, with 10 456 observations. It is concluded that all four techniques are able to reduce the impact of noise, as well as capture the overall trend and main fluctuations, however, for the selected dataset, the singular spectral analysis (SSA) showed a better signal-to-noise ratio.

Keywords: Decomposition models · Time series · Price · Agriculture · Corn

1 Introduction

The time series analysis of agricultural commodities is essential to understand trends, fluctuations, and patterns in agricultural commodity prices [1]. This paper presents four decomposition models applied to these time series, to analyze the inherent complexity of the data and provide a deeper understanding of the dynamics of the time series. Time series decomposition allows for the separation of the different components present in the data, such as long-term trends, seasonalities, and irregular variations, which is essential for informed decision making in the agricultural and financial sectors.

This study focuses on comparing and contrasting several decomposition models, including the Empirical Wavelet Transform (EWT), Empirical Modal Decomposition (EMD), Singular Spectral Analysis (SSA), and Variational Mode

© The Author(s), under exclusive license to Springer Nature Switzerland AG 2024
N. D. Duque-Méndez et al. (Eds.): CCC 2024, CCIS 2208, pp. 234–248, 2024.
https://doi.org/10.1007/978-3-031-75233-9_17

Decomposition (VMD). Each of these models presents unique and distinctive approaches to analyze and separate the different components present in agricultural price time series, offering a diverse range of analytical tools for researchers and professionals in the industry.

To contextualize the effectiveness of these models, a detailed data set is used that covers the closing prices of corn traded on the Chicago Board of Trade (CBOT) [2]. This dataset, which spans January 1980 to August 2021, provides a broad time perspective and a robust set of observations to assess the predictive and explanatory abilities of the different decomposition models. In general, this study provides a comprehensive and rigorous exploration of the decomposition models applied to time series of agricultural commodities. Given the limitations of existing methods, more useful methods for decomposing time series need to be explored.

This article is structured as follows. Section 2 describes related works. Section 3 gives a detailed explanation about the stages of the proposed computer vision system. In Sect. 4 show experimental results. Discussions regarding the obtained results and comparisons with different state-of-the-art methodologies are also reported. Finally, conclusions and future work are given in Sect. 5.

2   Background and Related Works

This section provides a comprehensive review of the existing literature on time series decomposition models. Previous research that has addressed similar issues is explored, highlighting the methodological approaches used and relevant findings.

2.1   Time Series

In general, a time series is an equispaced sequence of points in time [3]. Therefore, time series are discrete or equivalent time data sets. The forecasting responsibility of a time series is to forecast $xt + 1$ one step ahead in the significance of the feature at time $t + 1$. This task is accomplished by historical time series datasets [4].

The objective of analyzing a time series is to understand the behavior, nature, and mechanisms that generate the time series, through the application of probabilistic models that represent the data as accurately as possible and allow predictions to be made. The main objective of these models is to capture the underlying structure of the time series, using statistical and probabilistic techniques. This involves identifying patterns, trends, seasonality, and other factors that influence the data over time. In developing a time series model, the aim is to find the best representation of the observed data to make future predictions and better understand the behavior of the series.

Several concepts are defined below, necessary for understanding the various models used in time series decomposition.

Stationarity. The concept of stationarity can be visualized as a form of statistical equilibrium [5]. A time series $\{X_t, t = 0, \pm 1, \dots\}$ is stationary if it has statistical properties similar to those of a time-deferred series $\{X_{t+h}, t = 0, \pm 1, \dots\}$ for each integer h. Let $\{X_t\}$ be a time series with $E(X_t^2) < \infty$. The mean of $\{X_t\}$ is $\mu_x(t) = E(X_t)$. The covariance of $\{X_t\}$ is $\gamma_x(r, s) = \text{Cov}(X_r, X_s) = E[(X_r - \mu_x(r))(X_s - \mu_x(s))]$ for all integers r and s. $\{X_t\}$ is (weakly) stationary if $\mu_x(t)$ is independent of t and if $\gamma_x(t + h, t)$ is independent of t for every h. Furthermore, we define the autocovariance function (ACVF) of $\{X_t\}$ as $\gamma_x(h) = \text{Cov}(X_{t+h}, X_t)$ and the autocorrelation function (ACF) of $\{X_t\}$ as $\rho_x(h) \equiv \frac{\gamma_x(h)}{\gamma_x(0)} = \text{Cor}(X_{t+h}, X_t)$. The latter is used in comparing the correlation in the regular component of a time series [4].

2.2   Empirical Wavelet Transform Decomposition Method

As proposed by EWT [4] is a signal decomposition technique based on the adaptation of wavelets to the intrinsic structure of the signal. This approach pro-build adaptive wavelets capable of extracting AM-FM components from a signal [6]. The key idea is that these AM-FM components have compact support of the Fourier spectrum. Separating the different modes is equivalent to segmenting the Fourier spectrum and applying filtering corresponding to each detected support. It is demonstrated that the wavelet formalism can be adapted by considering different Fourier supports, thereby enabling the construction of a set of functions that form an orthonormal basis. Based on this construction, an empirical wavelet transform (and its inverse) is presented to analyze a signal. The main steps of EWT are as follows [7].

Step 1: The Fourier spectrum of the original series is adaptively divided into N continuous segments.

Step 2: Construct the filter bank based on the wavelet transform and take the Amplitude Modulation (AM) and Frequency Modulation (FM) components of each segment. The empirical wavelet function $\psi_n(\omega)$ and the empirical scale function $\varphi_n(\omega)$ can be defined as Eqs. 1 and 2. Where the transition function $\beta(x)$ is defined in Eq. 3.

Step 3: The approximate coefficients Wf (0, t) and the detailed coefficients Wf (n, t) are calculated in Eqs. 4 and 5. Where $\varphi_1(\tau - t)$ and $\psi_n(\tau - t)$ represent the complex conjugate of $\varphi_1(\tau - t)$ and $\psi_n(\tau - t)$.

Step 4: Calculate the components of the original series; the first component and the nth component are given by set of Eqs. 6 and 7. Where $*$ is the convolution operation.

$$\phi_n(\omega) = \begin{cases} 1, & |\omega| \leq \omega_1 - \tau_1 \\ \cos\left(\frac{\pi}{2}\beta\left(\frac{1}{2\tau_1}(|\omega| - \omega_1 + \tau_1)\right)\right), & \omega_1 - \tau_1 < |\omega| \leq \omega_1 + \tau_1 \\ 0, & \text{otherwise} \end{cases} \quad (1)$$

$$
\psi_n(\omega) = \begin{cases}
1, & \omega_n + \tau_n < |\omega| \le \omega_{n+1} + \tau_{n+1} \\
\cos\left(\frac{\pi}{2}\beta\left(\frac{(1)}{2\tau_{n+1}}(|\omega| - \omega_{n+1} + \tau_{n+1})\right)\right), & \omega_{n+1} - \tau_{n+1} < |\omega| \le \omega_{n+1} + \tau_{n+1} \\
\sin\left(\frac{\pi}{2}\beta\left(\frac{1}{2\tau_n}(|\omega| - \omega_n + \tau_n)\right)\right), & \omega_n - \tau_n < |\omega| \le \omega_n + \tau_n \\
0, & \text{otherwise}
\end{cases}
\tag{2}
$$

$$
\beta(x) = \begin{cases}
0, & x \le 0 \\
x^4(35 - 84x + 70x^2 - 20x^3), & 0 < x < 1 \\
1, & x \ge 1
\end{cases}
\tag{3}
$$

$$
W_f(0,t) = \int f(\tau)\overline{\varphi_1(\tau - t)}\, d\tau
\tag{4}
$$

$$
W_f(n,t) = \int f(\tau)\overline{\psi_n(\tau - t)}\, d\tau
\tag{5}
$$

$$
f_1(t) = W_f(0,t) * \varphi_1(t)
\tag{6}
$$

$$
f_n(t) = W_f(n,t) * \psi_n(t)
\tag{7}
$$

2.3   Variational Mode Decomposition Model

The variational mode decomposition model can decompose a complex series into a set of modes around the center frequency. Decomposing complex signals into AM-FM signals with K preset layers using preset scales [1,8]. The IMF is defined as Eq. 8.

$$
U_K(t) = A_k(t)\cos[\Phi_k(t)]
\tag{8}
$$

where $A_k(t)$ is the time amplitude, $\Phi_k(t)$ is the instantaneous phase, and $\Phi_k(t)$ is a monotonically decreasing function. In the VMD decomposition process, if each IMF component has limited bandwidth and different center frequencies, they will continuously alternately and iteratively update each other and adaptively decompose the signals to obtain K IMFs. It requires each IMF to have a minimum bandwidth accumulation sum.

Compared to WT and EMD, VMD has a superior denoising capability and can separate tones of similar frequencies for better signal characterization. The essence of VMD is the construction and solution of the corresponding variational problem. To construct a variational problem, IMFs are first defined as amplitude and frequency modulated (AM-FM) signals and then, in turn, subjected to a series of mathematical inequality of the Hilbert transform, frequency mixing and heterodyne demodulation. Studies by [9] show that the decomposition of VMD is more comprehensive and that the accuracy of the final prediction is higher.

In [10] propose decomposing the wind power data using VMD to obtain components with different fluctuation characteristics [8]. These components are divided into high-, intermediate- and low-frequency components according to their fluctuation characteristics. Then a set of characteristics is established that contains historical wind power data and meteorological factors, and a set of characteristics is selected for each component [1].

In [10] propose decomposing wind power data using VMD to obtain components with different fluctuation characteristics [8]. These components are divided into high-, intermediate- and low-frequency components according to their fluctuation characteristics. Then, a set of features is established that contains historical wind power data and meteorological factors, and a set of features is selected for each component [1]. Similarly, [11] a traffic flow prediction model based on CEEMDAN and VMD, demonstrating that data decomposition improves accuracy by handling nonlinear complexity and reducing noise. This approach outperformed other methods in precision and stability.

2.4   Empirical Modal Decomposition Method

The adaptive time-frequency signal processing method known as EMD was first proposed by [12]. Huang and his colleagues developed this method as a technique for decomposing nonlinear and nonstationary signals into intrinsic components called 'modes'. These modes are extracted by an iterative process of finding local minima and maxima in the original signal.

Each mode represents a signal with a compact supporting Fourier spectrum. This method has been popular in the study of signals in the last decade, mainly because of its ability to separate stationary and non-stationary components of signals. The EMD has some drawbacks, such as the lack of an accurate mathematical model, the choice of interpolation used to perform the interpolation between peak points to obtain the envelope curve, and the sensitivity to both noise and sampling [13].

EMD decomposes the input data into multiple and different intrinsic mode functions (IMF), where each IMF sequence represents a different implicit characteristic of the original time series data. Some studies suggest discarding IMF1 as it can cause significant forecast errors. The IMF1 can be trained using a deep learning approach [14]. The EMD process is called the sifting process and the steps are as follows [3]

1. For a given time series $X = \{x_1, x_2, x_3, \ldots, x_n\}$, find all extreme points (maximum and minimum points) in X.
2. Use the cubic spline interpolation method to fit the upper envelope $U(t)$ of the maximum point set and the lower envelope $L(t)$ of the minimum point set, respectively. Calculate the average upper and lower envelope value $m(t) = \frac{U(t)+L(t)}{2}$.
3. Calculate $H(t) = X - m(t)$ and judge whether $H(t)$ is an IMF according to the evaluation criterion.

4. If $H(t)$ is not an IMF, set $X = H(t)$. Then repeat the above steps until $H(t)$ meets the IMF criterion. Then extract $H(t)$ as $IMF_k(t)$, $k = 1, 2, 3, \ldots, n$.
5. Subtract the sum of all IMF sequences from the original data X to obtain the error sequence r_n.

After the above calculation, the original signal X can be decomposed into multiple IMF subsequences and the remaining error sequence (see Eq. 9).

$$x = \sum_{i=1}^{N} c_i(t) + r_n(t) \tag{9}$$

Here, $c_i(t)$ represents $IMF_k(t)$ and $r_n(t)$ is the residual sequence.

As time series characteristics are generally one of the most important factors to consider when modeling and forecasting future values, EMD is receiving increasing interest in data analysis for forecast purposes [15]. For example, [14] propose a neural decomposition based on EMD to forecast spot prices for crude oil. Likewise, [16] presents a method inspired by the EMD method from a robust mathematical formulation; its performance does not depend on numerical parameters such as the number of changes or the stopping criterion, which seem to have an important effect on the original EMD method; it is less sensitive to noise disturbances.

Other research, such as that presented by [17] developed a hybrid EEMD-NAGU model that uses Ensemble Empirical Mode Decomposition to process time series data of soybean future prices, improving the accuracy of the model by reducing noise and handling complex nonlinearities. This approach outperformed thirteen other prediction models, demonstrating the efficacy of EEMD in improving machine learning algorithms for time series predictions.

2.5   Single Spectral Analysis

SSA is a time series decomposition technique used to separate the signal and noise components present in a time series. SSA is useful for the analysis of non-stationary and stationary signals.

The formula for SSA involves several steps, the most fundamental being the Singular Value Decomposition (SVD) applied to the trajectory matrix. The basic formula for SVD is related in Eq. 10.

$$X = U \Sigma V^T \tag{10}$$

where:

X is the trajectory matrix of the time series.
U is a unitary matrix containing the singular vectors on the left.
Σ is a diagonal matrix that contains the singular values.
V^T is the transpose of a unitary matrix containing the right singular vectors.

The SSA process uses this decomposition to identify the principal components of the time series. After obtaining U, Σ, and V^T, the principal components (also known as eigenvector components) can be calculated from the Eq. 11.

$$C_i = \sqrt{\lambda_i} \cdot \mathbf{u}_i \cdot \mathbf{v}_i^T \tag{11}$$

where:

C_i is the i-th principal component.
λ_i is the i-th singular value (contained on the diagonal of Σ).
$\mathbf{u}_i$ is the i-th singular vector left.
$\mathbf{v}_i^T$ is the i-th transposed right singular vector.

Matrix decomposition: A SVD technique is applied to the trajectory matrix. This involves decomposing the matrix into singular components (singular vectors) and associated singular values. These components can then be used to reconstruct the original time series or to analyze the characteristics of the different components in terms of their contribution to the total signal.

3   Methodology

The methodology aims to compare the effectiveness of different decomposition models in handling data from agricultural commodity price time series, ultimately identifying the most suitable approach for this type of analysis.

Step 1: Decomposition methods (i.e., SSA, VMD, EMD, and EWT) are applied to the time series data to break it down into its constituent components.
Step 2: The decomposed components are reconstructed to form the signal, involving the selecting relevant components and combining them to create a cleaner and more interpretable version of the original time series.
Step 3: The reconstructed signals are evaluated to assess their quality and effectiveness. This involves analyzing the signal noise and evaluating the performance of the reconstruction methods in reducing noise and retaining important features of the original time series.

Figure 1 illustrates the steps involved in the methodology used for this study.

Data Description

The analysis is based on daily data on closing corn prices from the Chicago Board of Trade (CBOT). The dataset ranges from January 1980 to August 2021, encompassing a total of 10,456 observations from the website https://www.cmegroup.com/. These data points were obtained from a reliable financial data source, providing a comprehensive and detailed view of corn price fluctuations over the specified period. The analysis aims to utilize this extensive dataset to evaluate the performance of various decomposition models in handling and interpreting agricultural commodity price time series.

Data Preparation

The data were processed in Google Colaboratory, which allows running Python code. The Python libraries used for data processing include Pandas for data manipulation, NumPy for numerical operations, Matplotlib and Seaborn for data visualization, and the MATLAB programming platform.

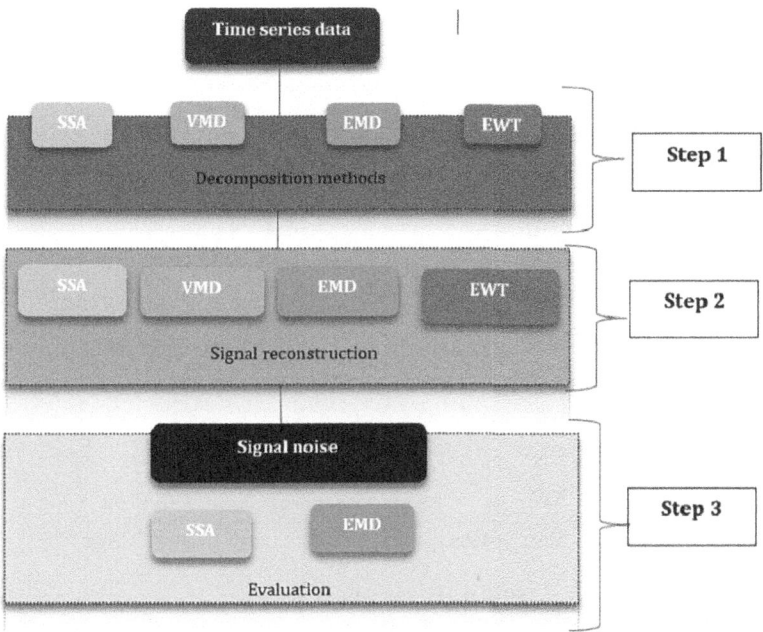

Fig. 1. Flowchart of the proposed methodology.

4   Results

This section details the results obtained by applying the decomposition techniques in terms of the signal-to-noise ratio and the precision to represent the fluctuations and trends in corn prices.

4.1   Step 1 - Decomposition of the Original Series Using the SSA-EWT-EMD VMD Techniques

Decomposition of the Original Series Using the SSA Techniques: For SSA, 30 decompositions were achieved. The noise-removed series was reconstructed by selecting the four principal components. All other decompositions contributing less than 0. 1% were considered non-results of the decomposition and following the SSA procedures,g SSA procedures [18], the first four components were selected and reconstructed. Mode 1 captured the main long-term trend, eliminating rapid fluctuations, while Modes 2 and 3 reflected seasonal patterns and economic cycles, respectively. Mode 4 complemented these results by capturing significant medium-term fluctuations. This decomposition allowed for the identification of both general trends and cyclical and seasonal oscillations, significantly improving the ability to analyze and predict corn prices. Figure 3 illustrates the methodology used in this study, highlighting the steps involved in the decomposition and reconstruction process (Fig. 2).

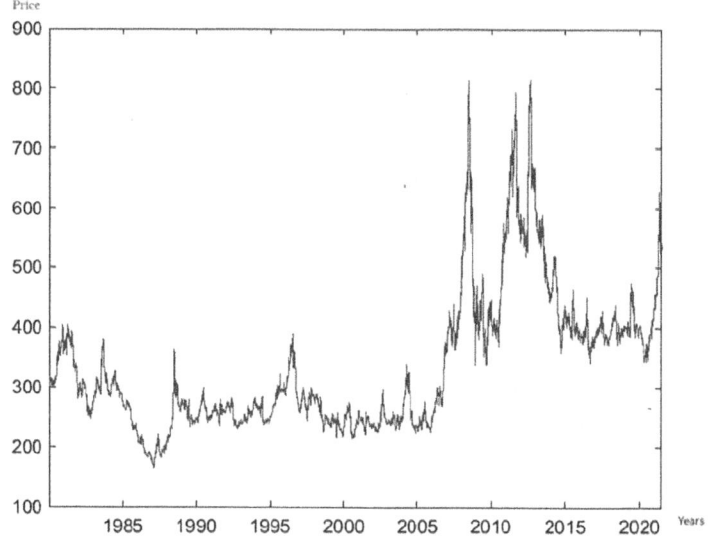

Fig. 2. Daily wheat prices.

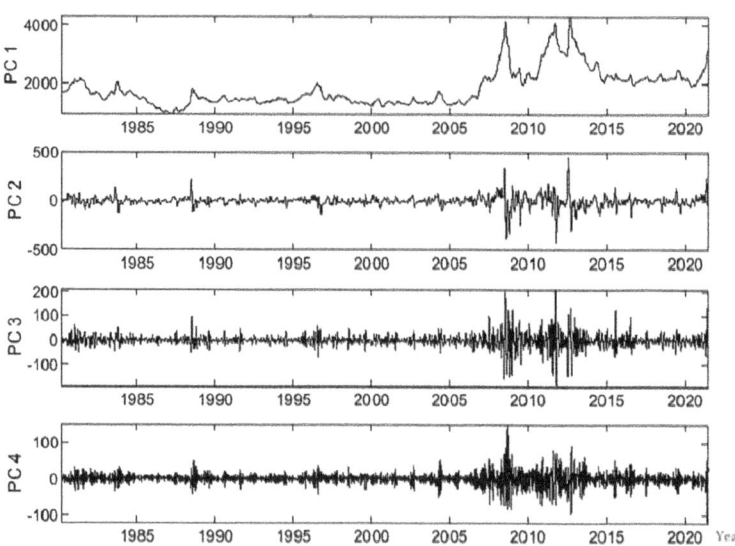

Fig. 3. Decomposition results with SSA.

Decomposition of the Original Series Using EWT Techniques

The Empirical Wavelet Transform (EWT) decomposes the original signal into 12 bands, allowing for isolation of variations in different frequency ranges. The high-frequency components (Modes 1 to 4) capture rapid oscillations and noise, useful for understanding short-term volatility, although they are not significant for long-term trend analysis. The medium frequency components (Modes 5 to 8) reflect seasonal and cyclical patterns, useful for identifying seasonal or economic cycles that affect corn prices. The low-frequency components (Modes 9 to 12) capture long-term trends and smoother variations, crucial for analyzing general market trends. content presented in Fig. 4. This decomposition allows the original series to be separated into several parts, each representing different characteristics or modes present in the data.

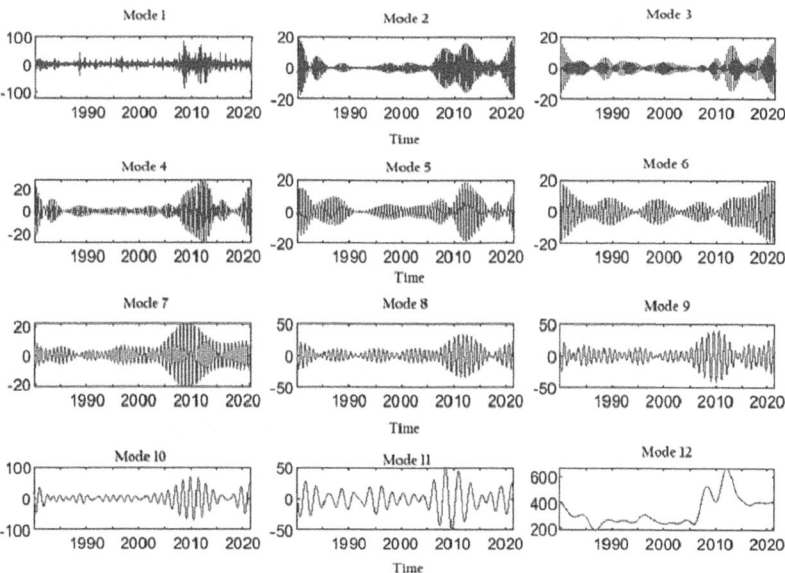

Fig. 4. Decomposition results with EWT.

Decomposition of the Original Series Using VMD Techniques. A critical parameter, the number of modes K, must be confirmed before the VMD technique decomposes the original series. By iteratively analyzing the results of the center frequency in relation to the different modes of various components, a center frequency aliasing is evident in the frequency spectrum of the tenth mode when K equals 10. Consequently, it is chosen to select the number of K modes as 9. The results of the decomposition are presented in Fig. 5.

For VMD, we decompose into nine components, the highest frequency component is summed to obtain the noise-free series.

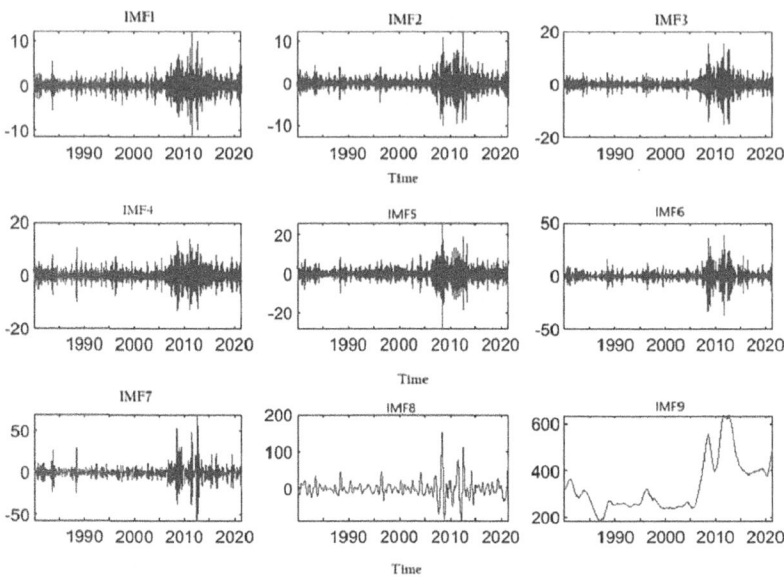

Fig. 5. Decomposition results with VMD.

Decomposition of the Original Series Using the EMD Technique. The EMD decomposition of the corn price time series produced five IMFs and a residual term. IMF 1 contains the highest frequency oscillations and the most noise, capturing rapid variations that generally represent market noise. IMFs 2 and 3 represent intermediate frequency oscillations, capturing cyclical and seasonal market patterns. IMFs 4 and 5 contain low-frequency oscillations, representing long-term trends and cycles. Finally, the residual term shows the underlying long-term trend in corn prices, eliminating faster oscillations and providing a clear view of long-term variation.

4.2   Step 2 - Reconstruction of the Series

In Step 2, the series reconstruction is performed using the components obtained from the decomposition in the previous step. This process involves appropriately combining the decomposed components to recreate the original series or to obtain an approximation of it. The reconstruction may vary in complexity and accuracy depending on the specific application and the decomposition method used. Figure a shows the results of the reconstruction with SSA, Figure b presents the results of the reconstruction with EMD, Figure c presents the results of the reconstruction with VMD and Fig. 7d shows the results of the reconstruction with EWT (see Fig. 7) (Fig. 6).

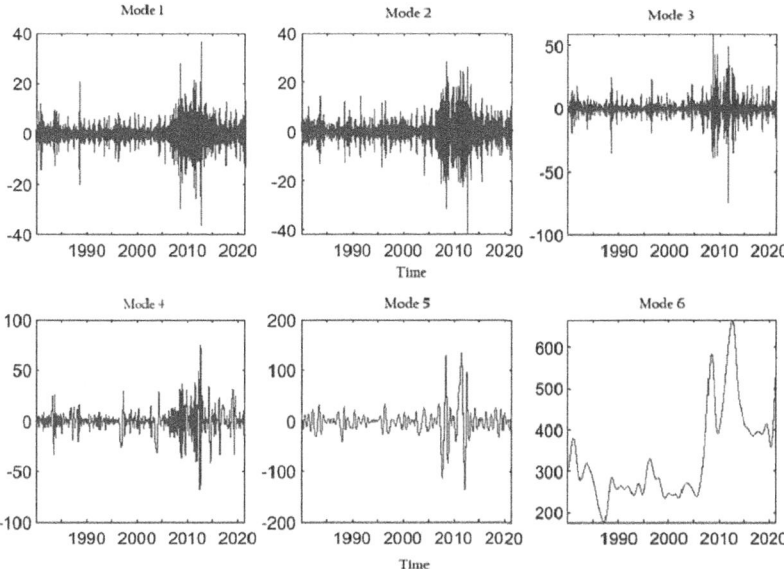

Fig. 6. Decomposition results with EMD.

4.3   Step 3 - Signal-to-Noise Ratio

Therefore, the methods process signals efficiently, but the effectiveness of the methods at different frequencies is different. The performance of the methods is measured by calculating the signal-to-noise ratio (SNR). The signal-to-noise ratio (SNR) is a measure that compares the power of a desired signal to the background power or noise surrounding it. The general formula for calculating the signal-to-noise ratio (SNR) is expressed in Eq. 12.

$$SNR = \frac{P_{\text{señal}}}{P_{\text{ruido}}} \tag{12}$$

P_{signal} is the power of the desired signal. P_{noise} is the power of the noise present in the system. For this study, the signal-to-noise ratio in decibels (dB) is used, which is calculated using the formula shown in Eq. 13

$$SNR_{\text{dB}} = 10 \log_{10} \left(\frac{P_{\text{señal}}}{P_{\text{ruido}}} \right) \tag{13}$$

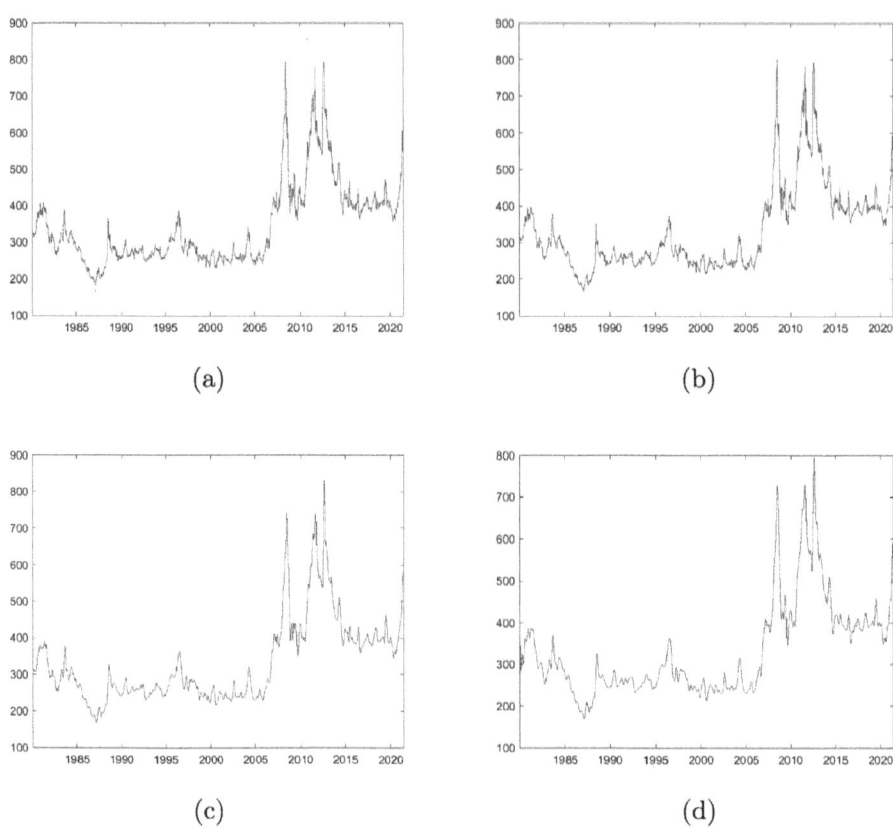

Fig. 7. Reconstruction models.

This converts the SNR to a logarithmic scale, which is usually more convenient to represent large ranges of SNR values which are presented in Table 1.

Table 1. Comparison of SNR for Different Decomposition Methods

	Decomposition method	SNR
0	EMD	31.17581
1	VMD	37.492635
2	EWT	29.480222
3	SSA	37.794697

5 Conclusions

The decomposition methods compared achieved a significant improvement in signal-to-noise ratio (SNR) compared to the original signal. In particular, the VMD and SSA methods show the largest improvements in SNR, with values of 37.49 and 37.79 respectively, indicating a high efficiency in separating the desired signal from the noise for this data set. It is important to note that the choice of the decomposition method may depend on the type of signal and the specific context of the application. In addition to SNR, other criteria, such as computational complexity, interpretability of decomposed components, and robustness to different input conditions, may influence the selection of the most appropriate method.

All four techniques are capable of reducing the impact of noise and capturing the overall trend and major fluctuations. It has been demonstrated that these methods decompose the input signal into various frequency modes and that the modes can be used to further process the data. Furthermore, their performances are analyzed by applying them to the same signal.

The comparison of decomposition models applies not only to the specific time series used in the study, but also to a wide variety of time series. Methods such as EMD, SSA, and VMD are versatile and can be used in different contexts due to their ability to handle complex characteristics such as trend, seasonality, and noise. Although the results may vary depending on the particularities of each dataset, the methodology to evaluate and select the most suitable decomposition method remains crucial to optimize data quality and prediction accuracy in various applications.

6 Future Research Lines

In future work, it is proposed that time series decomposition models be explored as a fundamental strategy to optimize the process of constructing machine learning models that utilize these series. The evaluation and selection of the best decomposition method is essential. Comparison of methods such as EMD, SSA, and VMD allows identifying which provides the best signal-to-noise ratio and most effectively captures important trends and fluctuations in the data. An optimal decomposition method ensures that the essential characteristics of the data are preserved and noise is minimized, thereby improving the quality of the data used to train the models. This facilitates the design of algorithms specialized in specific patterns, optimizing the time and resources needed for data preprocessing, and allowing ML models to focus on clearer and more predictable patterns, thus enhancing prediction accuracy. Previous studies have shown that combining decomposition techniques with ML algorithms significantly improves prediction accuracy, supporting the effectiveness of this strategy in optimizing the construction of robust and efficient predictive models.

References

1. He, X., Luo, J., Li, P., Zuo, G., Xie, J.: A hybrid model based on variational mode decomposition and gradient boosting regression tree for monthly runoff forecasting. Water Resour. Manage **34**, 865–884 (2020)
2. Group, C.: Corn overview (2024). Accessed 02 June 2024
3. Yang, Y., Fan, C.: Efficient and robust time series prediction model based on REMD-MMLP with temporal-window. Expert Syst. Appl. **207**, 117979 (2022)
4. Mohammadi, H.A., Ghofrani, S., Nikseresht, A.: Using empirical wavelet transform and high-order fuzzy cognitive maps for time series forecasting. Appl. Soft Comput. **135**, 109990 (2023)
5. Hyndman, R.J., Athanasopoulos, G.: Introduction to Time Series and Forecasting, 2nd edn. Springer, Cham (2016)
6. Gilles, J.: Empirical wavelet transform. IEEE Trans. Signal Process. **61**, 3999–4010 (2013)
7. Zeng, L., Ling, L., Zhang, D., Jiang, W.: Optimal forecast combination based on PSO-CS approach for daily agricultural future prices forecasting. Appl. Soft Comput. **132**, 109833 (2023)
8. Liu, Z., Chai, T., Tang, J., Yu, W.: Signal analysis of mill shell vibration based on variational modal decomposition. In: 2020 39th Chinese Control Conference (CCC), pp. 1168–1173. IEEE (2020)
9. Hou, T.Y., Shi, Z.: Adaptive data analysis via sparse time-frequency representation. Adv. Adapt. Data Anal. **3**, 1–28 (2011)
10. Zhang, G., Xu, B., Liu, H., Hou, J., Zhang, J.: Wind power prediction based on variational mode decomposition and feature selection. J. Mod. Power Syst. Clean Energy **9**(6), 1520–1529 (2021)
11. Yu, Q., Li, L., Zhao, H., Teng, L., Mualla, Y.: The short-term traffic flow prediction based on an improved data decomposition integration strategy. In: 2023 China Automation Congress (CAC), pp. 4376–4381. IEEE (2023)
12. Huang, N.E., et al.: The empirical mode decomposition and the hilbert spectrum for nonlinear and non-stationary time series analysis. In: Proceedings of the Royal Society of London. Series A: Mathematical, Physical and Engineering Sciences, vol. 454, pp. 903–995 (1998)
13. Yang, H.-F., Chen, Y.-P.P.: Hybrid deep learning and empirical mode decomposition model for time series applications. Expert Syst. Appl. **120**, 128–138 (2019)
14. Yu, L., Wang, S., Lai, K.K.: Forecasting crude oil price with an EMD-based neural network ensemble learning paradigm. Energy Econ. **30**, 2623–2635 (2008)
15. Lahmiri, S.: Comparing variational and empirical mode decomposition in forecasting day-ahead energy prices. IEEE Syst. J. **11**(3), 1907–1914 (2017)
16. Park, J.-M., Lee, J.-H., Ko, J.-H., Lee, S.-H.: Stock price prediction using machine learning algorithms. Int. J. Inf. Technol. Decis. Making **10**(4), 641–657 (2011)
17. Liu, J., Zhang, B., Zhang, T., Wang, J.: Soybean futures price prediction model based on EEMD-NAGU. IEEE Access **11**, 99328–99338 (2023)
18. Zeng, L., Luo, H.: Optimal forecast combination based on PSO-CS approach for daily agricultural future prices forecasting. Appl. Soft Comput. **122**, 109833 (2022)

Cyber Security and Information Security

Feature Selection in Machine Learning-Based IDS Performance

Jose Albeiro Montes Gil[1]([✉]) [iD], Néstor Darío Duque Méndez[1] [iD],
Gustavo Adolfo Isaza[2] [iD], Fabián Alberto Ramírez[2] [iD], and Jeferson Arango López[2] [iD]

[1] Universidad Nacional de Colombia, Bogotá, Colombia
joamontesgi@unal.edu.co
[2] Universidad de Caldas, Manizales, Colombia

Abstract. Computer security faces many challenges, including the detection of attacks generated by various intrusions. Intrusion detection systems (IDS) have different approaches, and those based on supervised learning have high capabilities to predict different types of attacks. As in other cases, machine learning algorithms supporting these tools benefit from dimensionality reduction. In this paper we present the results of applying different algorithms to obtain subsets of features that maintain good performance in classification tasks and apply the ensemble operations strategy to obtain the final features that are fed to the classifiers to determine the outputs. The results show that in the different classifiers all the metrics go down if the number of features is decreased in high degree, but that with the operation union of the subsets of the features obtained from the importance of the features, the number of attributes obtained is significantly lower and the good performance of the classifier is maintained.

Keywords: IDS · feature selection · dimensionality reduction

1 Introduction

In a world increasingly interconnected through computer networks and other devices, security in terms of privacy, reliability and availability is fundamental. The availability of systems is affected, among several situations, by various attacks from outside or inside the organization, including denial of service attacks, DoS, which reduce the availability of systems and even leave them completely inactive. Among the alternatives that have been implemented to deal with this situation are intrusion detection systems (IDS), which are systems that can determine whether security has been compromised at any time by permanently examining frame traffic. There are several approaches to IDS and several techniques that have been implemented to obtain a good performance of these, among them, artificial intelligence, and machine learning techniques, which in recent times have demonstrated the great capabilities that ensure systems with high performance achieving quite satisfactory metrics.

IDSs based on supervised learning have shown their great possibilities and capabilities to predict these situations with high performance Figs, but on many occasions,

N. D. Duque-Méndez et al. (Eds.): CCC 2024, CCIS 2208, pp. 251–268, 2024.
https://doi.org/10.1007/978-3-031-75233-9_18

with a high false positive rate. For training, a high volume of data is required and for the real-time execution phase, with already trained models, algorithms with acceptable computational cost must be implemented and high dimensionality negatively affects response times.

High dimensionality is recognized as one of the situations that can negatively influence machine learning algorithms. Reducing dimensionality under approaches such as relevant feature selection is a promising way, but it is not a trivial problem since the selected attributes change with different algorithms. For the case CICIDS-2017 data set, with 79 features originally and a class that reflects different types of attacks, it is very important to define a minimum number of features that allow a satisfactory classification or prediction, which forces to use computational methods to get those attributes with good performance while reducing the computational cost.

The work presented in this paper aims to run different algorithms that allow the selection of features and face the fact of obtaining non-homogeneous results in different cases, which maintains uncertainty. As a strategy, the subsets are integrated using the Union operator or the Intersection operator, seeking to determine the final features to be selected.

For the performance evaluation, different algorithms are applied to those that generate the subsets of features and are fed with the total of the attributes and with the features obtained with union and intersection. The results show that the best output is obtained, for the different metrics and algorithms, with the total features, but with the subset obtained with union the output drops very little in performance. The results of this work will be exploited in an IDS deployed in the cloud and for public use.

The rest of the paper is organized as follows: The next section is devoted to collect some concepts of interest for the paper, then some related work is discussed. Section 4 presents the methodology and Sect. 5 presents the validation and finally the Conclusions and Future work are included.

2 Theoretical Framework

2.1 Denial of Service Attack (DoS)

A Denial of Service (DoS) attack is defined as the massive sending of requests to a service or equipment on the network, which generates a collapse in the receiver's resources, partially or totally affecting the availability of the service or equipment [1].

2.2 Distributed Denial of Service Attack (DDoS)

Distributed Denial of Service attacks are requests coming from various sources, as shown in Fig. 1. Unlike DoS attacks, in a DDoS attack common users can be infected under the concept of "zombie computers", which generates a greater number of requests, which fall on the receiver, causing response times to increase or the definitive collapse for an indeterminate period [1].

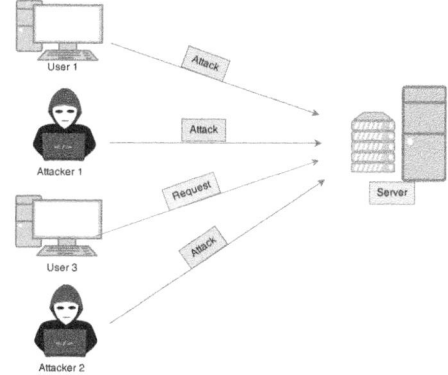

Fig. 1. DDoS Attack. **Source:** Own elaboration.

2.3 Dataset: CICIDS-2017

The CICIDS-2017 Dataset was created by the University of New Brunswick in 2017 at the faculty of Computer Science. For 5 days a traffic analysis was performed on 25 users making use of HTTP, HTTPS, FTP, SSH and email protocols, which allowed the capture of $2'827,876$ records. The analysis yielded 8 different types of attacks, including Brute Force FTP, Brute Force SSH, DoS, Heartbleed, Web Attack, Infiltration, Botnet and DdoS [2, 3].

The records were divided by days into files in CSV format, and DoS/DDoS attacks were used for this research. In total there are 1448366 rows and 79 columns, before preprocessing.

2.4 Intrusion Detection System

An Intrusion Detection System is defined in [4] like a tool that can detect attacks both inside and outside the network, given its ability to monitor traffic and apply classification algorithms. An IDS can be classified into two categories: signature-based IDS and anomaly-based IDS. The operation of signature-based IDS relies on logs stored in databases, while anomaly-based IDS compares network patterns and behavior.

2.5 Feature Selection

Attribute selection is a strategy often used in some research to reduce dimensionality by eliminating irrelevant or redundant attributes. As mentioned in [5], non-essential features negatively affect classifier performance and computational resource requirements.

Attribute selection is a strategy often used in some research to reduce dimensionality by eliminating irrelevant or redundant attributes. As mentioned in, non-essential features negatively affect classifier performance and computational resource requirements.

Since the techniques that allow attribute selection can vary in performance depending on the nature of the data, the general procedure for feature selection is defined in [5].

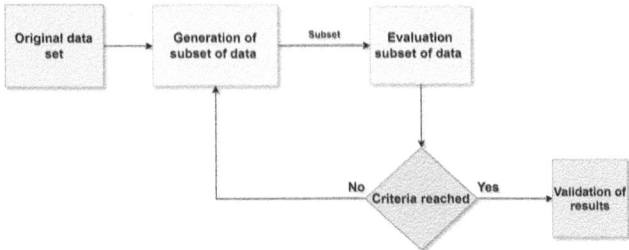

Fig. 2. General Attribute Selection Procedure. *Source: From* [5].

There, 4 stages are mentioned that can be applied to the CICIDS-2017 dataset, as shown in Fig. 2.

There are several strategies and algorithms for feature selection, including Boruta, feature importance, principal component analysis (PCA), Chi-Square, among others.

2.6   Decision Tree (DT)

A Decision Tree is a non-parametric classification technique that uses rules based on the grouping of criteria [6]. Its graphical representation allows to clearly represent the events that are generated according to a decision [7].

2.7   Extreme Gradient Rise – XGBoost (XGB)

As mentioned in [8], XGBoost is a technique developed by Tianqi Chen which combines different decision trees that improve their performance with respect to those previously defined. In [9], the use of the XGBoost algorithm is highlighted as an approach successfully used in different models for intrusion detection.

As mentioned in, XGBoost is a technique developed by Tianqi Chen which combines different decision trees which improve their performance with respect to those previously defined. The use of the XGBoost algorithm is highlighted as an approach successfully used in different models for intrusion detection.

2.8   RandomForest (RF)

RandomForest is defined in [10] as a non-parametric classification technique, supported on trees that generate predictions based on random vectors. Its application is established in classification and regression tasks. In [10], the advantage of RandomForest is highlighted, given its capacity for attribute selection and its resistance to overfitting.

2.9   K Nearby Neighbors – KNN

The K Nearest Neighbors (KNN) algorithm is a nonparametric supervised learning technique that classifies by means of data proximity, since it performs its prediction by calculating the Euclidean distance between a value and the nearest records [11].

2.10 Multilayer Perceptron – MLP

It is an artificial neural network which is composed of several hidden layers, which in turn constitute the input of another layer of neurons. An MLP neural network has a set of input neurons, a layer of hidden neurons and the neurons that constitute the prediction [12].

3 Related Research

The following are some papers presenting strategies for feature selection in data sets like the one exploited in this work.

In [2] the authors applied Boruta and Permutation Importance for Feature Selection of the CICIDS-2017 dataset. Different Supervised Learning techniques were applied to analyze the accuracy of each of the classifiers, among which RandomForest and Decision Trees are highlighted. The authors mention the importance of continuing to perform analyses with other types of classifiers and feature selectors.

In [13] a proposal was presented based on Feature Selection using Filter and Wrapper based techniques on the NSL-KDD dataset. Subsequently, the authors built an Intrusion Detection System using Machine Learning techniques with the features obtained by Chi-Square and Correlation. The results obtained reflect a higher accuracy of the Artificial Neural Network compared to the Support Vector Machine. The authors emphasize the importance of continuing research in the field of intrusions, given the false positive rates that are still occurring.

In [14] the CSE-CIC-IDS2018 dataset was used for the application of Machine Learning and Feature Selection techniques with the implementation of Chi-Square and Spearman's Correlation Coefficient. The model proposed by the authors was validated using 7 different classifiers, however, as future work, the need to advance studies under other hyperparameters and different types of Artificial Neural Networks is proposed.

In [15] a scheme for an IDS divided in 2 stages was built. Stage 1 consisted of data pre-processing and Feature Selection, and stage 2, the use of the classifier supported by Light Gradient Boosting Machine. Comparisons were made with other techniques such as Support Vector Machines, RandomForest, Convolutional Neural Networks, Multilayer Perceptron, among others.

In [16] the authors used the NSL-KDD dataset with random features to reduce the number of attributes, training times and model complexity. They implemented 8 Machine Learning techniques, which showed a good behavior in DoS attacks, while in U2R attacks their performance was inferior. As future work, the importance of implementing different methods for Feature Selection to reduce the computational complexity of IDS proposals is mentioned.

In [17] Chi-Square was applied as a feature selection technique for the UNSW-NB15 dataset. The performance of the 5 Machine Learning classification algorithms was analyzed in terms of accuracy, false positive rate, F1 score and mean square error. The analysis was performed under two proposals; the first one considered the dataset considering the selection technique, while the second one, did not consider the feature selection technique. This paper concludes that the Random Forest classifier was the best performing classifier compared to Naive Bayes, Logistic Regression and KNN.

In [18] the authors conducted a review of papers around the field of IDS, which were implemented under Machine Learning and Deep Learning techniques. Twenty-eight papers were analyzed, defining which is the most analyzed dataset and the most frequent learning technique. As a conclusion, the authors highlight the importance of performing validations with different data sets, given that most of the research is based on the KDD Cup'99 and NSL-KDD data set. Additionally, it is recommended to find mechanisms through feature selection to reduce the dimensionality of the problems and increase the performance of the classification models.

In [19], a comparison is made between normalization and feature selection using the UNSW-NB15 dataset, with the aim of determining which aspect is more relevant when analyzing the performance of a model. In this paper, the authors conclude that the evaluated models performed better when trained with recent datasets like UNSW-NB15 compared to older ones, such as NSL-KDD.

In [20] a critical review of other studies was carried out in terms of the most analyzed data sets, feature selection techniques and evaluation of the proposals in terms of parameters such as complexity, correlation, and performance. The importance of including techniques that allow the selection of features that can make significant contributions, considering the number of variables to be analyzed, is highlighted.

The review of research articles in the field of intrusion detection conducted in [21] includes 29 papers that use different mechanisms for Feature Selection and Classification algorithms supported by Machine Learning. It also highlights the popularity of Artificial Neural Networks in their different classifications.

While feature reduction can generate a reduction in the complexity of a model, in [22] a higher performance in terms of accuracy is observed when the number of attributes increases. In [23] the performance of some algorithms applying feature selection was analyzed, from there it is concluded that the accuracy of the models can be increased in some cases as the number of features increases.

The literature review conducted allows observing that feature selection in computer security related datasets is a field of interest and can bring improvements to the performance of classification models, however, a unique feature subset that delivers the best results in different classifiers is not determined.

Table 1 below presents a summary of the focus of each of the papers that make up the state-of-the-art review.

The following convention is used:

- Meets criterion (x).
- Does not meet criterion (-).

As shown in Table 1, the related works do not reflect a tendency to determine the most important characteristics in the generation of DoS/DDoS attacks, in addition to the absence of proposals that seek to evaluate the results from subsets of data defined by different classifiers. None of the works present the different characteristics selected by different methods used and a strategy to determine the final subset.

Table 1. Summary of research approach.

Research	Define Importance of each attribute	Dataset CICIDS 2017	Presents subassembly integration strategy	Comparison between experiment using all attributes and using attribute selection
[2]	–	X	–	X
[9]	–	–	–	–
[10]	X	–	–	X
[11]	–	–	–	–
[13]	–	–	–	–
[24]	–	–	–	–
[25]	–	–	–	X
[26]	–	–	–	–

Source: Own elaboration.

4   Metodology

This work was developed following a strategy divided into 3 stages, as shown in Fig. 3. With the proposed segmentation by stages, it is expected that it will be easier for future interested parties to replicate the present research. The stages defined in the present methodology are supported by what is mentioned in [2] and [5] for the definition of the criteria in the preprocessing and the stop condition to perform the validation of the results obtained in the feature selection.

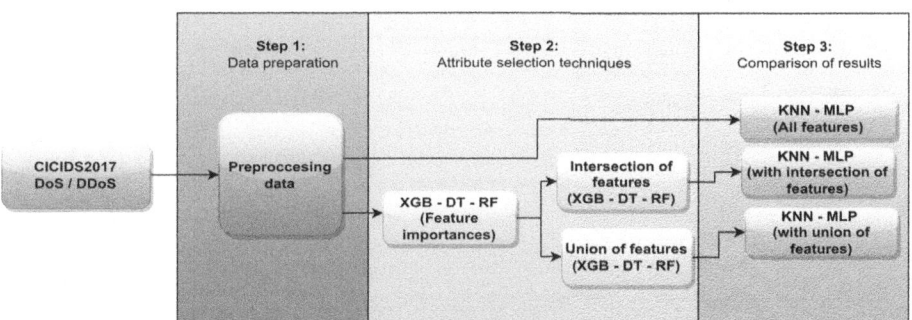

Fig. 3. Methodology by Stages. **Source:** Own elaboration

4.1   Stage 1 - Data Preparation

Data preparation is the stage in which the aim is to manipulate the records in such a way that they are useful for further analysis. The objective is to eliminate noise, subtract outliers, complete missing records, among others [27]. Based on the CICIDS-2017 data set, records with trends to infinity, non-numerical (Not a Number), blank spaces, a repeated column, replacing attribute names with values from 0 to 78 were eliminated, as shown in Table 2.

Table 2. CICIDS2017 Columns renamed.

Number	Column CICIDS2017	Number	Column CICIDS2017	Number	Column CICIDS2017
1	dst_port	27	bwd_iat_mean	53	pkt_size_avg
2	flow_duration	28	bwd_iat_std	54	fwd_seg_size
3	tot_fwd_pkts	29	bwd_iat_max		_avg
4	tot_bwd_pkts	30	bwd_iat_min	55	bwd_seg_size
5	totlen_fwd_pkts	31	fwd_psh_flags		_avg
6	totlen_bwd_pkts	32	bwd_psh_flags	56	fwd_byts_b_avg
		33	fwd_urg_flags		
7	fwd_pkt_len_max	34	bwd_urg_flags	57	fwd_pkts_b_avg
8	fwd_pkt_len_min	35	fwd_header_len		
		36	bwd_header_len	58	fwd_blk_rate_avg
9	fwd_pkt_len_mean	37	fwd_pkts_s		
10	fwd_pkt_len_std	38	bwd_pkts_s	59	bwd_byts_b_avg
11	bwd_pkt_len_max	39	pkt_len_min		
12	bwd_pkt_len_min	40	pkt_len_max	60	bwd_pkts_b_avg
13	bwd_pkt_len_mean	41	pkt_len_mean	61	bwd_blk_rate_avg
14	bwd_pkt_len_std	42	pkt_len_std		
15	flow_byts_s	43	pkt_len_var	62	sub-flow_fwd_pkts
16	flow_pkts_s	44	fin_flag_cnt		
17	flow_iat_mean	45	syn_flag_cnt	63	sub-flow_fwd_byts
18	flow_iat_std	46	rst_flag_cnt		
19	flow_iat_max	47	psh_flag_cnt	64	sub-flow_bwd_pkts
20	flow_iat_min	48	ack_flag_cnt		
21	fwd_iat_tot	49	urg_flag_cnt	65	sub-flow_bwd_byts
22	fwd_iat_mean	50	cwe_flag_count		
23	fwd_iat_std	51	ece_flag_cnt	66	init_fwd_win_byts
24	fwd_iat_max	52	down_up_ratio		
25	fwd_iat_min			67	init_bwd_win_byts
26	bwd_iat_tot				
				68	fwd_act_data_pkts
				69	fwd_seg_size_min
				70	active_mean
				71	active_std
				72	active_max
				73	active_min
				74	idle_mean
				75	idle_std
				76	idle_max
				77	idle_min
				78	Label

Source: Own elaboration.

Additionally, the values were normalized. Table 3 shows the amount of data before and after the preprocessing stage.

Table 3. Variations in data with preprocessing.

Item	CICIDS-2017 (DoS – DDoS) without preprocessing	CICIDS-2017 (DoS – DDoS) with preprocessing
Rows	1448355	1447279
Columns	79	78

Source: Own elaboration.

Table 4 shows the coding process to which the classes were subjected, where the outputs are represented by numerical values between 0 and 5.

Table 4. Class Categorization.

Label	Class	Records
0	Benigno	1067540
1	DDoS	128027
2	DoS GoldenEye	10293
3	DoS Hulk	230124
4	DoS Slowhttptest	5499
5	DoS Slowloris	5796
Total		**1447279**

Source: Own elaboration.

4.2   Stage 2 - Application of Techniques for Attribute Selection

Figure 4 shows that stage 2 consisted of identifying the percentage of importance of each of the attributes in the dataset. To determine the importance of the features, the XGB, RF and DT classifiers were defined using the well-known free software library for machine learning Scikit-Learn.

To avoid overfitting, the data set was divided into 75% of the records for the training process and 25% for the tests. These values were defined from experience in previous work and guidelines in the literature.

To determine the importance of each of the features, the attribute "feature_importances_" of the Scikit-Learn library is used, as recommended in [28, 29] y [30].

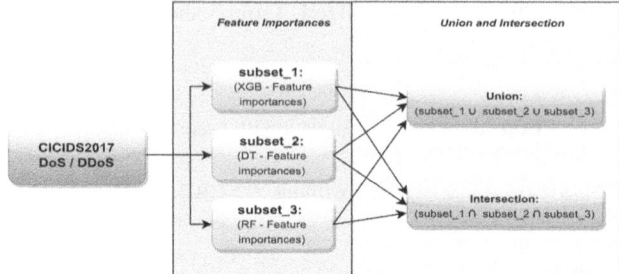

Fig. 4. Attribute selection stage. **Source:** Own elaboration.

Figure 5 shows the performance of the classifiers in terms of precision, F1 score, accuracy and sensitivity. It is possible to observe that the 3 techniques obtained significant metrics, when the model was validated with 25% of the data.

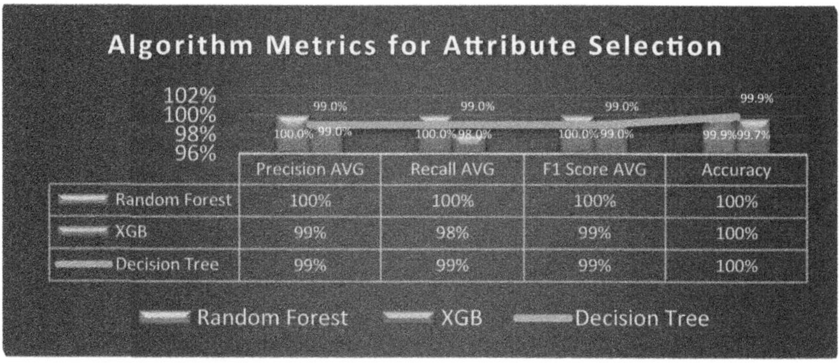

Algorithm Metrics for Attribute Selection

	Precision AVG	Recall AVG	F1 Score AVG	Accuracy
Random Forest	100%	100%	100%	100%
XGB	99%	98%	99%	100%
Decision Tree	99%	99%	99%	100%

Random Forest   XGB   Decision Tree

Fig. 5. Classifier Performance - Attribute Importance. **Source:** Own elaboration.

The chosen classifiers are run on the preprocessed data set and give a weighting of the weight of each attribute in obtaining the classifier results. Tables 5, 6 and 7 show the percentage of importance of each attribute according to each classification technique. Since "feature_importances" determines in percentage the importance of each attribute, the cumulative sum of the attributes was considered for the creation of each subset until the percentage of importance reached 80%.

Table 5. Feature Importance XGB.

Feature	% Importance	% of Importance (cumulative)
3	16,00%	16%
35	8,67%	25%
36	6,42%	31%
38	6,35%	37%
4	6,16%	44%
25	5,26%	49%
0	3,85%	53%
7	3,29%	56%
42	3,20%	59%
6	2,93%	62%
67	2,89%	65%
75	2,79%	68%
11	2,48%	70%
22	2,27%	73%
24	2,18%	75%
65	2,05%	77%
14	1,78%	79%
23	1,76%	80%

Source: Own elaboration.

Table 6. Feature Importance DT.

Feature	% Importance	% of Importance (cumulative)
75	22,50997%	22,50997%
42	17,87567%	40,38564%
0	13,88353%	54,26917%
36	13,20165%	67,47083%
71	8,34455%	75,81537%
6	4,38788%	80,20326%

Source: Own elaboration.

Table 7. Feature Importance RF.

Feature	% Importance	% of Importance (cumulative)
36	5,366%	5,366%
75	4,782%	10,148%
73	4,638%	14,786%
76	4,436%	19,222%
0	4,382%	23,604%
1	4,138%	27,742%
22	3,864%	31,606%
42	3,615%	35,221%
65	3,533%	38,754%
15	3,461%	42,215%
23	3,299%	45,514%
4	3,064%	48,578%
13	2,761%	51,339%
14	2,584%	53,923%
21	2,352%	56,275%
20	2,220%	58,495%
62	1,957%	60,452%
17	1,779%	62,231%
16	1,758%	63,989%
18	1,747%	65,735%
34	1,727%	67,462%
69	1,650%	69,112%
5	1,632%	70,744%
71	1,611%	72,355%
6	1,584%	73,940%
10	1,479%	75,419%
72	1,401%	76,819%
27	1,389%	78,208%
25	1,373%	79,581%
9	1,363%	80,943%

Source: Own elaboration.

Table 8 presents the characteristics obtained with each of the classifiers.

To compare the performance of the model using all the features and attribute selection, it was defined to perform the intersection and union between the 3 subsets of features obtained by each classifier, obtaining 5 features in the intersection (Table 9) and 37 with the union (Table 10).

Table 8. Attribute subsets for each classifier.

Subset	Features
subconjunto_RF	36, 75, 73, 76, 0, 1, 22, 42, 65, 15, 23, 4, 13, 14, 21, 20, 62, 17, 16, 18, 34, 69, 5, 71, 6,10, 72, 27, 25, 9
subconjunto_XGB	3, 35, 36, 38, 4, 25, 0, 7, 42, 6, 67, 75, 11, 22, 24, 65, 14, 23
subconjunto_DT	75, 42, 0, 36, 71, 6

Source: Own elaboration.

Table 9. Intersection of attributes.

Subset	Features
subset_intersección	0, 36, 6, 42, 75

Source: Own elaboration.

Table 10. Union of attributes.

Subset	Features
subset_union	0, 1, 3, 4, 5, 6, 7, 9, 10, 11, 13, 14, 15, 16, 17, 18, 20, 21, 22, 23, 24, 25, 27, 34, 35, 36, 38, 42, 62, 65, 67, 69, 71, 72, 73, 75, 76

Source: Own elaboration.

4.3 Stage 3 - Comparison of Results

This stage is defined with the objective of comparing the metrics of the KNN classifiers and the NN-MLP multilayer perceptron neural network with the subsets defined in the feature selection stage, as shown in Fig. 6.

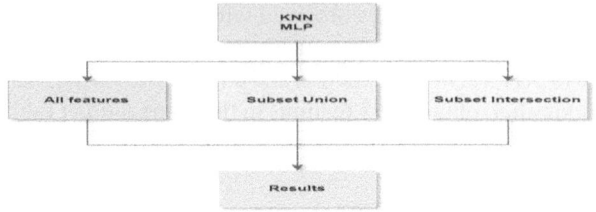

Fig. 6. Application of subsets to KNN and MLP. **Source:** Own elaboration.

Stage 3 ends with the analysis of the KNN and MLP classifier metrics trained on the intersection, union and full-featured dataset subsets.

5 Validation with Feature Subset

All calculations were performed using the Google Colab environment under Python 3.7.15. The code is available at https://acortar.link/c9PTFm.

5.1 Evaluation of KNN and MLP Models Using Subsets

To validate the relevance of the features selected in the previous phase they build the models with KNN and MLP. Figure 7 shows the summary of the metrics in terms of precision, accuracy, F1 score and sensitivity.

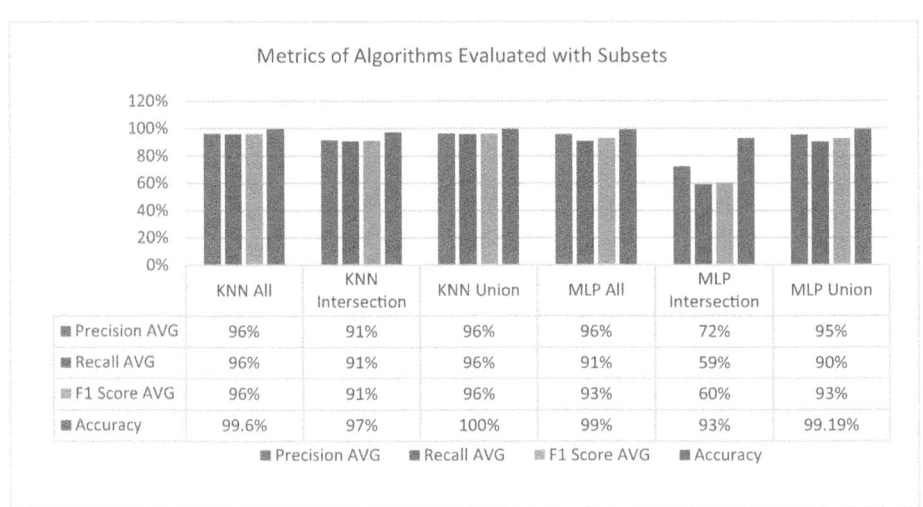

Fig. 7. Evaluation of Metrics with Subsets and all Attributes. **Source:** Own elaboration.

As shown in Fig. 7, the KNN model yielded significant metrics in the 3 moments in which it was evaluated (see Figs. 8, 9 and 10); however, its performance improved as the number of attributes increased. On the other hand, the MLP model did not perform as well (see Figs. 11, 12 and 13) and as the number of attributes decreased, the quality of the classification decreased. The KNN model was able to group the different DoS/DDoS attacks and the traffic identified as benign, as can be seen in the different confusion matrices (Figs. 8, 9 and 10).

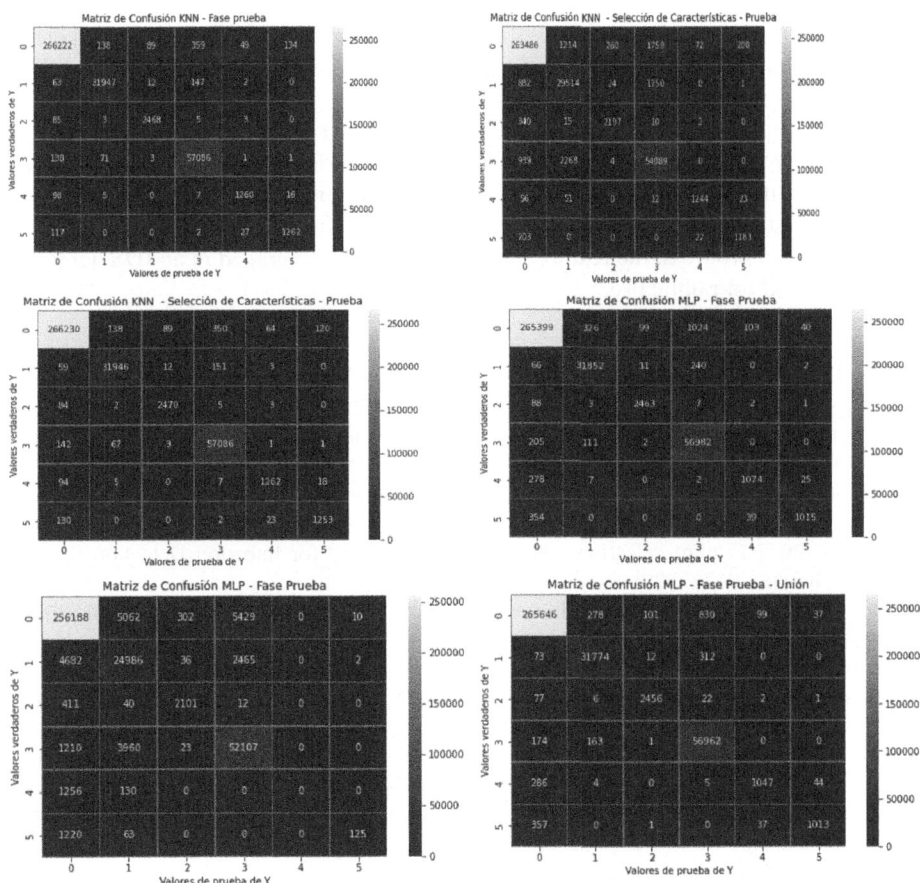

Fig. 8. Confusion matrices. **Source:** Own elaboration.

6 Conclusions and Future Work

In this paper, an experimental study was conducted on the classification of different types of DDoS/DoS attacks defined in the CICIDS2017 data set. Given the large number of attributes and instances in the dataset, a proposal was made to evaluate different classification techniques using all attributes and analyzing the metrics when a reduction to the dimensionality of the data is performed. The attribute selection algorithm yielded different subsets of features with different classifiers. In order to determine the behavior of other classifiers with the obtained features, they were run with the total features, intersection and union.

The results show that when having few features (those obtained with the intersection) the performance is reduced. But the union, which yields 37 attributes, has a performance very close to the one obtained with the total of features (78), in the 2 classifiers with which the validation was performed and for all metrics.

This result, which is not conclusive, is a contribution that can improve the effectiveness of intrusion detection systems aimed at improving computer security by inspecting significant features in the classification process.

Future work is expected to validate the results obtained with other data sets, including those obtained with actual traffic. Tests will also be performed with variations of the models with different hyperparameters and other classification techniques, as well as including data obtained from DDoS attacks generated with new signatures.

As a work in progress, the results of this work will be leveraged in an IDS deployed in the cloud and for public use.

References

1. Obaid, H.S.: Denial of service attacks: tools and categories. Int. J. Eng. Res. **V9**(03), 631–636 (2020). https://doi.org/10.17577/ijertv9is030289
2. Ortiz Martínez, E.M., Arguijo, P., Hiram Vázquez López, A., Ángel, R., Armenta, M.: Selección de características con método wrapper para un sistema de detección de intruso: caso CICIDS-2017 Feature Selection with a Wrapper Method for Intrusion Detection System: Case CICIDS-2017 (2020)
3. University of New Brunswick. Intrusion Detection Evaluation Dataset (CIC-IDS2017). https://www.unb.ca/cic/datasets/ids-2017.html. Accessed 28 Sept 2021
4. Mohammadi, S., Mirvaziri, H., Ghazizadeh-Ahsaee, M., Karimipour, H.: Cyber intrusion detection by combined feature selection algorithm. J. Inf. Secur. Appl. **44**, 80–88 (2019). https://doi.org/10.1016/j.jisa.2018.11.007
5. Kumar, V., Sonajharia, M.: Feature selection: a literature review. Smart Comput. Rev. **4**(3) (2014). https://doi.org/10.6029/smartcr.2014.03.007
6. Mahbooba, B., Timilsina, M., Sahal, R., Serrano, M.: Explainable artificial intelligence (XAI) to enhance trust management in intrusion detection systems using decision tree model. Complexity **2021** (2021). https://doi.org/10.1155/2021/6634811
7. Layme Fernández, C., et al.: Application of decision trees in the identification of fraudulent websites. Rev. Innov. Softw. **3**(1) (2022)
8. Song, Y., Li, H., Xu, P., Liu, D.: A Method of intrusion detection based on WOA-XGBoost algorithm. Discrete Dyn. Nat. Soc. **2022**, 1–9 (2022). https://doi.org/10.1155/2022/5245622
9. Bedi, P., Gupta, N., Jindal, V.: I-SiamIDS: an improved siam-IDS for handling class imbalance in network-based intrusion detection systems (2021)
10. Choubisa, M., Doshi, R., Khatri, N., Hiran, K.K.: A simple and robust approach of random forest for intrusion detection system in cyber security. In: 2022 International Conference on IoT and Blockchain Technology, ICIBT 2022, Institute of Electrical and Electronics Engineers Inc. (2022). https://doi.org/10.1109/ICIBT52874.2022.9807766

11. González, H., Santos, G., Campos, F., Morell Pérez, C.: Evaluación del algoritmo KNN-SP para problemas de predicción con salidas compuestas evaluation of KNN-SP algorithm for multi-target prediction problems. Rev. Cubana Ciencias Inform. **10**(3), 119–129 (2016). http://rcci.uci.cu

12. Yigit, Y., Bal, B., Karameseoglu, A., Duong, T.Q., Canberk, B.: Digital twin-enabled intelligent DDoS detection mechanism for autonomous core networks. IEEE Commun. Stand. Mag. **6**(3), 38–44 (2022). https://doi.org/10.1109/MCOMSTD.0001.2100022

13. Taher, K.A., Jisan, B.M.Y., Rahman, M.M.: Network intrusion detection using supervised machine learning technique with feature selection. IEEE (2019)

14. Qusyairi Ridho, S.F., Kalamullah, R.: Implementation of ensemble learning and feature selection for performance improvements in anomaly-based intrusion detection systems (2020)

15. Hua, Y.: An efficient traffic classification scheme using embedded feature selection and LightGBM (2020)

16. .Iram, A., Zahrah, A., Faheem, M., Alwi, B.M.: A machine learning approach for intrusion detection system on NSL-KDD dataset (2020)

17. .Kocher, G., Kumar, G.: Analysis of machine learning algorithms with feature selection for intrusion detection using UNSW-NB15 dataset. Int. J. Netw. Secur. Appl. **13**(1), 21–31 (2021). https://doi.org/10.5121/ijnsa.2021.13102

18. Ahmad, Z., Shahid Khan, A., Wai Shiang, C., Abdullah, J., Ahmad, F.: Network intrusion detection system: a systematic study of machine learning and deep learning approaches. Trans. Emerg. Telecommun. Technol. **32**(1) (2021). https://doi.org/10.1002/ett.4150

19. Albarka Umar, M., Chen, Z., Shuaib, K., Liu, Y.: Effects of feature selection and normalization on network intrusion detection. Commun. Netw. Broadcast Technol. 1–27 (2024). https://doi.org/10.36227/techrxiv.12480425.v3

20. di Mauro, M., Galatro, G., Fortino, G., Liotta, A.: Supervised feature selection techniques in network intrusion detection: a critical review. Eng. Appl. Artif. Intell. **101** (2021). https://doi.org/10.1016/j.engappai.2021.104216

21. Kalimuthan, C., Arokia Renjit, J.: Review on intrusion detection using feature selection with machine learning techniques. In: Materials Today: Proceedings, pp. 3794–3802. Elsevier Ltd. (2020). https://doi.org/10.1016/j.matpr.2020.06.218

22. Nazir, A., Khan, R.A.: A novel combinatorial optimization based feature selection method for network intrusion detection. Comput. Secur. **102** (2021). https://doi.org/10.1016/j.cose.2020.102164

23. Mahmood, R.A.R., Abdi, A.H., Hussin, M.: Performance evaluation of intrusion detection system using selected features and machine learning classifiers. Baghdad Sci. J. **18**, 884–898 (2021). https://doi.org/10.21123/bsj.2021.18.2(Suppl.).0884

24. Liu, Z., Yin, X., Hu, Y.: CPSS LR-DDoS detection and defense in edge computing utilizing DCNN Q-learning. IEEE Access **8**(3), 42120–42130 (2020). https://doi.org/10.1109/ACCESS.2020.2976706

25. Xiao, S., Tong, W.: Prediction of user consumption behavior data based on the combined model of TF-IDF and logistic regression. J. Phys.: Conf. Ser. (2021). https://doi.org/10.1088/1742-6596/1757/1/012089

26. Kanimozhi, P., Aruldoss Albert Victoire, T.: Oppositional tunicate fuzzy C-means algorithm and logistic regression for intrusion detection on cloud. Concurr. Comput. **34**(40) (2022). https://doi.org/10.1002/cpe.6624

27. Mishra, P., Biancolillo, A., Roger, J.M., Marini, F., Rutledge, D.N.: New data preprocessing trends based on ensemble of multiple preprocessing techniques. TrAC - Trends Anal. Chem. **132** (2020). https://doi.org/10.1016/j.trac.2020.116045

28. Wang, J., Chang, X., Wang, Y., Rodríguez, R.J., Zhang, J.: LSGAN-AT: enhancing malware detector robustness against adversarial examples. Cybersecurity **4**(1) (2021). https://doi.org/10.1186/s42400-021-00102-9

29. Leevy, J.L., Hancock, J., Zuech, R., Khoshgoftaar, T.M.: Detecting cybersecurity attacks across different network features and learners. J. Big Data **8**(1) (2021). https://doi.org/10.1186/s40537-021-00426-w

30. Sharma, N.V., Yadav, N.S.: An optimal intrusion detection system using recursive feature elimination and ensemble of classifiers. Microprocess Microsyst. **85** (2021). https://doi.org/10.1016/j.micpro.2021.104293

Exploring Advanced Deep Learning Methods for Image Steganography: New Approaches to Concealing Videos Within Videos

Omar Esteban Vargas Salamanca[(⊠)]

Universidad de los Andes, Bogotá, Colombia
o.vargas@uniandes.edu.co

Abstract. Steganography is the art of concealing information within another medium, traditionally involving the embedding of text within images. However, an increasingly relevant application is the concealment of one image within another, known as image-in-image steganography. Extending this concept further, this research introduces the innovative approach of video-in-video steganography, which applies image-in-image techniques to conceal entire video sequences within other videos. By leveraging the capabilities of Deep Learning, specifically convolutional neural networks, improving the undetectability and efficiency of our steganographic methods. Focus on refining how visual data is concealed within other multimedia content leads to more structured and cohesive embedding strategies. The models were trained using instance the popular ImageNet dataset [1], showcasing proficiency in embedding secret images and video sequences within host media, undetectably. This work evaluated the performance of our models using Structural Similarity Index Measure (SSIM) to assess their capacity to encode and decode secret data effectively. The primary contribution of our study is the development of a system capable of performing steganography between two videos of the same length, thereby offering a novel perspective within the field of traditional steganography. This exploration not only pushes the boundaries of conventional techniques but also paves new pathways for secure visual data transmission, highlighting the transformative potential of convolutional neural networks in revolutionizing steganography.

Keywords: Steganography · Convolutional Neural Networks · Deep Learning · Encoding · Decoding

1 Introduction

Steganography, the art of hiding messages within a medium to ensure they remain unnoticed, involves complex layers of analysis and processing [2]. Traditionally, a hidden message is embedded within a carrier—often an image—forming what is known as a stego medium or container, described by the equation:

N. D. Duque-Méndez et al. (Eds.): CCC 2024, CCIS 2208, pp. 269–282, 2024.
https://doi.org/10.1007/978-3-031-75233-9_19

$$\text{stego medium} = \text{hidden message} + \text{carrier} \tag{1}$$

This process, though simple in concept, has evolved significantly due to technological advancements, expanding the range of potential carriers-often referred to as covers in other research-to include digital media such as images, audio, and video [3,4]. Innovations have particularly enhanced the method of image-into-image steganography, where one image is concealed within another, ensuring both undetectability and reconstructibility [5].

Inspired by these advancements and the pioneering work by Wani, M. A and Sultan, B. in *Deep learning based image steganography: A review. WIREs Data Mining and Knowledge Discovery* [6], this research introduces a novel approach: video-into-video steganography. This method utilizes Convolutional Neural Networks (CNNs) to encrypt frames of two videos of identical length, applying and extending principles from image-in-image techniques to video sequences. By determining optimal placements for the secret information and efficiently encoding and decoding it, CNNs enhance the utility of traditional methods for dynamic media contexts. Leveraged the diverse image sizes and qualities available in an instance of the popular dataset ImageNet [1] to develop and train models that proficiently handle high-quality video content with minimal perceptual distortion. This approach not only challenges the traditional boundaries of steganography but also significantly expands upon the foundational concepts introduced in the use of Machine Learning in Steganography. It specifically applies these principles to color videos, which are effectively sequences of images. For a detailed illustration of how these components interact within our system, see Fig. 1.

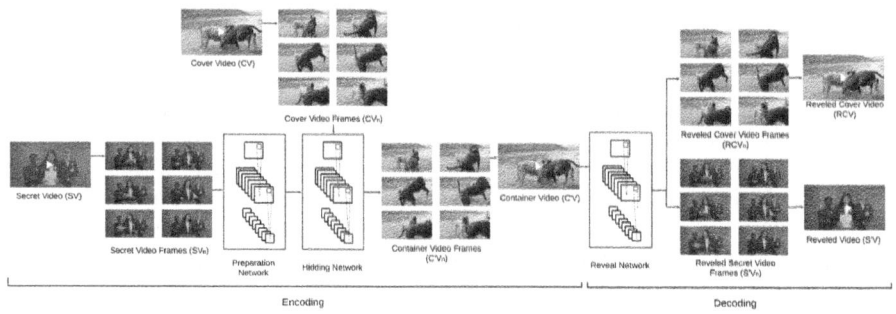

Fig. 1. Three components of the full CNN system.

To evaluate the effectiveness of our models, this study employs Structural Similarity Index Measure (SSIM), which is crucial for assessing the integrity and quality of stego videos [5]. Our results demonstrate that our CNN-based models not only achieve high rates of data embedding but also maintain the fidelity of the cover video, significantly improving upon traditional steganographic techniques.

This study's contributions are significant, introducing a robust framework for video-in-video steganography that utilizes CNNs to seamlessly embed large volumes of data within videos. By exploring the trade-offs between embedding capacity and visual quality, our work offers viable solutions for real-world applications where security and quality are paramount, thereby expanding the boundaries of traditional steganography and setting the stage for future secure digital communication research.

2   Related Work

This section reviews the existing literature on steganographic techniques, with a particular focus on the evolution from traditional methods to advanced applications involving deep learning techniques and multimedia data.

First of all, it is pertinent to point out that the main studies involving the relationship between steganography and deep learning are based on classifiers. For example, the most known steganography techniques such as LSB of images are used and deep learning models and traditional ML techniques are used to detect if given an image contains any message or type of hidden content. On the other hand, the objective of this research is to find and explore models based on deep learning that are alternatives to the most common steganography techniques in which integrity is not the priority [2].

On the other hand, the use of deep learning models for image steganography are mainly based on GANs and in general show acceptable results in terms of metrics such as Peak Signal to Noise Ratio (PSNR) and Structural Similarity Index (SSIM), one of the limitations they may have refers to amount of computational resources and data and training time required to get a good model. On the other hand, the architecture proposed in this research approaches to propose a simpler encoder-decoder based convolutional network architecture that minimizes both training and prediction time of the image with steganography [7].

Returning to contemporary times, the digital revolution has significantly strengthened the field of steganography, particularly when combined with deep learning techniques. This combination aims to encrypt messages within containers more efficiently and undetectably. For instance, in a recent study, Hashemi (2022) [8] explains how to create image steganography utilizing deep learning, specifically convolutional autoencoders combined with ResNet architecture. This method addresses the limitations of traditional steganography techniques, such as low capacity and robustness, by employing a reverse ResNet architecture to extract the hidden image from the stego image. Similarly, Hayes, J., & Danezis, G. (2017) [9] explore ResNet-based architectures to extract stego images, reporting an SSIM of 0.98, indicating that the obtained container images are nearly identical to the cover image without encrypted secrets, which is a significant indicator of the success rate of these models. This sets a goal for us to achieve and even surpass in our research. It is worth mentioning that advances in steganography using machine learning have been developing for years, as can be seen in the case of HUGO, a highly popular method in the field of steganography. HUGO is specialized in embedding for spatial-domain digital images and was proposed by Pevný, T., Filler, T., & Bas, P. (2010) [10].

With the rise of deep learning, researchers have explored the use of convolutional neural networks to create innovative ways of embedding information in containers with multiple digital formats. Notably, Baluja's work [6] on image-in-image steganography using deep learning set a new benchmark for the field, inspiring further research in the field of encoding digital images inside of other digital images in an undetectable way [11]. Finally, it is noteworthy to highlight recent studies in the state of the art regarding the application of steganography within videos to encode messages in text format, as explained by Kunhoth, J., Subramanian, N., Al-Maadeed, S., & Bouridane, A. (2023) [12]. However, there is limited information regarding the utilization of the computational capabilities mentioned in this section for video-into-video steganography. This lack of information motivates us to generate this knowledge and lead our own research to contribute to this field.

3   Architecture of the Model and Application Explanation

In this project, an autoencoder architecture tailored for steganography tasks is proposed. The model consists of three primary components: the Preparation Network, Hiding Network (Encoder), and Reveal Network. The overarching objective is to encode information from a secret image (S) into a cover image (C), thereby generating C', which closely resembles C, while retaining the capacity to decode information from C' to reconstruct the secret image (S').

The Preparation Network is responsible for preprocessing data from the secret image to be merged with the cover image input for the Hiding Network. Subsequently, the Hiding Network transforms this combined input into the encoded cover image (C'). Finally, the Reveal Network decodes the hidden information from C' to generate the decoded secret image (S') (Fig. 2).

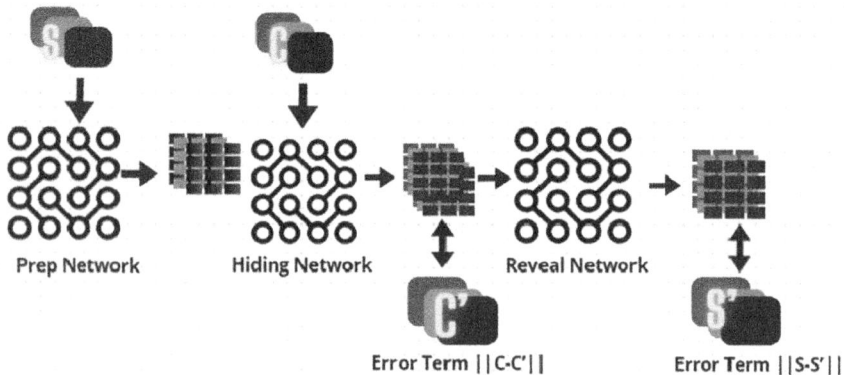

Fig. 2. Architecture of Steganography Autoencoder Model.

To enhance stability during decoding, noise is introduced prior to the Reveal Network, as referenced in literature. Although specific architectural details for the three networks were not originally delineated by the paper's author, a structure with 5 layers comprising a total of 65 filters, distributed across 3×3, 4×4, and 5×5 filter sizes, is implemented. Notably, the Preparation Network is simplified to consist of only 2 layers following the same filter distribution.

The loss model used in this study was trained over 100 epochs, with a redefined loss function aimed at comparing both the generated S' and C' images, using a value of $\beta = 1.0$.

For the reveal network, the loss function was formulated as $\beta \times |S - S'|$, where S represents the ground truth secret image and S' denotes the predicted secret image.

For the full model, encompassing both preparation and hiding networks, the loss was computed as the sum of two terms: $|C - C'|$ to evaluate the similarity between the cover images, and $\beta \times |S - S'|$ to assess the fidelity of the secret images.

3.1   Justification for Excluding Component Analysis

While techniques such as PCA or ICA could potentially reduce the complexity of the problem, they were not included in this model. This decision is based on several key considerations:

Integrity of Reconstructed Image: Techniques like PCA or ICA can reduce dimensionality, but they may also lead to the loss of crucial information necessary for accurately reconstructing the secret image. In steganography, ensuring a faithful representation of the input data is essential for the accurate recovery of the hidden information. The model prioritizes maintaining the full integrity of the secret image, even if it means not reducing the complexity beforehand.

Representation in the Input: The fidelity of the input representation is vital for the effectiveness of the steganographic process. By not applying dimensionality reduction techniques, the Preparation Network ensures that all relevant information from the secret image is retained and utilized in the encoding process. This comprehensive representation helps the Hiding Network encode the secret image into the cover image more effectively.

Robustness and Visual Quality: The autoencoder is designed to be robust against variations and noise in the input data, enhancing its stability and performance without the need for prior dimensionality reduction. Additionally, maintaining the visual quality of the encoded cover image (C') is critical to avoid detection. Techniques like PCA or ICA might alter the visual characteristics of the cover image, increasing the risk of detection.

3.2 Usage of the Model for Developing a Video Steganography Application

Once the steganography model was properly serialized and ready for being used freely, the next step was to test it with video input formats. Firstly, as the model only receives as input two images of 124×124 pixels at a time, it was necessary to come up with a way for the app to process two entire videos of any length and resolution. The process of processing the video can be seen in Fig. 4 (Fig. 3).

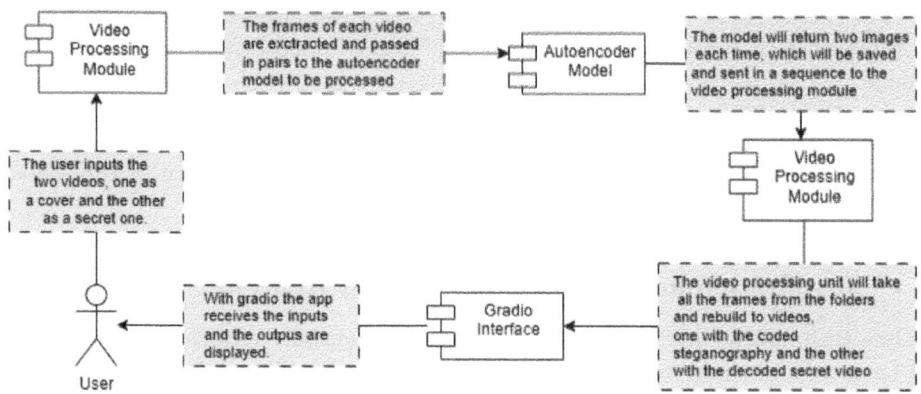

Fig. 3. Diagram that shows the process for processing a video.

As a general approach to this process, the app takes each of the videos, and by using the python library "cv2" their frames are extracted one by one and stored under the name "framex.jpg" (where x represents the number of the frame) in two different folders, one for each video. Then, the app takes the original folder (that is the one with the frames of the cover video) and iterates in all of its elements, taking each one of the iterations and pairing them with its respective frame counterpart (that comes from the secret video). Each of these pairs is passed to the autoencoder model (that was previously loaded using TensorFlow) and the new frames for the videos are obtained. The frames are stored independently in different folders in order to re-build a video from them when the process is completed. Lastly, the library "cv2" is used again for re-building a video from the processed frames folders. The videos obtained from this process always come with a resolution of 124 × 124 pixels, meaning that it reduces significantly the native resolution from the videos. Even though this is not optimal, due to computational limitations it was impossible to train the model with images that match the standard video resolution used (ranges from 1280 × 720 px to 3840 × 2160 px).

It is also very important to note that when the results of the processing of each frame are obtained, the Structural Similarity Index Measure (SSIM) value between the original frame (previously rescaled to 124 × 124 px) and the processed frame is calculated. This metric is really useful for evaluating steganography, as the SSIM takes into account changes in structural information, luminance and contrast. It is designed to more closely model human visual perception.

The framework used for deploying the application is Gradio. The selection of this technology over others was due to the great documentation that exists from it and the accessibility when trying to deploy the model, even in environments like Colab (from Google).

The application also supports the processing of individual images, taking two as an input and returning the processed images and a small graph that shows the noise between the original images and the processed ones.

4   Experimentation and Results

In this section, it describes the training process of the model using a dataset consisting of over 6000 images [1], resized to 124 × 124 pixels. It is crucial to emphasize that ensuring a diverse dataset is paramount for guaranteeing optimal performance of both the encoder and decoder components of the model. This diversity helps mitigate the risk of overfitting and ensures that the model can generalize well to unseen data.

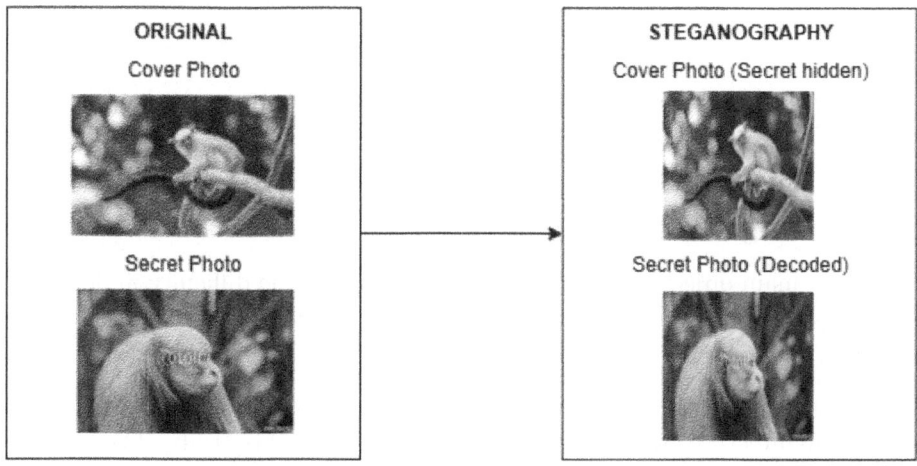

Fig. 4. Results of steganography with some example images.

Once satisfactory results are obtained for the decoder component, the model can be effectively utilized with any pair of images, and even videos. Remarkably, to the human eye, the differences between the original and encoded images can be challenging to discern. It is worth noting that in steganography, the model exhibits a remarkable capability to transfer or combine colors in some images, highlighting the CNN's ability to extract relevant features from the images.

The training process involved iteratively optimizing the model's parameters using backpropagation and gradient descent techniques. Data augmentation strategies, such as random rotations, flips, and adjustments to brightness and contrast, were employed to enhance the model's robustness and generalization capabilities. Additionally, early stopping and learning rate scheduling techniques were utilized to prevent overfitting and improve convergence speed.

Overall, the training process aimed to strike a balance between model complexity and generalization ability, ensuring that the model could effectively encode and decode information while maintaining fidelity to the original images. The successful training of the model on a diverse dataset underscores its potential for various applications in image and video processing, with steganography (Fig. 5).

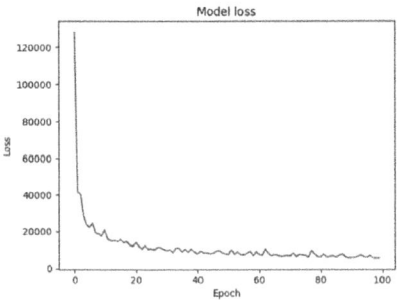

Fig. 5. Loss Metrics in a Training Steganography Autoencoder Model

4.1   Challenges in Detecting Encoder-Generated Predictions

One of the main objectives of this study was to assess the difficulty in detecting predictions generated by the encoder, particularly those involved in steganographic processes. Several methodologies were explored to tackle this challenge, each employing distinct machine learning architectures.

Initially, an attempt was made to leverage a network fashioned after the decoder architecture of the autoencoder proposed in this study. The encoder weights were frozen and amalgamated with a dense network housing a sigmoid neuron at the output layer. However, a significant obstacle was encountered when endeavoring to input a solitary image into the model, given the decoder's expectation of dual images. Various attempts were made to address this predicament, including padding the absent tensors with arrays of zeros or ones. Unfortunately, these endeavours yielded suboptimal outcomes.

Subsequent investigations delved into the utilization of pretrained models such as VGG16, MobileNetV2, and a customized variant of VGG16. Despite meticulous training attempts encompassing diverse configurations, satisfactory results remained elusive across all trials. The models exhibited either pronounced overfitting tendencies or achieved performances akin to random guessing (0.5) (Figs. 6 and 7).

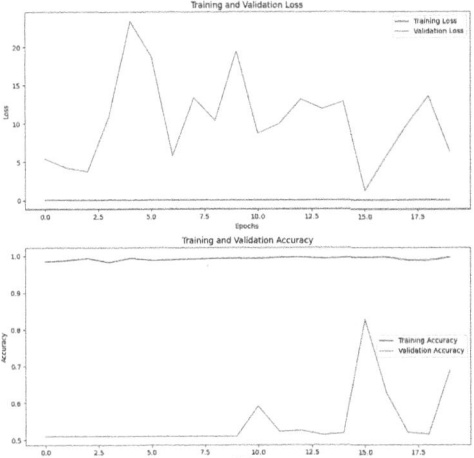

Fig. 6. Metrics Binary Classifier Based on MobileNetV2

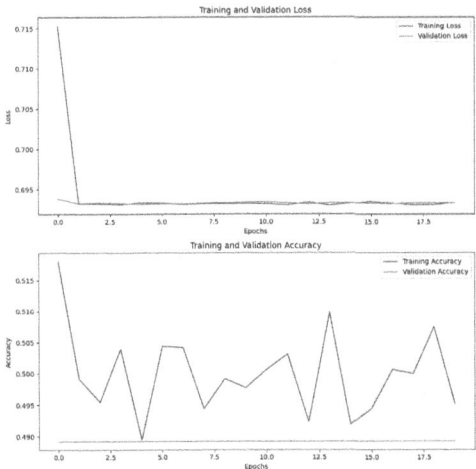

Fig. 7. Metrics Binary Classifier Based on VGG16

Confronted with these challenges, an alternative approach was pursued employing a siamese network architecture. This network was engineered to accept both the original image and the image synthesized by the encoder as input. However, formidable hurdles were once again encountered, as models grounded on convolutional or dense networks struggled to attain satisfactory performance levels. Even after exhaustive experimentation involving data augmentation techniques and fine-tuning of hyperparameters like learning rates, the results failed to meet the desired criteria.

Additionally, a convolutional siamese network was also employed, where the inputs were the original image and the one generated by the encoder. Despite these efforts, satisfactory results were not achieved.

The inherent difficulty in detecting predictions generated by the encoder lies in the complex and subtle alterations made to the input image. Unlike traditional generative models, which often introduce noticeable artifacts, the encoder aims to embed information seamlessly within the cover image, making it challenging for classification models to discern between original and encoder-generated predictions. Additionally, the nature of steganographic processes involves encoding information in imperceptible ways, further complicating the task of detection.

4.2   Video Experimentation

Regarding the results with video steganography, the main metric used was SSIM. Even though things like the noise and absolute pixel difference could be had into account, these metrics wouldn't be different from the ones obtained when originally training and testing the model. The results obtained for both the cover video and the secret video are pretty good, getting mostly values close to one. This means that for the human eye, there isn't much perceivable difference between the original content and the one that has steganography on it. The charts 10 and 11 show the results obtained when testing with two videos of 243 frames.

This graph 10 shows the results for the SSIM in the original cover video and the processed cover video. The range of values doesn't go under 0.98, and it oscillates between hundredths. This result is superb, meaning that the results for the usage of this model are great when hiding videos inside other videos.

For this other graph 11, the results show the SSIM values for the original secret video against the decoded secret video. With the secret video, the results are also interesting, but a little worse than the ones obtained in the cover video. The oscillation of the range goes under 0.86 and up to more than 0.93. This means that the model does a better job hiding the images than recovering them, yet giving an understandable output when recovering the inputs.

The model was trained on 3000 samples and took approximately 1 h and 20 min to achieve good results over 100 epochs. The training was conducted in a TPU v4 environment provided by Google Colab, and results might vary with different hardware configurations. In terms of prediction time, the model can reveal steganographic images of size 124×124 pixels in under 2 s per image. However, these times can vary depending on the hardware used and potential quantization (Figs. 8 and 9).

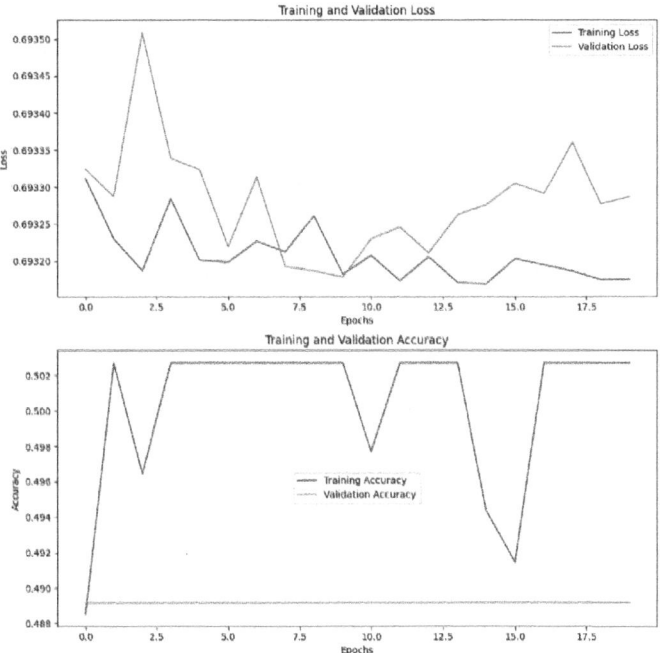

Fig. 8. Metrics Binary Classifier Based on Encoder-Weights

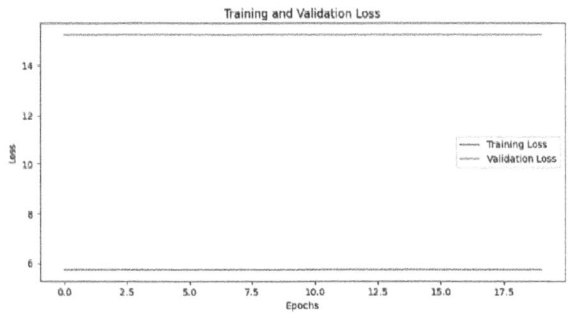

Fig. 9. Metrics Binary Classifier Based on Siamese Network

5   Future Work

It is worth mentioning that during the experimentation period, the tool StegExpose was utilized. Created by Boehm (2014) [13], StegExpose is designed to
detect whether an image contains a hidden text message using steganography.
The tool evaluates the ability of models to evade detection through parameters
such as Sample Pairs, Chi-Squared Attack, RS Analysis, and Primary Sets. However, this tool did not demonstrate any consistent results regarding its use in
image-in-image steganography, and thus, it was not included in our investigation.

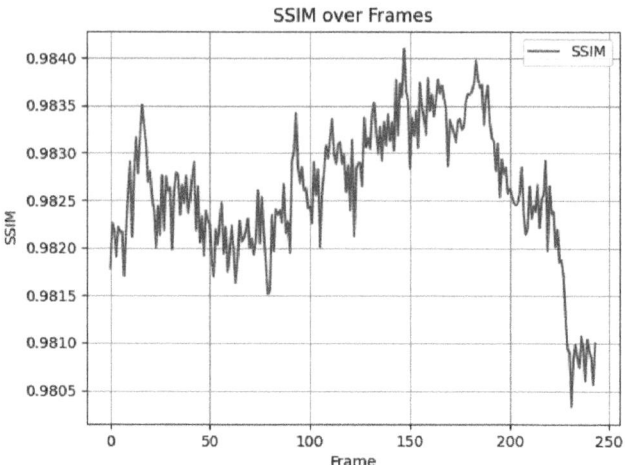

Fig. 10. Results for SSIM in the cover video.

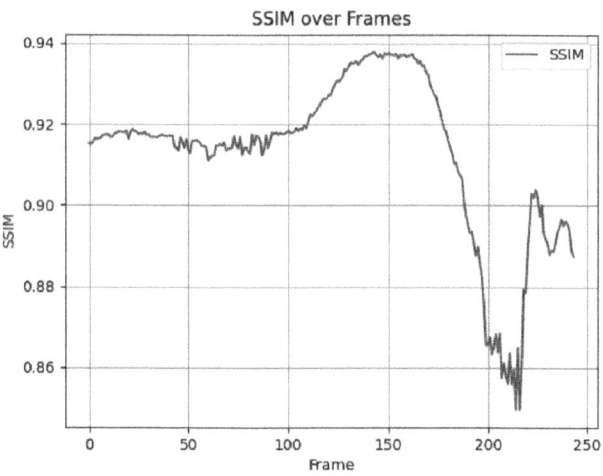

Fig. 11. Results for SSIM in the secret video.

Similarly, our research into the state of the art in this field led us to a study conducted by Zhang, Cuesta-Infante, Xu, and Veeramachaneni (2019) [14]. They introduced an innovative concept called SteganoGAN, which leverages Generative Adversarial Networks. Their results were promising, indicating the potential of models to hide high-quality images in imperceptible ways. Therefore, SteganoGAN is an architecture worth exploring in future work to enhance the quality of videos subjected to video-in-video steganography.

Future research should focus on improving the fidelity of the recovered content and exploring advanced techniques for detecting steganographic content.

Additionally, further investigation into the use of GANs, such as SteganoGAN, could lead to significant advancements in the field, enabling more effective and undetectable methods of video steganography.

6   Conclusions

The results obtained from the experiments in video steganography demonstrate the effectiveness of using Convolutional Neural Networks (CNNs) for embedding and extracting hidden information in videos. The primary metric used for evaluation was the Structural Similarity Index Measure (SSIM), which indicated a high level of similarity between the cover and stego videos. The SSIM values for the cover video consistently remained above 0.9805, suggesting that the modifications introduced by the steganographic process were imperceptible to the human eye. This is a significant achievement, as it confirms the model's ability to conceal information without degrading the visual quality of the cover video.

For the secret video, the SSIM values ranged from 0.86 to 0.93. Although these values are slightly lower than those for the cover video, they still indicate a reasonable level of fidelity in the recovered content. This discrepancy suggests that while the model excels at embedding information, there is room for improvement in the extraction process to ensure higher fidelity of the recovered secret video.

The robustness of the model was further validated by the use of various data augmentation techniques during the training process, such as random rotations, flips, and adjustments to brightness and contrast. These techniques enhanced the model's ability to generalize and perform well on diverse datasets, which is crucial for real-world applications.

In summary, the results of this study contribute significantly to the field of video steganography by demonstrating the viability of using CNNs for embedding and extracting hidden information. The high SSIM values achieved in both cover and secret videos affirm the model's capability to maintain visual quality while ensuring secure communication. However, further research is needed to improve the fidelity of the recovered content and to explore advanced techniques for detecting steganographic content.

When considering validation with other types of multimedia content, the challenge lies in finding a reliable metric to validate the training process. For instance, since this study focused on images, Mean Squared Error (MSE) per pixel was a suitable option. However, if the goal is to hide content like text, other metrics would need to be analyzed to ensure a good training process.

References

1. Imagenet-a   Dataset.   https://github.com/omar-vargas/imagenet-a-steganography-. Accessed 2024/06/25
2. Khan, A.A., et al.: IMG-forensics: multimedia-enabled information hiding investigation using convolutional neural network. IET Image Process. **16**, 2854–2862 (2022). https://doi.org/10.1049/ipr2.12272

3. Kunhoth, J., Subramanian, N., Al-Maadeed, S., et al.: Video steganography: recent advances and challenges. Multimed. Tools Appl. **82**, 41943–41985 (2023). https://doi.org/10.1007/s11042-023-14844-w
4. Liu, Y., Liu, S., Wang, Y., Zhao, H., Liu, S.: Video steganography: a review. Neurocomputing **335**, 238–250 (2019)
5. AlKhodaidi, T., Gutub, A.: Refining image steganography distribution for higher security multimedia counting-based secret-sharing. Multimed. Tools Appl. **80**, 1143–1173 (2021). https://doi.org/10.1007/s11042-020-09720-w
6. Wani, M.A., Sultan, B.: Deep learning based image steganography: a review. WIREs Data Min. Knowl. Discov. **13**(3), e1481 (2023). https://doi.org/10.1002/widm.1481
7. Zhang, K.A., et al.: SteganoGAN: high capacity image steganography with GANs. arXiv preprint arXiv:1901.03892 (2019)
8. Hashemi, S.H.O., Majidi, M.H., Khorashadizadeh, S.: Color image steganography using deep convolutional autoencoders based on ResNet architecture. arXiv preprint arXiv:2211.09409 (2022)
9. Hayes, J., Danezis, G.: Generating steganographic images via adversarial training. In: Advances in Neural Information Processing Systems, vol. 30 (2017)
10. Pevný, T., Filler, T., Bas, P.: Using high-dimensional image models to perform highly undetectable steganography. In: Information Hiding: 12th International Conference, IH: Calgary, AB, Canada, 28–30 June 2010, Revised Selected Papers 12, pp. 161–177. Springer (2010)
11. Wu, P., Yang, Y., Li, X.: Image-into-image steganography using deep convolutional network. In: Advances in Multimedia Information Processing-PCM 2018: 19th Pacific-Rim Conference on Multimedia, Hefei, China, 21–22 September 2018, Part II 19, pp. 792–802. Springer (2018)
12. Kunhoth, J., Subramanian, N., Al-Maadeed, S., Bouridane, A.: Video steganography: recent advances and challenges. Multimed. Tools Appl. **82**(27), 41943–41985 (2023)
13. Boehm, B.: Stegexpose-A tool for detecting LSB steganography. arXiv preprint arXiv:1410.6656 (2014)
14. Zhang, K.A., Cuesta-Infante, A., Xu, L., Veeramachaneni, K.: SteganoGAN: High capacity image steganography with GANs. arXiv preprint arXiv:1901.03892 (2019)

Automated Machine Learning Prototype for Detecting Phishing, Deepfakes, and Fraudulent Audio Patterns: A Systematic Literature Mapping

Juan Andres Torres Camargo[(✉)], Rodrigo Andres Martinez Mellizo, and Juan Jose Caiza Narvaez

Facultad de Ingeniería Institución Colegio Mayor del Cauca, Popayán, Colombia
`{jtorres,rmartinezm,juanjosecaiza}@unimayor.edu.co`

Abstract. The pervasiveness of cybercrime poses a significant threat to individuals and organizations worldwide, particularly in the realm of audio-based interactions. To combat this evolving threat landscape, this study delves into the application of artificial intelligence (AI), specifically machine learning techniques [1], to develop a prototype for mitigating cybercrime [2]. The prototype aims to address emerging threats such as phishing, deepfakes [3], and social engineering [4], which exploit audio signals and compromise personal data and security.

Keywords: Artificial intelligence · Machine learning · Cybercrime · Phishing · Deepfakes · Social engineering · Mobile devices

1 Introduction

The arrival of sophisticated audio-based fraud techniques, particularly in the context of phishing and social engineering, poses a significant challenge to cybersecurity. In Colombia, this issue has gained particular urgency, as the number of individuals falling victim to such scams has shown a concerning upward trend [5]. These fraudulent practices often involve the manipulation and deception of users through techniques like phishing, which employs coercive persuasion strategies, or in other words, manipulation, deception, and other identity impersonation techniques, to extract sensitive information or conduct fraudulent transactions [6]. The prevalence of these audio-based scams necessitates the development of robust countermeasures to protect individuals and safeguard their digital assets.

Given the rising incidence of fraud and the excessive daily use of these devices or platforms; scams pose a serious threat to the security and privacy of users, as well as to trust in telecommunications services. The current initiative has the potential to provide an additional layer of security and protection for users, helping to mitigate the risks associated with the trend of fraudulent activities by cybercriminals employing AI-powered audio analysis techniques for threat identification [7].

© The Author(s), under exclusive license to Springer Nature Switzerland AG 2024
N. D. Duque-Méndez et al. (Eds.): CCC 2024, CCIS 2208, pp. 283–297, 2024.
https://doi.org/10.1007/978-3-031-75233-9_20

Artificial intelligence: The ability of a machine or computer system to solve problems that require human intelligence. This is achieved using algorithms and mathematical models. On the other hand, there exists a more advanced model known as Machine learning: Consists in teaching a computer to learn on its own, similar to how humans learn from experience. It's a technology that helps us automate tasks and solve complex problems more efficiently and intelligently. This is one of the main reasons it is very popular in the Cybercrime battle. Crimes committed through digital networks or computers, such as identity theft, stealing personal information, fraud, extortion [19]. Nowadays, we are all a potential target with all this modern tech and scammers have also developed modern techniques. Phishing: When someone pretends to be a trusted entity or person in order to obtain personal information. Similar but not the same as the next technique. Deepfakes: They are realistic digital illusions where, with the help of artificial intelligence, fake videos or audios can be created in which a person appears to say or do something they never actually did. It's as if someone took your image or voice to make you "act" in a movie without your consent. There's also another lesser-known technique that consists of hacking people rather than hacking their computer. Social engineering: Could be described as a psychological trick in the digital world where, instead of hacking computers, scammers attempt to "hack" people by manipulating them into taking actions they shouldn't.

To gain a comprehensive understanding of the current landscape and identify potential research gaps, this study conducted a thorough literature review. This involved meticulously searching for relevant academic articles, research papers, and reports from reputable databases and information repositories. The search terms encompassed various aspects of AI-powered cybersecurity, including threat detection, machine learning techniques, and audio processing. The goal was to gather a broad spectrum of insights and perspectives on the intersection of AI and cybercrime mitigation, particularly in the context of audio-based threats.

2   Research Protocol

To investigate the application of artificial intelligence in combating cybercrime on audio-processing devices, a research protocol was followed based on the one proposed by Kitchenham et al. [8]. This protocol was adapted to the specific characteristics of the research and was structured into four main approaches: research questions approach (QGM), search definition and selection strategies, review execution, and review report preparation.

Each of these approaches was implemented in a rigorous and systematic manner, enabling a thorough understanding of the subject matter and the identification of pertinent findings. The outcomes of this investigation will serve as a robust foundation for the development of an effective prototype and the formulation of valuable conclusions and recommendations in the domain of artificial intelligence against cybercrime.

3 First Section: GQM Approach

Following the first approach, the research objectives and questions that guided the systematic mapping were established. These objectives focused on analyzing the application of AI (Fig. 1).

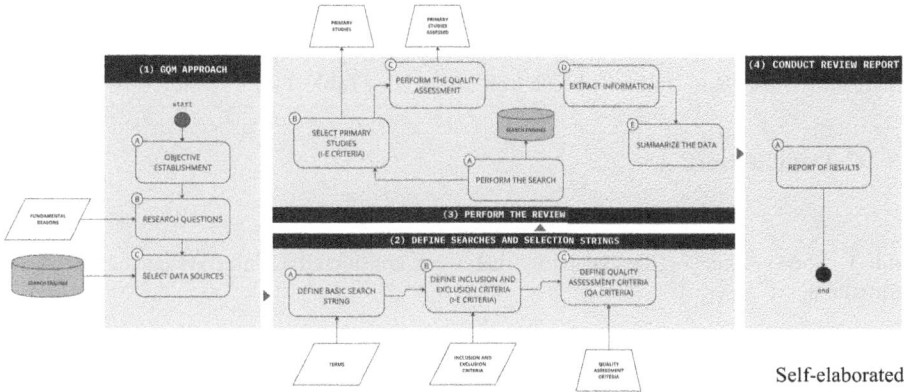

Self-elaborated

Fig. 1. Visual representation of the research protocol proposed by Kitchenham et al. [8] for conducting a systematic mapping study. The original phases and activities have been adapted to align with the specific focus of the research.

3.1 A. Objective Establishment

- OB1: Characterize AI techniques used to combat fraudulent activity patterns, considering their ability to address current limitations and improve effectiveness.
- OB2: Identify the impact of AI techniques on the development of solutions to prevent and mitigate fraudulent patterns, analyzing their contribution to user security and protection.
- OB3: Evaluate the potential of AI in creating adaptive and intelligent solutions to prevent and mitigate fraudulent activity in audio samples.
- OB4: Provide an updated guide for researchers and professionals interested in applying AI to combat fraudulent activity patterns, synthesizing findings and providing recommendations for future research.

3.2 B. Research Questions (RQ)

- RQ1: How many primary studies have been published from 2020 to 2024?
- RQ2: What kind of publications are the most frequent considering the type?
- RQ3: What are the artificial intelligence techniques applied to the detection of fraudulent patterns in audio samples?

Research questions (RQ1, RQ2, and RQ3) are intricately linked to Objectives (OB1, OB2, OB3 and OB4), forming a cohesive framework for the study. RQ1 delves into the identification of publication trends or interest in the proposed topic between 2020 and 2024. RQ2 sheds light on the most prolific authors contributing to the field, while RQ3 unravels the predominant variety of publications within the research context.

The alignment of RQ1, RQ2, RQ3 and RQ4 demonstrates a well-structured and cohesive research approach. Each research question contributes to the fulfillment of the objectives, providing a comprehensive understanding of the topic and its associated trends and publication types. This synergy between research questions and objectives ensures a focused and impactful research endeavor (Table 1).

3.3   C. Data Sources

Table 1. Shows some of the databases that were considered for the systematic search for information.

Databases	Area
Google Scholar	Interdisciplinary
IEEE Xplore	Engineering
SpringerLink	Engineering

4   Second Section: Search Definition and Selection Strategies

4.1   A. Search String Definition

It is defined as a structured set of keywords related to a specific topic of interest, designed to optimize the search for elements in different scientific and academic databases.

Table 2 showcases five of the most relevant crafted search strings that unearthed a wealth of information directly with AI, audio and phishing. These strings were constructed by strategically nesting technical terms and leveraging logical operators like "AND" and "OR" to refine the search results. These searches were carried out in both English and Spanish in order to achieve a wider scope.

4.2   B. Definition of Inclusion and Exclusion Criteria

Establishing well-defined inclusion and exclusion criteria is fundamental for ensuring the relevance and rigor of the research material. Table 3 outlines the inclusion criteria, while Table 4 presents the exclusion criteria, both of which were essential for qualifying the primary studies based on the defined search strings.

Table 2. Most relevant search strings in the research process.

No.	Search String
1	"deepfake" AND ("deep learning" OR "machine learning") AND "cybersecu rity" AND ("audio" OR "voice")
2	"threat detection" AND "artificial intelligence" AND "voice" AND "phishing"
3	"detección de amenazas" AND "inteligencia artificial" AND "voz"
4	("Framework" OR "Prototype") AND ("Audio" OR "Voice" OR "Deepvoice") AND ("Risk" OR "Threat" OR "Attacks") AND ("Cybersecurity" OR "Information Security") AND ("ARTIFICIAL INTELLIGENCE" OR "IA") AND ("Machine Learning" OR "Artificial Intelligence")
5	("Framework" OR "Prototype") AND ("Risk" OR "Threat" OR "Attacks") AND ("Cybersecurity" OR "Information Security") AND ("ARTIFICIAL INTELLI GENCE" OR "IA") AND ("Machine Learning" OR "Artificial Intelligence")

Table 3. Inclusion criteria

Id	Inclusion Criteria (IC)
IC1	Studies centered intersecting with the following key areas: Artificial Intelligence, Machine Learning, Threat Detection, Deep Learning, Phishing, Deepfakes, Deep Voice, Audio Analysis, Voice Recognition and Cybersecurity
IC2	Studies within the four-year timeframe (2020–2024)
IC3	Studies in English or Spanish
IC4	Studies complemented with high-quality images, graphics, or other helpful visual elements

Table 4. Exclusion Criteria

Id	Exclusion Criteria (EC)
EC1	Duplicate Studies
EC2	Studies Unrelated to the Research Topic
EC3	Studies with Paid, Restricted, or Denied Access
EC4	Studies that were mainly addressing website phishing

4.3 C. Definition of Quality Assurance Criteria (QA)

Table 5 presents 11 quality criteria that were meticulously considered in the selection of primary studies, following the grouping defined in [41]. This comprehensive evaluation framework ensured a more understandable assessment and improved classification of the studies.

To measure quality, a threshold or range between -11 and 11 was assigned. In the quality assessment, a "Yes" corresponds to a value of 1, a "No" corresponds to a value of -1, and in this case, a "Partially" corresponds to a value of 0. The quality level of each study was determined based on its overall score.

To categorize the quality, we considered the levels or labels of high, medium, moderate, and low. It is important to note that if an article had a low score, it did not mean its exclusion from the total selected studies. On the contrary, it, along with the grouping of the 12 criteria, provided a way to create an order of relevance for other types of research and analysis in the future.

5   Third Section: Review

5.1   A. Search Execution

A total of 38 primary studies were meticulously selected based on the four proposed inclusion criteria (IC1, IC2, IC3 and IC4). Among these, the majority (37 studies) fell under IC1, followed by IC2 and IC3 with 34 studies both. Last but not least IC4 with 27 studies. This collection represents 65.5% of the initial 58 studies. For those seeking a deeper dive into each study, a comprehensive list of the primary studies, along with their assigned references, is presented in Table 6.

On the other hand, an analysis of the 53 studies excluded based on the four proposed exclusion criteria (EC1, EC2, EC3,and EC4) revealed that the majority of non-selected studies (27 studies) fell under EC2, followed by EC4 with 15 studies, EC3 with 8 studies, and EC1 with 3 studies. Within Table 5, Eleven essential quality benchmarks are established.

Table 5. Quality assurance criteria

Area	ID	Quality Assurance Criteria (QA)	+1	NEUTRAL	−1
Introduction	QA1	The study presents a clear description of the reasons for the proposed research	Yes ‖ Partially ‖ No
Planification	QA2	The study includes a structure or process plan used in the development of the research	Yes ‖ Partially ‖ No
Planification	QA3	The study focuses on artificial intelligence techniques applied to the detection of phishing through audio samples	Yes ‖ Partially ‖ No

(continued)

Table 5. (*continued*)

Area	ID	Quality Assurance Criteria (QA)	+1	NEUTRAL	−1
Tematic	QA4	The study provides a clear description of the problem it addresses	Yes ‖ Partially ‖ No
Tematic	QA5	The study provides a clear description of the proposed technological solution	Yes ‖ Partially ‖ No
Tematic	QA6	The study materializes the proposed technological solution in models, prototypes, applications, etc.	Yes ‖ Partially ‖ No
Solution	QA7	The study offers a good description of the technologies implemented in the proposed solution	Yes ‖ Partially ‖ No
Solution	QA8	The proposed solution in the study has been implemented or tested in real-world environments	Yes ‖ Partially ‖ No
Difficulties	QA9	The study describes limitations or difficulties in the research and/or analysis of the results	Yes ‖ Partially ‖ No
Difficulties	QA10	The study describes issues tied to the proposed solution	Yes ‖ Partially ‖ No
Results	QA11	The study exhibits in a detailed and clear manner the results obtained	Yes ‖ Partially ‖ No

Table 6. Primary Studies

ID	Title
A1	Publicación: Prevención de fraudes en el sector financiero colombiano a través de controles que usan el aprendizaje automático [9]
A2	Sistema de detección de patrones de fraude con redes neuronales en la Provincia De Los Rios y su incidencia en la telefonía celular [10]
A3	Estudio de fraudes basados en la técnica de Ingeniería Social [4]
A4	Desarrollo de un modelo predictivo para detectar casos de fraude interno en una institución bancaria [11]
A5	A comprehensive survey of AI-enabled phishing attacks detection techniques [12]
A6	Investigation of Phishing Susceptibility with Explainable Artificial Intelligence [13]
A7	Phishing Detection Implementation using Databricks and Artificial Intelligence [14]
A8	Digital Deception: Generative Artificial Intelligence in Social Engineering and Phishing [15]

(*continued*)

Table 6. (*continued*)

ID	Title
A9	Evaluación de clasificadores de voz para Smart Personal Assistants [16]
A10	DeepDetection: Privacy-Enhanced Deep Voice Detection and User Authentication for Preventing Voice Phishing [17]
A11	Deepfake audio detection and justication with Explainable Articial Intelligence (XAI) [18]
A12	Method for identifying cyberattacks based on the use of social engineering over the phone [7]
A13	Are ChatGPT and Deepfake Algorithms Endanger
A14	Audio Stream Analysis for Deep Fake Threat Identification [8]
A15	Voice Detection Using Convolutional Neural Network [20]
A16	Deep neural networks in acoustic model [21]
A17	Detecting telecommunication fraud by understanding the contents of a call [22]
A18	Deep Learning and Artificial Intelligence Framework to Improve the Cyber Security [23]
A19	Deep Learning-Based Analysis of a Real-Time Voice Cloning System [24]
A20	The efficacy of Deep Learning and Artificial Intelligence Framework in Enhancing Cybersecurity, Challenges and Future Prospects [25]
A21	Human Perception of Audio Deepfakes [26]
A22	Study of artificial intelligence in cyber security and the emerging threat of ai-driven cyber attacks and challenge [27]
A23	Automated Machine Learning Enabled Cybersecurity Threat Detection in Internet of Things Environment [28]
A24	Inteligencia artificial y el aprendizaje automático en la ciberseguridad [40]
A25	Tecnologías de defensa frente a inteligencia de amenazas y ciberataques [2]
A26	Impacto de la inteligencia artificial en la ciberseguridad empresarial: unanálisis crítico de la evolución de amenazas y medidas preventivas [29]
A27	Threat Hunting basado en técnicas de Inteligencia Artificial [1]
A28	Inteligencia artificial en la seguridad de la información en una organización [30]
A29	Desarrollo de un chatbot cognitivo capaz de dar respuesta a diferentes incidentes de ciberseguridad [31]
A30	Aplicación de machine learning a un modelo tradicional de prevención y detección de fraude [32]
A31	The Human Factor in Cybersecurity: Addressing Social Engineering and Insider Threats [33]
A32	Phishing Attacks: Unraveling Tactics, Threats, and Defenses in the Cybersecurity Landscape [34]

(*continued*)

Table 6. (*continued*)

ID	Title
A33	A Comprehensive Survey: Evaluating the Efficiency of Artificial Intelligence and Machine Learning [35]
A34	Deepfake Speech Detection: A Spectrogram Analysis [36]
A35	Data driven: AI Voice Cloning [37]
A36	Deepfake Audio Detection via MFCC Features Using Machine Learning [3]
A37	Automatic Multispeaker Voice Cloning [38]
A38	Applicability of deepfakes in the field of cyber security [39]

5.2 B. Information Extraction

Throughout the process of selecting and evaluating primary studies, information extraction was conducted simultaneously, focusing on crucial aspects to address the research questions. These aspects included the study title, publication year, source, country of origin, publisher, study genre, abstract, findings, relevance to diverse stakeholders, and other essential elements. This methodology ensured a deep and precise understanding of each study, facilitating an exhaustive and meaningful interpretation of the data.

5.3 C. Data Synthesis

The compilation and organization of data were meticulously executed, adhering to the elements outlined in the information extraction section and applied to the 38 selected studies. Consequently, this comprehensive information has been condensed into the approach presented below.

6 Four Section: Review Reports

6.1 A. Results Report

Insights from each research question (RQ) in approach 1.

1. **How many primary studies have been published from 2020 to 2024?**

A total of 38 primary studies were chosen for in-depth analysis after excluding the first five, which provided foundational context on the "origin" of the issue and documented early cases, allowing us to point out how this problem has evolved. Here's the breakdown of how the studies were distributed by year and country: 2.6% (1/38) for the year 2015, 5.3% (2/38) for the year 2016, 2.6% (1/38) in 2018, 2.6% (1/38) in 2019, 2.6% (1/38) in 2020, 2.6% (1/38) in 2021, 13.2% (5/38) in 2022, 42.1% (16/38) in 2023, and finally the year 2024 with 26.3% (10/38). The primary studies originated from 27 countries, with Spain contributing the highest number of studies.

2. **What types of publications are the most frequent considering the type?**

In Table 7, the classification of the 38 primary studies according to the type of publication is shown.

Table 7. Distribution of study types

Type/Topic	Percentage
Article	71,05(27/38)
Proyect	7,9(3/38)
Masters-Thesis	21,05(8/38)

3. What are the artificial intelligence techniques applied to detecting fraudulent patterns in audio samples?

Table 8. Artificial intelligence techniques employed

Techniques	Study
Machine Learning (ML)	[A2], [A4], [A10], [A11], [A15], [A16], [A30], [A22]
Deep Learning (DL)	[A9], [A10], [A11], [A35], [A36], [A37], [A38], [A21]
Natural Language Processing (NLP)	[A10], [A17], [A22]

Table 9. Applied artificial intelligence architectures

Architecture	Study
Support Vector Machine (SMV)	[A4], [A10], [A11], [A36], [A17]
Deep Neural Network - (DNN)	[A2], [A4], [A25], [A27], [A32], [A21]

To ensure the robustness and reliability of the studies included in this analysis, a rigorous selection process was implemented. 19 studies meeting strict criteria of relevance and quality were identified, specifically those receiving a positive assessment (+1 / Yes) or neutral (0/Partially) evaluation in both the criteria for describing the data handling (QA3) and model architecture (QA7). This selection allows for a meticulous focus on studies that provide a clear and comprehensive description of the methodology used, including data handling and model architecture—crucial aspects for evaluating the validity and replicability of findings. This provides a solid foundation for understanding the application of AI and machine learning in fraud prevention (Tables 8, 9 and 10).

Table 10. Input data processing techniques

Processing technique	Study
Mel-Frequency Cepstral Coefficients(MFCCs)	[A7], [A9], [A25], [A27], [A29]
Non-Negative Matrix Factorization(NMF)	[A11], [A37]
Self-Organizing Maps (SOM)	[A2], [A22]
WAVLM MODEL	[A35]
RTVC	[A38]
Automatic Speech Recognition (ASR)	[A37], [A38], [A36], [A16]

7 Overview of the Research

7.1 A. General Observations

The objective of this systematic mapping is to identify studies or research related to the application of artificial intelligence (AI) techniques in mitigating cybercrime, specifically in cases involving the use of audio to compromise personal security. After analyzing the obtained results, the following observations can be highlighted:

Firstly, there is a significant interest in applying autonomous systems for detecting, classifying, and scoring cyber threats that exploit audio signals, such as phishing, deepfakes, and social engineering techniques. This interest is enhanced by leveraging modern technologies like AI and the availability of portable intelligent systems.

Secondly, in most of the proposed solutions in the 59 reviewed studies, it was identified that the creation of these tools or models involved a thorough analysis of audio characteristics (voice). Many of these solutions focus on processing and evaluating audio to identify and mitigate cybercrime threats. These solutions typically utilize machine learning (ML) and deep learning (DL) techniques, along with signal processing through spectral analysis, such as the use of Mel-Frequency Cepstral Coefficients (MFCC).

Lastly, within most of the studies, the importance of non-invasive, affordable, and easily accessible systems is emphasized. This is crucial since many end users, such as individuals and small organizations, may face socioeconomic and logistical challenges in implementing advanced security solutions. These AI-based solutions would allow for greater protection against cyber threats without requiring costly or complex infrastructure, offering an additional layer of security that can be used continuously and effectively in various environments.

Therefore, these solutions would help reduce the burden on individuals and organizations by providing advanced tools for detecting and mitigating cybercrime, facilitating the protection of sensitive data, and enhancing overall security in the digital realm.

7.2 B. Limitations of the Systematic Mapping

Some of the limitations refer to the exclusion of studies in languages other than English and Spanish, as some studies with significant contributions to this research may not have been considered. Similarly, this applies to the selected time interval and studies based on undergraduate, graduate projects, or those that are not accessible.

7.3 C. Relevance of Research and Practice

Considering the obtained results, it can be affirmed that these will have great importance for future research related to the application of systems or models involving AI techniques in addressing cybercrime that exploits audio signals. These results will contribute information related to different techniques, application methods, architectures, data management, preprocessing, and types of solutions, based on the objectives of the proposed model and solution, which theoretically should address a specific problem or area. Thus, this study provides an initial insight into the research and practical interest in leveraging modern technologies alongside machine learning systems for their application in security-related environments, more specifically, given the results, in the context of cybercrime threats such as phishing, deepfakes, and social engineering.

8 Conclusions

In the ever-evolving landscape of cybercrime, audio-based interactions mostly through mobile devices have become prime targets for malicious actors. Phishing scams, deepfakes, social engineering tactics, and other sophisticated techniques present significant threats to global security and privacy. To combat this growing menace, we must harness the transformative power of artificial intelligence (AI) and machine learning (ML) to develop innovative solutions that effectively protect users.

Integrating cutting-edge machine learning algorithms allows us to effectively detect and mitigate phishing attempts, identify deepfakes, and neutralize social engineering tactics. This prototype aims to provide users with protection measures, safeguarding the community from pervasive audio-based cyber threats that have been taking place recently.

While AI and ML-based solutions for cybercrime prevention have demonstrated their potential, they still require further maturity to address complex real-world scenarios promptly. It is crucial to train models with diverse data, evaluate them in complex situations, and continuously optimize and update them. Addressing these challenges, along with interdisciplinary collaboration, an ethical legal framework, and user awareness, will enable us to fully harness the potential of AI and ML to build a safer future in the realm of cybersecurity.

9 Future Investigations

As a future endeavor, it would be interesting to conduct further research on the management and analysis of voice signals for the interpretation and identification of specific characteristics related to speech pathologies. With this knowledge of audio preprocessing, we can now take some of the models already described and proposed in the studies analyzed in this systematic mapping and recreate them using datasets from a Spanish -

speaking population. This will allow us to measure the ease of process replication and the effectiveness of the proposed models within a specific context. English is the global language so it's weird but not impossible to find these datasets in Spanish. Nevertheless, the team has been training a model with real and fake audios of some USA presidents.

References

1. Lozano, M.A.: Threat Hunting basado en técnicas de Inteligencia Artificial (2024). https://riunet.upv.es/handle/10251/204427. Accessed 21 June 2024
2. InnDev, L.Q.: Tecnologías de defensa frente a inteligencia de amenazas y ciberataques. QuevedoInnDev (2024). revistas.itecsur.edu.ec. Accessed 21 June 2024
3. Hamza, A., et al.: Deepfake audio detection via MFCC features using machine learning. IEEE Access (2022). https://ieeexplore.ieee.org/abstract/document. Accessed 21 June 2024
4. Alzas Hernandez, J.: Estudio de fraudes basados en la técnica de Ingeniería Social (2023). Accessed 20 June 2024
5. Audio Stream Analysis for Deep Fake Threat Identification (2024). https://doi.org/10.31648/cetl.9393
6. Ciberespacio, D.Q.-F.-R.: La ciberseguridad y la ciberdefensa frente a los factores de inestabilidad económicos y sociales. esdegrevis-tas.edu.co (2022). https://doi.org/10.25062/2955-0270.4767
7. Lysenko, S., Bokhonko, O., Vorobiyov, V.: Method for identifying cyberattacks based on the use of social engineering over the phone (2024). ceur-ws.org. Accessed 21 June 2024
8. Kitchenham, B., Pearl Brereton, O., Budgen, D., Turner, M., Bailey, J., Linkman, S.: Systematic literature reviews in software engineering–a systematic literature review. Inf. Softw. Technol. **51**(1), 7–15 (2009). Accessed Jan 2008
9. Prada, G.L., Meneses, C.H.: Prevención de fraudes en el sector financiero colombiano a través de controles que usan el aprendizaje automático (2024). Accessed 20 June 20
10. Conectividad, M.E., et al.: Sistema de detección de patrones de fraude con redes neuronales en la Provincia De Los Rios y su insidencia en la telefonia celular, año 2015 (2015). Accessed 20 June 2024
11. Jurado, D.G.: Desarrollo de un modelo predictivo para detectar casos de fraude interno en una institución bancaria (2016). https://repositorio.uchile.cl/handle/2250/143695. Accessed 21 June 2024
12. Basit, A., et al.: A comprehensive survey of AI-enabled phishing attacks detection techniques. Telecommun. Syst. **76**(1), 139–154 (2021). https://doi.org/10.1007/s11235-020-00733-2
13. Gritti, C., Litou, I., Fan, Z., Li, W., Laskey, K.B., Chang, K.-C.: Investigationof phishing susceptibility with explainable artificial intelligence. Future Internet (2024). https://doi.org/10.3390/fi16010031
14. Kalla, D., Samaah, F., Smith, N.B., Kuraku, S., Smith, N.: Phishing detection implementation using databricks and artificial Intelligence. Int. J. Comput. Appl. **185**(11), 975–8887 (2023). https://doi.org/10.5120/ijca2023922764
15. Schmitt, M., Flechais, I.: Digital Deception: Generative Artificial Intelligence in Social Engineering and Phishing (2023). http://arxiv.org/abs/2310.13715. Accessed 21 June 2024
16. A. Pablo Ríos Goytre Dirigido por Gregorio López López Roberto Gesteira Miñarro: Evaluación de clasificadores de voz para Smart Personal Assistants (2023). Accessed 21 June 2024
17. Kang, Y., Kim, W., Lim, S., Kim, H.: Deep-detection: privacy-enhanced deep voice detection and user authentication for preventing voice phishing. Appl. Sci. (2022)

18. Govindu, A., et al.: Deepfake audio detection and justification with Explainable Artificial Intelligence (XAI) (2023). https://doi.org/10.21203/rs.3.rs-3444277/v1

19. Dash, B., et al.: Are ChatGPT and deepfake algorithms endangering the cybersecurity industry? A review. Int. J. Eng. Appl. Sci. (2023). https://doi.org/10.31873/IJEAS.10.1.01

20. Vishniakou, U.: Voice detection using convolutional neural network. https://cyberleninka.ru/article/n/voice-detection-using-convolutional-neural-network. Accessed 21 June 2024

21. Camacho, O., Tejedor, A., Bartosz, Z., Advisor, J., Adrián, Fonollosa, R.: Deep neural networks in acoustic model (2016). https://upcommons.upc.edu/bitstream/handle/2117. Accessed 21 June 2024

22. Zhao, Q., Chen, K., Li, T., Yang, Y.: Detecting telecommunication fraud by understanding the contents of a call. 1(1) (2018). https://doi.org/10.1186/s42400-018-0008-5

23. Ghillani, D.: Deep learning and artificial intelligence framework to improve the cyber security. Am. J. Artif. Intell. (2022). https://doi.org/10.22541/au.166379475.54266021/v1

24. Isaaka, N., Emmanuel, L.: Deep learning-based analysis of a real-time voice cloning system. Emmanuelijisrt.com, vol. 8, no. 7 (2023). Accessed 21 June 2024

25. I. N.-I. C. S. Journal and undefined 2021, "The efficacy of Deep Learning and Artificial Intelligence Framework in Enhancing Cybersecurity, Challenges and Future Prospects. innovatesci-publishers.comI NaseerInnovative Computer Sciences Journal, 2021 innovatesci-publishers.com. https://innovatesci-publishers.com/index.php/ICSJ/article/view/1. Accessed 21 June 2024

26. Müller, N., et al.: Human perception of audio deepfakes. In: Proceedings of the 1st International Workshop on Deepfake Detection for 2022 dl.acm.org, vol. 22, pp. 85–91 (2022). https://doi.org/10.1145/3552466.3556531

27. S. H.-A. at S. 4652028 and undefined 2023, "Study of Artificial Intelligence in Cyber Security and The Emerging Threat of AI-Driven Cyber Attacks And Challenge. papers.ssrn.com SMUH Hassan Available at SSRN 4652028, 2023 papers.ssrn.com, vol. 43, pp. 1557–1570 (2023). https://papers.ssrn.com/sol3/papers.cfm?abstractid=4652028. Accessed21 June 2024

28. Alrowais, F., et al.: Automated machine learning enabled cybersecurity threat detection in internet of things environment. Computer Systems Science Engineering (2023). academia.edu. Accessed 21 June 2024

29. Cortez, N.A.: Impacto de la inteligencia artificial en la ciberseguridad empresarial: un análisis crítico de la evolución de amenazas y medidas preventivas (2024). http://190.15.129.146/handle/49000/15738. Accessed 21 21 June 2024

30. Aldair Villacorta Vidal, C., Nacional de Trujillo, U., Elvis Steve, P., Alberto Carlos Mendoza de los Santos, P.: Inteligencia artificial en la seguridad de la información en una organización: Artificial intelligence in information security in an organization. revistas.unu.edu.pe CAV Vidal, ESO Centurion, ACM de los SantosInvestigación Universitaria UNU, 2023 revistas.unu.edu.pe, vol. 13, pp. 1046–1063. http://revistas.unu.edu.pe/index.php/iu/article/view/120. Accessed: 21 June 2024

31. Mata Suñe, E.: Desarrollo de un chatbot cognitivo capaz de dar respuesta a differentes incidentes de ciberseguridad (2023). https://upcommons.upc.edu/handle/2117/392841. Accessed 21 June 2024

32. Acuña, L.Q.: Aplicación de Machine Learning a un modelo tradicional de Prevención y detección de fraude en entidad financiera proyectado periodos trimestrales (2023). https://ciencia.lasalle.edu.co/maestanaliticainteligencianegocios/7/. Accessed 21 June 2024

33. Kasowaki, L., Yusef, O.: The Human Factor in Cybersecurity: Addressing Social Engineering and Insider Threats (2023). https://easychair.org/publications/preprintdownload/wDQQ. Accessed 21 June 2024

34. Kheruddin, M.S., et al.: Phishing Attacks: Unraveling Tactics, Threats, and Defenses in the Cybersecurity Landscape (2024). https://doi.org/10.22541/au.170534654.48067877/v1

35. Ozkan-Ozay, M., et al.: A comprehensive survey: evaluating the efficiency of artificial intelligence and machine learning techniques on cyber security solutions. IEEE Access (2024). https://ieeexplore.ieee.org/abstract/document/10403908/
36. Firc, A., Malinka, K.: Deep-fake speech detection: a spectrogram analysis. In: Proceedings of the 39th ACM/SIGAPP Symposium on Applied Computing, pp. 1312–1320 (2024). https://doi.org/10.1145/3605098.3635911
37. S. LA Luca CAGLIERO Doc Moreno QUATRA Doc Lorenzo VAIANI Candidate Alessandro Emmanuel PECORA, Data driven: AI Voice Cloning (2023). Accessed 21 June 2024
38. Jemine, C.: Automatic Multispeaker Voice Cloning (2019). Accessed 21 June 2024
39. A. F.-B. U. of Technology, F. of Information, and undefined 2021. Applicability of Deepfakes in the Field of Cyber Security. A FircBrno University of Technology, Faculty of Information Technology, Brno (2021). https://doi.org/10.1080/13683500.2020.1738357
40. "Inteligencia Artificial y el Aprendizaje Automático en la Ciberseguridad. Accessed 22 June 2024
41. Usman, M., Bin Ali, N., Wohlin, C.: A quality assessment instrument for systematic literature reviews in software engineering. E-Informatica Softw. Eng. J. **17**(1) (2023). https://doi.org/10.37190/E-INF230105

Entropy Wall for Symmetric Cryptographic Key Generation Through Frame Processing

Reinaldo Toledo[✉] and Jorge E. Camargo

UnsecureLab, Universidad Nacional de Colombia, Bogotá, Colombia
{rtoledo1,jecamargom}@unal.edu.co

Abstract. The entropy wall concept plays a vital role in ensuring the high degree of randomness and unpredictability necessary for robust cryptographic systems. Natural sources of entropy, such as the dynamic and non-reproducible properties of physical phenomena, provide a superior foundation for generating cryptographic keys. By leveraging such sources, one can effectively defend against potential vulnerabilities and emerging threats, including those posed by artificial intelligence and quantum computing.

This paper introduces a method for generating encryption keys for AES 128 bits algorithm by utilizing entropy barriers from real-world sources, specifically video frames of juice being continuously shaken. By exploiting the inherent randomness of these video frames, the proposed approach aims to enhance cryptographic security by producing keys that provide an additional layer of protection to conventional encryption algorithms. The analysis includes an evaluation with the performance based on different key sizes and entropy evaluation.

Keywords: Entropy wall · Randomness · Cryptography · Key generation · Information security · Cybersecurity

1 Introduction

In the current digital era, data has become one of the most valuable and critical resources. Consequently, cybersecurity, the field responsible for protecting this data, has gained increasing importance due to the constant threats seeking unauthorized access to information. From trivial data to industrial secrets, digital information is the driving force behind the contemporary economy and technology, making its protection essential.

Cybersecurity encompasses the protection of computer systems, networks, and data against attacks and unauthorized access. As technology advances, the amount of data generated and shared increases exponentially, thereby making the protection of this information ever more vital. Furthermore, the evolution of cyber threats has transformed the landscape: what were once attacks by individual hackers have now become a highly sophisticated industry, with attackers seeking both valuable information and the hijacking of computer systems to extort businesses and governments.

Moreover, nation-states have increasingly involved themselves in cybersecurity issues. It is not uncommon to witness state-sponsored cyberattacks with objectives such

© The Author(s), under exclusive license to Springer Nature Switzerland AG 2024
N. D. Duque-Méndez et al. (Eds.): CCC 2024, CCIS 2208, pp. 298–312, 2024.
https://doi.org/10.1007/978-3-031-75233-9_21

as destabilizing the infrastructure networks of other countries or obtaining governmental secrets. Consequently, this puts individuals' privacy and national security at constant risk, highlighting the growing importance of cybersecurity in protecting various data sources.

One of the pillars of cybersecurity is prevention. Preventing attacks before they occur is crucial, as the damage from a successful attack can be significant and sometimes irreversible. A compromised system can become unusable, information can be stolen or erased, and the recovery of this data can be costly and, in some cases, impossible.

To mitigate these risks, various security measures are implemented, such as firewalls, data encryption, and two-factor authentication. Preventive measures are critical because once an attack has occurred, the damage is already done and can be very significant. For instance, a system might become unusable, the company's information might be stolen or deleted, leading to substantial economic impacts due to the time and money required to recover these data. In some cases, the data may never be recoverable.

In this context, this project focuses on strengthening cybersecurity by generating random encryption keys, thus making it more difficult for attackers to decrypt the data using known algorithms. By implementing diverse security measures, such as firewalls, data encryption, and two-factor authentication, this approach aims to alter the data in such a way that it becomes exceedingly difficult for attackers to decipher it. Furthermore, as digital information ranges from trivial data to industrial secrets, it constantly exposes itself to various threats seeking unauthorized access, thereby emphasizing the need for robust protection measures.

The rest of the paper is organized as follows. In Sect. 2 related works are presented. In Sect. 3 there is the description of the methodology followed in order to carry out the project and experiments. Section 4 describes the communication scheme implementation, test suit of messages between sender-receiver and the 128 bits key generation process including the design of experiments for the respective analysis. Section 5 shows the results represented with data visualization as well as a discussion and future works in Sect. 6. Finally, Sect. 7 concludes the paper.

2   Related Works

Several approaches have been explored in the realm of secure key generation (SKG) for wireless communication. While traditional methods have encountered challenges, such as compromised channel reciprocity in Frequency Division Duplexing (FDD) systems, recent advancements propose innovative solutions to enhance Key Generation Rates (KGR) and overall security robustness. For instance, Chai et al. [1] introduce a lightweight SKG scheme leveraging the Random Forest (RF) algorithm and Multi-to-Single Mapping (M2SM) strategy, complemented by Moving Average (MA) and subtraction algorithms to mitigate noise and frequency-selective fading in FDD channels [1]. Their experimental results demonstrate an error-free KGR of 57.71 Kbps, underscoring the efficacy of their approach in achieving high-speed and secure SKG suitable for practical deployment in modern wireless networks. Moreover, the scheme exhibits low complexity, high flexibility, and ensures information-theoretic security, addressing critical shortcomings observed in conventional SKG methods.

Moreover, several methods have investigated the utilization of deterministic dynamical systems for pseudo-random number generation (PRNG). Traditionally, discrete chaotic systems such as maps have been widely studied, alongside recent investigations into continuous chaotic systems. However, a novel approach introduced by [2] utilizes a deterministic Brownian system to generate pseudo-random sequences of binary numbers. This system exhibits two positive Lyapunov exponents, distinguishing it from typical chaotic systems characterized by a single maximum positive Lyapunov exponent. The implementation employs fixed-point arithmetic with 64-bit precision, utilizing 5 bits for the integer part and 58 bits for the fractional part. Each variable from the third-order Brownian system contributes its eight least significant bits to form the random bit sequences. Rigorous statistical testing using the NIST and TestU01 suites confirms the generated sequences' randomness. This innovative approach underscores the system's potential for enhancing cryptographic security through robust random number generation.

One notable study [3] proposes a time-invariant mechanism using electroencephalogram (EEG) signals. The research employs Discrete Wavelet Transform and autoencoders to extract biometric features from EEG signals, which are then used to generate secure seeds for cryptographic keys. This approach ensures that each cryptographic key is unique to the current EEG signal received, eliminating the need to store previous keys or EEG data. Results indicate a robust security against random attacks, achieving a 0% False Acceptance Rate and generating seeds with high entropy (0.968) in less than 500 ms. This method represents a significant advancement in utilizing EEG biometrics for continuous and secure key generation, addressing challenges such as consistency, uniqueness, efficiency, and data privacy preservation.

Furthermore, a recent work [4] proposes an approach for cryptographic key generation based on iris biometrics. This method utilizes advanced techniques such as Ensemble Gradient Minimization and an edge-preserving filter to extract iris structure from images. Subsequently, iris texture data is normalized, and a consistent region is selected using a novel statistical method. Feature vectors are generated using ensemble local space-filling curve descriptors and their variants, followed by the creation of a discriminant feature vector through hybrid feature selection and neighborhood component analysis. Finally, an interval-based encoding scheme is employed to derive a key from the discriminant feature vector. This approach proves effective in dynamically generating keys from online iris images, with potential applications in data storage security, remote authentication protocols, and blockchain security systems. The experimental results confirm that the keys generated exhibit randomness and meet criteria for unlinkability and revocability, demonstrating robustness against brute-force attacks and JPEG compression.

The convergence of these studies underscores a common objective: to advance the field of SKG by developing reliable and efficient methods that can operate securely in diverse environments and applications. Continued research in these areas is essential to further enhance the security and scalability of cryptographic systems, ensuring they meet the evolving demands of modern digital ecosystems.

3   Methodology

To ensure the successful development of the proposed project, a comprehensive planning process was established, encompassing abstract modeling to detailed specification and implementation phases. This meticulous approach facilitates a coherent transition from conceptualization to execution, enabling a thorough understanding and precise application of each project component. The adopted methodology is explicitly goal-oriented, designed to achieve clearly defined objectives at each stage, thereby ensuring that all aspects of the project are addressed methodically and efficiently. This structured approach not only enhances the clarity and precision of the development process but also ensures that each phase is seamlessly integrated, contributing to the overall success of the project.

3.1   Establishment of the Cryptographic Model to be Developed

In this phase, we establish the cryptographic model to be utilized in the research. Two primary cryptographic models were considered: symmetric cryptography and public key cryptography. Symmetric cryptography involves the use of a single key for both encryption and decryption processes, whereas public key cryptography utilizes a pair of keys, one for encryption and another for decryption. Each model offers distinct advantages and challenges, which were meticulously analyzed to determine their suitability for the intended application.

After a thorough evaluation, the most appropriate cryptographic model was selected based on several criteria, including security requirements, computational efficiency, and ease of implementation. This decision was crucial to ensure that the chosen model aligns with the goals of the project and can be effectively tested under relevant conditions.

Subsequently, the specific cryptographic algorithm to be employed within the chosen model was identified. The selection process for the algorithm involved a comprehensive review of existing cryptographic techniques, focusing on their robustness, performance, and applicability to real-world scenarios. The algorithm had to meet stringent security standards to protect against potential threats while maintaining efficiency to facilitate practical use.

Once the cryptographic algorithm was defined, the next step involved planning and setting up the experimental framework. This framework was designed to simulate conditions that closely resemble everyday environments, ensuring that the experimentation reflects realistic and practical applications. This phase included configuring the necessary hardware and software, establishing secure channels for data transmission, and creating test cases that mimic common use cases.

In addition, the experimental conditions were meticulously documented to ensure reproducibility and accuracy of results. Parameters such as key lengths, encryption and decryption times, and throughput were recorded and analyzed. This detailed documentation is essential for comparing the performance of different algorithms and models under similar conditions.

By defining the cryptographic model and algorithm in this manner, the study aims to achieve a comprehensive understanding of their practical implications, ensuring that the results are both reliable and applicable to real-world cryptographic applications.

3.2 Establishment of Tools and Communication Sources

Under this phase, the methodology defines the connection methods between the involved devices in the cryptographic model. It entails a thorough analysis and selection of the most suitable connection protocols to ensure reliable and efficient communication. Criteria for selecting these methods focus on aspects such as reliability, low latency, and compatibility with the chosen cryptographic algorithms.

Moreover, the phase involves the identification and classification of communication channels as either secure or insecure. Secure channels utilize advanced encryption techniques and authentication protocols to offer high levels of protection against unauthorized access and eavesdropping. Conversely, insecure channels lack adequate security measures and may pose risks to data integrity and confidentiality.

Understanding the characteristics of these channels is crucial to ensure the efficacy of the cryptographic model. Secure channels enable the safe transmission of sensitive data, while awareness of insecure channels assists in developing strategies to manage data in less secure environments. This understanding ensures the cryptographic model can be implemented effectively across various real-world scenarios where the security of communication channels varies.

In essence, this phase plays a critical role in establishing a dependable and secure communication infrastructure for the cryptographic model. By carefully defining connection methods and identifying secure and insecure channels, the study ensures the model functions effectively in practical settings. This foundational work is essential for supporting subsequent cryptographic processes and achieving robust security outcomes.

3.3 Design and Implementation of the Entropy Wall Algorithm

Several methods for generating random keys from an entropy wall source are explored and evaluated in this phase. The initial step involves a comprehensive review of existing techniques and approaches in the field of random key generation, focusing on their applicability to the specific requirements of the study. Various algorithms and methodologies are considered, each selected for its potential to maximize entropy extraction and ensure robust key generation.

Following the identification of suitable approaches, the next critical task involves the creation and configuration of the entropy source itself. This process entails setting up a reliable framework that can consistently provide a high-quality stream of entropy, essential for generating secure and unpredictable cryptographic keys. Special attention is given to optimizing the entropy collection mechanisms to minimize bias and enhance randomness in the generated keys.

Simultaneously, the algorithmic implementation phase commences, where the selected approach is translated into a functional cryptographic algorithm tailored to the study's objectives. This involves coding the algorithm in a programming language suited for cryptographic applications, ensuring efficiency and correctness in key generation processes. Rigorous testing protocols are integrated during this phase to validate the algorithm's performance, assessing factors such as randomness, entropy distribution, and computational efficiency under varying conditions.

3.4 Identification and Design of Performance Test, Entropy Metrics Adopted and Analysis Sources

The final phase focuses on the identification and design of comprehensive performance tests, entropy metrics adopted, and sources of analysis. To begin with, a meticulous approach was adopted to document the experimental conditions thoroughly, ensuring the reproducibility and accuracy of results across all phases of the study. Critical parameters including key lengths, encryption and decryption times, and overall throughput were systematically recorded and rigorously analyzed. This meticulous documentation plays a pivotal role in facilitating comparative evaluations of various algorithms and models operating under similar conditions, thereby enabling researchers to discern nuanced differences in performance metrics and validate the efficacy of the cryptographic methods employed. Moreover, the establishment of standardized performance tests serves to enhance the reliability and robustness of findings, offering insights into the operational capabilities and limitations of the cryptographic systems under scrutiny. As part of this phase, the adoption of specific entropy metrics is pivotal, encompassing measures to quantify the randomness and unpredictability of generated cryptographic keys. These metrics include entropy estimation techniques such as Shannon entropy, Kolmogorov complexity, and NIST Special Publication 800-90B guidelines, which provide standardized frameworks for assessing the quality and security of random number generators. The integration of diverse sources of analysis further enriches the evaluation process, encompassing statistical tools, simulation environments, and real-world deployment scenarios to validate the resilience and efficacy of the cryptographic algorithms in practical applications. By employing a rigorous approach to performance testing and entropy analysis, this phase aims to substantiate the theoretical foundations with empirical evidence, thereby advancing the state-of-the-art in cryptographic research and reinforcing the reliability of cryptographic systems in safeguarding sensitive information.

4 Communication and Encryption Process

4.1 Cryptography Scheme

The starting point for the experimental execution is based on a symmetric cryptography model. Within this model, the essential components are identified as follows: the message denoted by m, the encrypted message represented as c and the key corresponding to k. The encryption process is denoted as E_k, which refers to encryption using the Advanced Encryption Standard (AES) with a 128-bit key length. Correspondingly, the decryption process is represented as D_k, indicating decryption with the same AES 128-bit key. The cryptographic scheme is illustrated as follows (Fig. 1).

To elaborate, the symmetric cryptography model operates on the principle that the same key is used for both encryption and decryption. This model ensures that a message m is transformed into an encrypted message c through the application of the encryption algorithm E_k, utilizing the key k. Subsequently, the original message m can be recovered from the encrypted message c by applying the decryption algorithm D_k with the same key k.

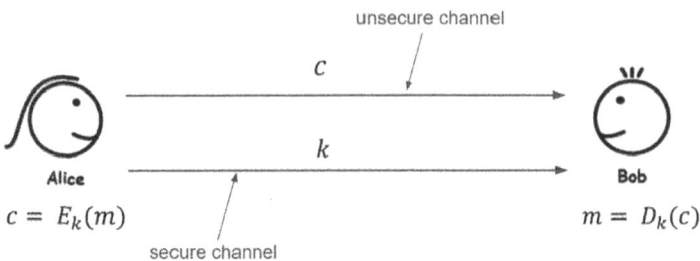

Fig. 1. Symmetric cryptography model taken into account for the shared key generation.

The selection of AES with a 128-bit key for this cryptographic model is driven by several compelling reasons. AES is a well-established and widely accepted encryption standard that has been thoroughly vetted for security and efficiency. The 128-bit key length, while being the shortest key length recommended under the AES standard, strikes a balance between security and performance. It offers robust protection against brute-force attacks due to the vast number of possible key combinations (approximately), which makes it computationally infeasible for an attacker to exhaustively search through all possible keys within a reasonable timeframe.

Furthermore, it is pertinent to clarify the workings of the chosen algorithm, AES-128, a subset of the Advanced Encryption Standard (AES) defined by the National Institute of Standards and Technology (NIST) [5], is widely recognized for its robust security and efficiency. The AES algorithm operates on a block size of 128 bits, employing a series of transformation steps, including substitution, permutation, and mixing of the input data, combined with key addition. These steps are repeated over 10 rounds in the case of AES-128, each round involving processes such as SubBytes, ShiftRows, MixColumns, and AddRoundKey. The final round omits the MixColumns step to complete the encryption.

Moreover, the implementation of AES-128 is efficient both in terms of computational resources and execution time. This efficiency is particularly crucial in environments with limited processing power and memory, such as embedded systems and mobile devices. Despite its shorter key length compared to AES-192 and AES-256, AES-128 provides a high level of security suitable for most applications while ensuring faster encryption and decryption processes.

4.2 Communication Scheme Implementation

The communication scheme: The identification of Alice is defined as a host machine running the Windows 10 operating system. On the other hand, the entity of Bob, the message receiver, is a virtual machine operating on Kali Linux. The transmission of information corresponding to the messages is carried out over a network with an Ethernet interface, establishing communication between the host machine and the virtual machine using the host-only Ethernet adapter integrated within VirtualBox, the virtualization tool of the physical device.

Communication between the host and the virtual machine is established via the Ethernet interface provided by VirtualBox. Messages are transmitted through ICMP (Internet Control Message Protocol) packets between both machines. The messages are loaded into the ICMP payload, which is the data field of the ICMP packet that carries the actual message content. This approach ensures a straightforward and efficient means of message transmission within a controlled environment.

The chosen communication model, operating within a single physical device, offers several advantages. Firstly, it simplifies the setup and management of the network environment, reducing the complexity and potential external threats associated with multi-device networks. The use of ICMP packets for message transmission provides a lightweight and efficient protocol for the exchange of messages, taking advantage of its simplicity and the widespread support across different operating systems.

Before being loaded into the ICMP packets, the messages are encrypted using the AES-128 algorithm implemented through the PyAES library. This step ensures that the data remains secure and protected from unauthorized access during transmission. The encryption process adds a layer of security, leveraging the robustness of AES-128 to safeguard the integrity and confidentiality of the messages. The encryption key itself is securely stored in a shared directory between the two machines, which acts as the secure channel of the proposed symmetric cryptographic model. In this shared directory, the host machine has read and write permissions because the constant generation of random keys is carried out in a parallel process. Meanwhile, the virtual machine has read-only permissions to ensure secure and controlled access to the keys.

Upon receipt of the ICMP packets, the virtual machine employs a Sniffer, implemented using the Scapy library, to capture the messages. The same library is utilized to send the messages from the host machine, ensuring consistency and reliability in the communication process. The use of Scapy provides a powerful and flexible framework for packet manipulation and network communication, facilitating the development of custom networking tools and enhancing the overall functionality of the communication scheme.

Implementing this communication model within a single device, using ICMP packets, allows for a controlled and isolated environment, minimizing external variables and potential security vulnerabilities. This controlled setup also facilitates the process of result analysis and performance evaluation, ensuring accurate and reliable testing conditions. The use of ICMP packets simplifies the message transmission process, reducing overhead and ensuring efficient delivery. Additionally, the encryption of messages with AES-128 before transmission guarantees that the data remains secure, protecting against potential eavesdropping and tampering.

The communication scheme described leverages the strengths of a single-device setup, ICMP packet transmission, and robust encryption methods to ensure secure and efficient message exchange between the host and virtual machine. The combination of VirtualBox networking capabilities, the PyAES library for encryption, and the Scapy library for packet handling provides a comprehensive and reliable framework for secure communication within this research context. Figure 2 depicts the communication scheme.

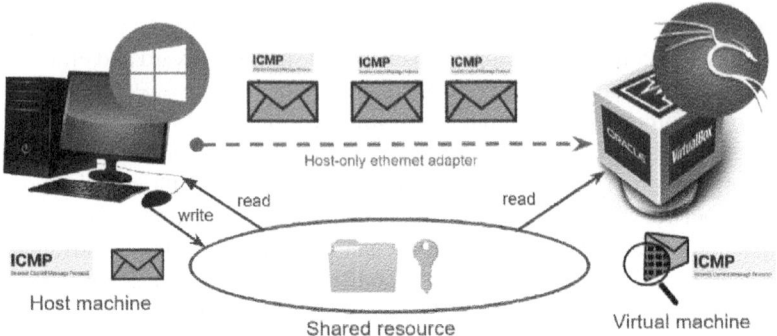

Fig. 2. Communication scheme implemented based on the symmetric cryptography model defined.

5 Key Generation

Regarding the selection of an entropy wall, a video recording captured a glass of lulo juice being stirred with a metal spoon, utilizing a frame rate of 180 frames per second for precise frame processing, as it is illustrated in Fig. 3. The process of generating a 16-byte key from an entropy wall frame unfolds through structured process.

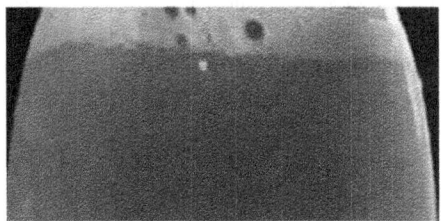

Fig. 3. Entropy wall represented as a random frame resulted from video recording

Initially, 16 random pixels denoted by P are meticulously selected from the frame, optimizing the selection process using numpy. This library, renowned for its efficiency in numerical computations and array operations, enables swift processing essential for handling large datasets and complex calculations in real-time applications.

Following pixel selection, the RGB components (Red, Green, Blue) of each chosen pixel are computed to derive a single numerical value representing the pixel's intensity and color composition represented by M. This calculation results in an array of 16 numeric values, each uniquely encapsulating the essence of the selected pixels.

To standardize the data range suitable for cryptographic key generation, a normalization function adjusts the numeric values to fall within a specified range, aligning them with ASCII character codes ranging from 33 to 126, the resulted array is denoted by N. The use of a uniform distribution for pixel selection ensures randomness and unpredictability, fundamental traits for generating strong cryptographic keys.

The resulting array of normalized integers is then converted into ASCII characters, facilitated by Python's inherent ASCII encoding capabilities. This transformation yields a coherent sequence of characters that collectively form a 16-byte random key.

The generated key is securely stored in a shared resource accessible to both the host and virtual machines, establishing a secure channel within our cryptography scheme for key exchange and communication.

This methodological approach offers distinct advantages. By harnessing NumPy, the computational efficiency required for processing vast amounts of pixel data is significantly enhanced. NumPy's optimized array operations expedite the pixel selection and processing phases, ensuring minimal computational overhead and maximal performance efficiency.

Furthermore, the use of OpenCV streamlines frame manipulation and pixel extraction, facilitating seamless integration into real-time video processing applications. This integration underscores the practical applicability of computer vision techniques in generating cryptographic keys from dynamic sources of entropy.

The synergy between NumPy and OpenCV empowers the generation of robust and secure cryptographic keys from video frames. This approach not only ensures randomness and unpredictability but also underscores the versatility and reliability of modern computational tools in cryptographic applications. The complete generation process is illustrated in Fig. 4.

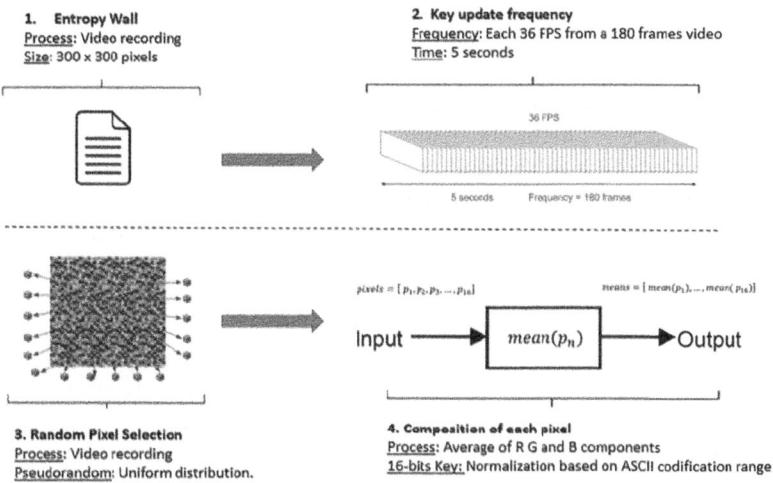

Fig. 4. Illustration of the key generation process based on frame processing

6 Analysis and Results

We conducted a comprehensive performance analysis of a random key generation system, focusing specifically on evaluating the computation time required for different key sizes. The study utilized Python's time library to measure the execution time for generating keys of varying sizes. To ensure accuracy, five runs were conducted for each key size, and the average times were calculated and stored in an array.

The approach adopted in the study is notable for its simplicity and effectiveness in using the time library, a well-established tool in Python, to measure execution times. This method provides a straightforward and precise means of assessing computational performance, which is crucial in cryptography where efficiency is as vital as security.

Advantages of this approach include its simplicity and clarity in measuring computation time compared to more complex methods used in related works. While other studies often employ multifaceted techniques to enhance Key Generation Rates (KGR) and security robustness, the authors' approach emphasizes a direct evaluation of computational performance. This simplicity is crucial for practical implementations where key generation speed is paramount.

For example, Chai et al. [1] present a method using the Random Forest (RF) algorithm and Multi-to-Single Mapping (M2SM) strategy, alongside advanced techniques like moving average and subtraction algorithms. Despite achieving a high KGR, their approach involves higher complexity. In contrast, the authors' method focuses on straightforward computation time evaluation, essential for applications requiring rapid and uncomplicated implementation.

Moreover, the study conducted by De La Fraga and Ovilla Martínez [2] introduces a deterministic Brownian system for generating pseudo-random sequences, employing fixed-point arithmetic with 64-bit precision. While innovative, this approach requires complex technical implementation. In contrast, the key generation by frames study simplifies performance evaluation through computation time measurement, offering a solid basis for efficiency assessments without intricate setups.

Additionally, advancements like the EEG signal-based approach [3] and iris biometric key generation [4] highlight significant progress in utilizing biometric features. However, these methods involve handling biometric data and complex signal processing techniques. The current project development addresses these challenges by providing a straightforward evaluation of computation time, crucial for real-time applications and resource-constrained environments.

Based on the visualization of computational performance, it is observed that the increase in key size, measured in bytes, correlates with a linear improvement in performance, see Fig. 5. This linear relationship underscores the scalability of the key generation system, indicating that larger key sizes not only enhance security through increased entropy but also do so efficiently in terms of computational overhead. This scalability is a significant advantage in scenarios where both robust security and operational efficiency are paramount considerations.

The frequency analysis conducted on a sample of 5 concatenated cryptographic keys provides valuable insights into the distribution and randomness characteristics of the generated keys. Each key, generated using the method described earlier, was concatenated to form a composite sample for analysis. The approach utilized Python's collections. Counter to ascertain the frequency of each byte within the concatenated keys, ensuring a comprehensive evaluation across the entire key space.

Upon analysis, it was found that the distribution of byte frequencies closely adhered to a uniform pattern, reflecting the effective randomness inherent in the key generation

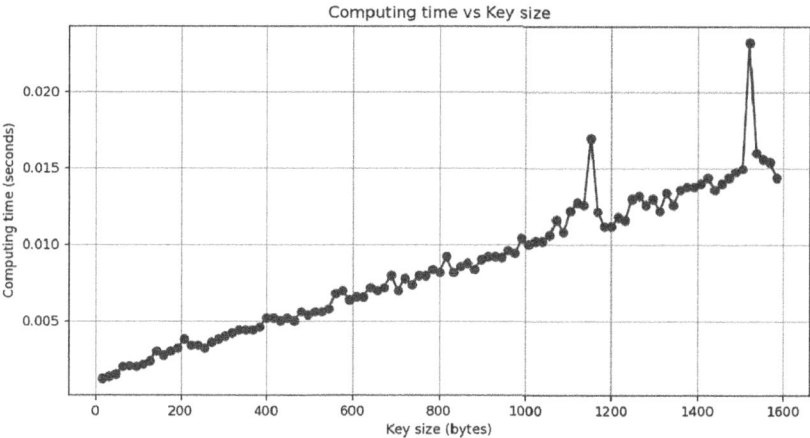

Fig. 5. Visualization of computing time for different key sizes generation

process. This observation is pivotal in cryptographic applications where unpredictability and uniformity are critical for robust security protocols. The calculated entropy per character of 5.0885 further corroborates the high degree of randomness achieved, surpassing conventional benchmarks and affirming the efficacy of the proposed method in generating secure cryptographic keys.

Comparatively, related studies often employ intricate algorithms and advanced techniques to bolster Key Generation Rates (KGR) and enhance security robustness. For instance, Chai et al. [1] utilize complex machine learning algorithms and multi-stage mapping strategies, emphasizing achieving high KGR alongside stringent security measures. Despite their advancements, these approaches may introduce additional computational overhead and implementation complexities, which contrasts with the straightforward evaluation of computational performance highlighted in this study.

Moreover, advancements in biometric-based key generation methods [3, 4], such as EEG signal analysis and iris biometric features, signify substantial strides towards leveraging physiological data for enhanced security. However, these approaches typically require specialized hardware and intricate signal processing techniques, which can limit their applicability in resource-constrained environments.

In contrast, the method presented in this study offers a streamlined approach to cryptographic key generation, focusing on fundamental principles of randomness and computational efficiency. By evaluating the frequency distribution and entropy characteristics of concatenated keys, this study establishes a robust foundation for assessing key generation methodologies in real-world scenarios. The linear scalability observed in computational performance with increasing key sizes further underscores the method's practicality and effectiveness for applications requiring both security and operational efficiency.

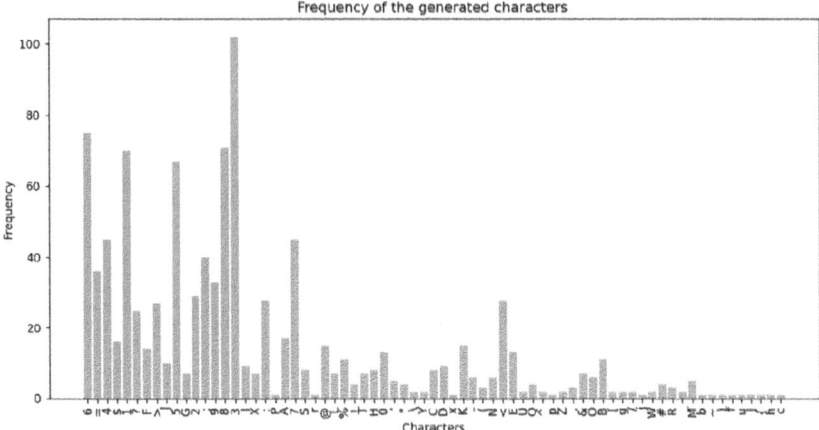

Fig. 6. Visualization of characters frequency through five generated size 100 keys from the same frame.

Furthermore, according to Fig. 6, the frequency analysis reveals a notable concentration of characters within a specific range, leading to the observation of a Poisson probability distribution in the visualization. This distribution pattern indicates that certain characters appear more frequently than others, aligning with the expected characteristics of Poisson distributions where events (or characters, in this case) occur independently at a constant average rate over a fixed interval. The observed distribution of character probabilities is influenced by the level of entropy present in the frames used for key generation. The entropy, quantified at 5.0885 bits per character, serves as a crucial determinant of the variability and unpredictability in character selection, thereby directly impacting the resultant cryptographic key's strength and randomness.

7   Discussion and Future Work

A promising avenue for future research lies in the development of sophisticated metrics for entropy analysis on frames, aiming to pinpoint sections within video frames that can lead to a more uniformly distributed character frequency. Current methodologies often rely on generalized entropy calculations, but tailoring these metrics to specific sections of frames could enhance the randomness and unpredictability of generated cryptographic keys. By identifying and optimizing these sections, researchers can potentially improve the overall security and reliability of cryptographic systems.

Another crucial area for future investigation involves a deeper exploration of conditions, constraints, and the effects of variables during random key generation. This entails examining factors such as frame resolution, color depth, sampling rates, and environmental conditions (e.g., lighting variations) to understand their impact on key generation efficiency and security. Developing algorithms that adaptively adjust key generation parameters based on these variables could lead to more efficient cryptographic systems capable of operating in diverse and dynamic environments.

Moreover, exploring the integration of machine learning and artificial intelligence techniques presents an exciting frontier. Techniques such as anomaly detection and pattern recognition could be leveraged to identify and mitigate potential biases or vulnerabilities in key generation processes. Furthermore, employing neural networks for real-time analysis of frame entropy could automate the optimization of key generation parameters, ensuring robust security without compromising computational efficiency.

8   Conclusion

The conclusion of this study underscore the importance of researching natural entropy methods compared to pseudo-random algorithms in the field of key generation and information security.

Research into natural entropy methods is crucial due to their ability to harness sources of truly unpredictable and non-deterministic randomness, such as physical or biological phenomena. This research is fundamental because cryptographic systems rely on the quality and quantity of entropy to generate secure keys. While this study used Python's time library to measure key generation time, future research could explore integrating natural entropy sources to enhance the security and efficiency of these systems.

The results obtained in this study provide a clear assessment of the computational performance of a key generation system by using runtime measurements with Python. The methodology employed identified differences in key generation time based on key size, which is essential for optimizing performance in applications where efficiency is critical. These findings underscore the importance of selecting key generation methods that are not only secure but also efficient in terms of execution time.

From an information security standpoint, this study highlights the need to balance security with efficiency. It is crucial to generate keys that are robust and difficult to predict to safeguard sensitive information. Simultaneously, it is vital to ensure that key generation processes are not overly slow, particularly in applications requiring swift and effective processing. The simplicity and effectiveness of the approach used in this study for measuring computation time provide a valuable tool for cybersecurity professionals seeking to evaluate and enhance the efficiency of their systems without compromising security.

References

1. Chai, Z., et al.: Lightweight and Fast Physical-Layer Key Generation in FDD Systems Using Random Forest, pp. 1–11. State Key Laboratory of Advanced Optical Communication System and Networks, Shanghai Jiao Tong University, Shanghai 200240, China (2023)
2. De La Fraga, L.G., Ovilla-Martínez, B.: Generating pseudo-random numbers with a Brownian system, pp. 1–4. Cinvestav, Computer Science Department, Mexico City, Mexico (2024)
3. Hernández Álvarez, L., et al.: KeyEncoder: a secure and usable EEG-based cryptographic key generation mechanism, 2nd edn. Institute of Physical and Information Technologies (ITEFI), Spanish National Research Council (CSIC), Madrid, Spain (2023)

4. Priyabrata, D., Pandey, F., Sarma, M., Samanta, D.: Efficient private key generation from iris data for privacy and security applications, pp. 1–10. Advanced Technology Development Centre, Indian Institute of Technology Kharagpur, Kharagpur, West Bengal, 721302, India (2023)
5. National institute of standards and Technology: Advanced Encryption Standard (AES), pp. 3–20 (2023)

Transparency and Security in Tourism: Exploring the Potential of Blockchain Technology

Juan Gutiérrez[1]([✉]) [ID], Martín Pantoja[1] [ID], and Marcelo López[1,2] [ID]

[1] Universidad Nacional de Colombia, sede Manizales, Manizales, Colombia
jgutierrezgi@unal.edu.co
[2] Universidad de Caldas, Manizales, Colombia

Abstract. This theoretical research addresses the current problem of the trust dilemma and presents a comprehensive approach to understanding the fundamental concepts of blockchain and the specific challenges in the context of tourism in the Caldas region (Colombia). To achieve this purpose A bibliometric and narrative analysis based on the scientific production from the Scopus and Web of Science (WoS) databases was rigorously conducted. The bibliometric, descriptive, and structural analysis, using a corpus of 2,156 documents analyzed with the R program, allowed for the construction of a solid theoretical framework on the topic. Additionally, the network data-based mapping analysis, carried out with VOSviewer, enabled the structuring and visualization of key thematic relationships and emerging trends in the field, identifying research clusters and predominant lines of study. The scope of this exploratory study includes the identification and analysis of trends and patterns in the scientific literature on the implementation of Blockchain technology in the tourism industry. It is also shown the significant implications of blockchain technology for tourism include improving the tourist experience, facilitating fast and hassle-free transactions, and providing financial inclusion for the unbanked. In this article, a general blockchain model for the tourism industry will be presented, specifically designed to integrate with the sector in the region of Caldas (Colombia), emphasizing the urgency of understanding and evaluating the positive impacts. The General Objective is Analyze the potential of Blockchain technology to transform traditional business models and create new opportunities for social and sustainable innovation in the tourism industry.

Keywords: Blockchain · Digitalization · innovation · smart tourism

1 Introduction

In the current digital era, blockchain technologies have emerged as potential catalysts to radically transform traditional business models. In a world where trust and security are paramount for the exchange of assets and data, conventional structures face significant challenges. The underlying problem lies in the lack of transparency, efficiency, and security inherent in traditional business models, often hindered by intermediaries and slow processes. This issue is exacerbated by globalization and the growing need for instant and secure transactions.

© The Author(s), under exclusive license to Springer Nature Switzerland AG 2024
N. D. Duque-Méndez et al. (Eds.): CCC 2024, CCIS 2208, pp. 313–327, 2024.
https://doi.org/10.1007/978-3-031-75233-9_22

Digitalization is expected to continue driving the travel experience towards a more fluid, frictionless, and high-quality experience, and in a way that contributes to the United Nations Sustainable Development Goals (SDGs) Tyan et al., (2021). The use of technologies such as the 'Internet of Things', location-based services, artificial intelligence, augmented and virtual reality, and blockchain technology has resulted in a more attractive, efficient, and inclusive tourism offering, which is also economically, socially, and environmentally sustainable. It has also facilitated innovation and the rethinking of processes, addressing challenges such as seasonality, overcrowding, and the development of smarter destinations.

The working hypothesis suggests that by harnessing the decentralized and transparent characteristics of blockchain technology, the Caldas region could establish itself as a more visible and attractive tourist destination on a global scale. Pioneering companies realized the competitive advantage of transparency, according to Saberi S. (2019) citing Ward (2017). By using this technology as management models to record and share information about tourist destinations, attractions, and services, an immutable and verifiable record is created. Travelers could access accurate and reliable information about itineraries, activities, and costs, which would help build greater trust in operators and the information provided.

In this context, the adoption of blockchain not only signifies technological evolution but also a fundamental shift that could position the Caldas region as a global tourism reference point, highlighting the importance of innovation in the industry to boost its competitiveness and appeal.

The theoretical methodology employed in this research is based on a bibliometric analysis using the R programming language, through the bibliometrix package. Timeline of analysis between 2016 and 2024. Additionally, the software VOSviewer was used, a tool that facilitates the creation of network-based maps from the extracted information. The combination of these two tools offers a comprehensive approach to exploring patterns, trends, and connections in the scientific literature, allowing for a deep and visually informative understanding of the research landscape on the topic addressed. This methodological approach strengthens the validity and thoroughness of the analysis by providing a broad and detailed view of the field of study from a bibliometric perspective.

2   Problem Statement

The issue addressed in this research stems from the recent evolution of the sustainable tourism paradigm towards a smarter approach, driven by the need to address the growing sustainability demands in data-driven tourist destinations. This shift is evidenced in works such as that of Liu et al. (2022), who highlight the transition towards sustainable and intelligent tourism. The interest in this new paradigm is supported by the increasing investment in blockchain by the tourism industry, with tourism being at the forefront of this adoption, as noted by J. Kwok et al. (2018).

Within the tourism industry in the Caldas region, there is a problem that impacts the improvement of the traveler's experience and sustainable development. One of the main challenges lies in establishing an identity as a tourist destination. The various interest groups have not worked together collaboratively around the use and appropriation of

ICT and knowledge for tourism. As mentioned by Trujillo et al. (2021). No research has been done to establish knowledge management or intellectual capital frameworks to guide this collaborative effort. Echeverri-Rubio et al. (2022) argue that a participatory tourism development approach should be formulated and implemented; however, this requires a complete overhaul of the sociopolitical, legal, administrative, and economic structures in many developing countries. This challenge is particularly relevant in the Caldas region, where such structural changes are crucial to fostering an environment conducive to sustainable and intelligent tourism.

Justification

This research aims to analyze the state of the art of blockchain technology literature to transform traditional business models and create new opportunities for social and sustainable innovation in the tourism industry in Caldas region involving different stakeholders such as hotels, airlines, travel agencies, tour operators, insurance companies, payment service providers, and others. The relevance of this research is justified from various perspectives: theoretical, practical, and methodological.

The need for this research is based on the current lack of specific studies addressing the implementation of blockchain technology in the tourism sector of Caldas region. As the tourism industry undergoes constant changes, there is an urgent need to understand how blockchain can meet the demands and challenges.

Objectives

General Objective: Analyze the potential of Blockchain technology to transform traditional business models and create new opportunities for social and sustainable innovation in the tourism industry in Caldas region. Identify the challenges and opportunities this presents and propose strategies for successful implementation.

Specific Objective 1: Conduct a literature review on the central concepts addressed in this study using the WOS and Scopus databases.

Specific Objective 2: Determine the potential of blockchain technology in creating opportunities for social and sustainable innovation in tourism.

Specific Objective 3: Propose a general model for implementing blockchain technology in transforming traditional businesses in the tourism industry for the region of Caldas (Colombia).

Methodology

The search equation utilized ("Blockchain technology" AND "tourism industry" OR "tourism sector"). This decision was supported by the equation's ability to cover relevant titles, abstracts, and keywords, resulting in a substantial collection of 2156 unique documents after removing duplicates from Scopus and Web of Science. The timeframe considered encompassed all available years. As a result of this process, a total of 1109 records from Scopus and 1577 records from Web of Science were collected. Subsequently, after removing 530 duplicate records, the corpus was reduced to 2156 unique documents using bibliometrix. Detailed information on the analyzed documents, once the information extracted from the databases was refined, is provided in the subsequent table (Table 1).

Table 1. Distribution of documents and references during the study period

Variable	
Analytical Period	2016–2024
Resources journals, articles, others	878
Documents	2156
Average publications per year	18.85
Average citations per document	5131
References	54917

Sources: Bibliometrix Package

Analyzing the different ways blockchain has to transform business processes was prioritized through network data-based mapping analysis carried out in VOSviewer by Van Eck & Waltman (2014) using the Scopus and Web of Science databases.

2.1 Blockchain in the Tourism Sector: Revolutionizing the Way We Travel

Theoretical Framework. Blockchain is a distributed database of records or shared public/private ledgers of all digital events executed and shared among blockchain participants Crosby et al., (2016). The technological features of blockchain, as described by Chen et al., (2018), include four main characteristics: decentralization (trust is established through mathematical methods rather than a central authority), traceability (each transaction is traceable through block information), immutability (once a transaction is conducted, it cannot be altered). Furthermore, these features result in technological advantages such as reliability, trust, security, and efficiency. Hannes T., et al., (2020).

Every node in the blockchain network serves as a distributed ledger, and the transaction information recorded becomes the distributed information that collectively forms a ledger, which in turn creates a decentralized distributed database. Each node stores accounting transactions. Simultaneously, any node connected to the blockchain network operates independently, holding the same status and viewing the ledger of the entire blockchain network. Liu Y. et al., (2022).

Decentralization is a crucial property of blockchain technology that prevents any tampering with information, therefore enhancing its validity. Deleting collectively maintained records is impractical, and participants can access verified transaction records through shared public or private ledgers Crosby et al. (2016). A centralized database is more vulnerable to hacking, corruption, or breakdowns.

The traceability function allows individuals to access previous data. Concerning products, traceability function assists people in observing how products are transported and produced step by step, along with the quality and brand of the product materials, paraphrasing Liu Y. et al., (2022).

Smart contracts employed by the blockchain network are a set of digitally defined commitments, including an agreement on which contract participants could implement these commitments. It is a contract that utilizes computer language instead of legal language to record the terms and is automatically executed by a program. Liu Y. et al., (2022).

Results

Table 2 provides a hierarchical view of the most cited documents within this bibliometric analysis. This ranking is based on the number of accumulated citations, showcasing the most influential works at the top of the list.

Table 2. Most cited authors

Autor's	Cites
Nakamoto S	451
Saberi	352
Kshetri N	254
Crosby M	230
Swan M	220
Wood G	214
Buterin V	206
Christidis K	199

Sources: Bibliometrix Package

Using the VOSviewer software in Fig. 1, an author coupling analysis has been conducted, a methodology that enables the identification and visualization of prominent collaborations among researchers in the field of study. This analysis process identifies patterns of co-authorship and connections between authors based on the frequency and depth of their collaborations.

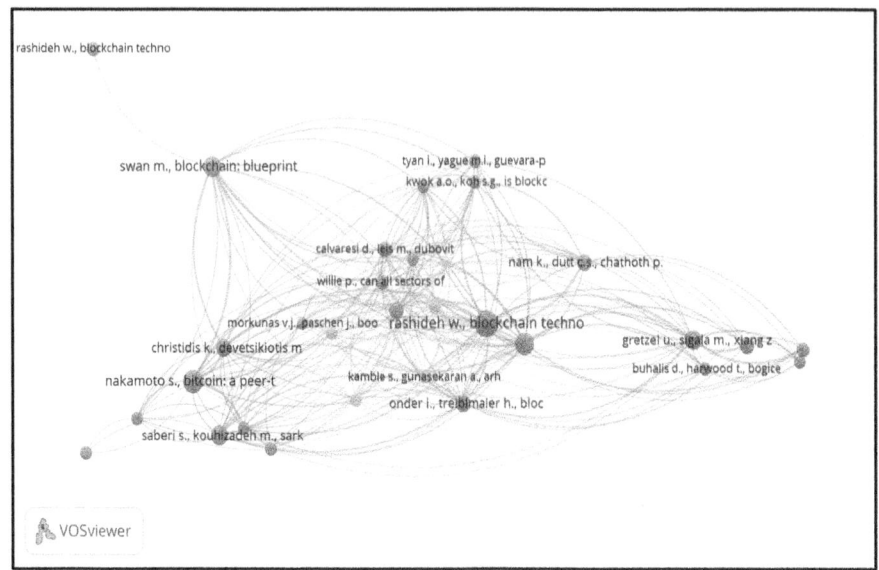

Fig. 1. Author Collaboration/ Own elaboration through Vosviewer.

Table 3 and Fig. 2 present the results obtained using the Bibliometrix package in R and the Vosviewer software (heatmap), highlighting information on the most productive countries in a particular research domain. China leads academic production with a significant total of 408 articles, representing 20.63% of the total contribution. India closely follows with 217 articles, accounting for 10.97%. Despite having fewer articles (153), the United States demonstrates significant impact, with a frequency of 7.74% and the highest ratio of citations per article (MCP_Ratio) among the highlighted countries (0.307).

Table 3. Information on the most productive countries.

Country	Articles	Freq	SCP	MCP	MCP Ratio
1 China	408	0.2063	317	91	0.223
2 India	217	0.1097	171	46	0.212
3 USA	153	0.0774	106	47	0.307
4 Italy	88	0.0445	72	16	0.182
5 United Kingdom	88	0.0045	56	32	0.364
6 Turkey	72	0.0364	43	29	0.403
7 Spain	68	0.0344	58	10	0.147
8 Australia	53	0.0268	31	22	0.415
9 Malaysia	48	0.0243	37	11	0.229
Taiwan	45	0.0228	34	11	0.244

Sources: Bibliometrix Package

Italy and the United Kingdom share the fourth position in terms of frequency (4.45%), but they show notable differences in the MCP_Ratio, highlighting the effectiveness of the publications from the UK with a ratio of 0.364 compared to Italy's (0.182). Turkey, Spain, Australia, Malaysia, and Taiwan complete the list, each with their own contribution to academic production, distinguished by their respective citation per article indexes. These results reveal interesting patterns in productivity and research impact among different countries, offering valuable insights into the geographic distribution of academic production in the analyzed area.

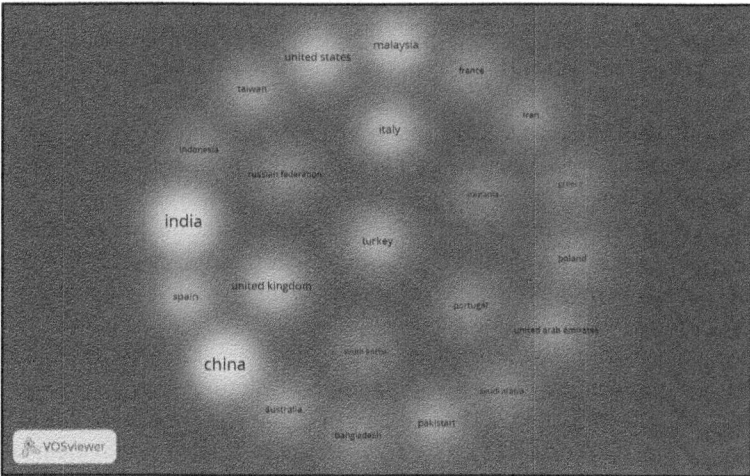

Fig. 2. Heat Map, Most Productive Countries/Source: Own elaboration through Vosviewer.

3 Final Discussion: The Blockchain Revolution in the Tourism Industry

The paradigm of sustainable tourism has transitioned towards smart and sustainable tourism, driven by the increasing need to address sustainability in data-driven tourist destinations. Liu Y, et al. (2022) citing Bibri, (2021). Currently, among industries, tourism leads the investment in blockchain J. Kwok et al. (2018).

Blockchain enables smart tourism with features like digital payments, credential management, inventory management, reservations, and ticketing, alongside the establishment of identity management and loyalty programs. Smart tourism is defined as tourism supported by technological efforts to collect, group, and analyze data in destinations. Liu Y, et al. (2022) citing Gretzel et al. (2015). Blockchain technology sets itself apart from most existing information system designs by incorporating four key characteristics: non-localization, decentralization, security, and auditability. It is considered that Blockchain technologies can be combined with multiple tourism service scenarios, such as reducing the rate of baggage loss, performing automatic compensations, eliminating overbooking of hotels, and allowing potential buyers to communicate directly with hotels and tour operators. (Kshetri Nir, 2018, p.85) Saberi et al. (2019) citing Steiner and Baker (2015).

The rise of blockchain technology could significantly disrupt the global economy. In paraphrasing J. Kwok et al. (2018), it is noted that small island economies are at the forefront of adopting this asset as digital technology. For instance, Caribbean economies are launching their first digital legal tender and Aruba is developing a blockchain platform to increase tourism revenue.

Eliminating commissions through blockchain technology can help reduce overall operating costs. Regarding revenue management, Blockchain presents several unique propositions. An example is provided by "LockChain," aiming to be the first Blockchain-based distribution platform exclusively for the accommodation sector. LockChain rebranded to "LockTrip" in April 2018 and operates with its own cryptocurrency called LOC token Willie (2019), citing LockTrip, (2018).

These significant implications are also echoed by Kizildag M, et al. (2020). When they mention that key applications of blockchain in this industry would relate to direct booking, online booking systems (i.e., airlines and online travel agencies), and digital identity-based check-in/out processes. The emphasis on online booking systems, utilized by airlines and travel agencies, highlights blockchain's ability to enhance transaction security and transparency, establishing a crucial trust framework in a digital environment Raluca-Florentina T., (2022).

The utilization of cryptocurrencies like Bitcoin in international tourism and addressing customer profiles through technology-intensive solutions and innovations will be beneficial for the tourism sector. Establishing a digital money system in tourism aims to resolve issues concerning external payments. Within this context, as payment challenges within the sector are addressed, discussions arise on attracting more foreign tourists and how cryptocurrencies can contribute to the economy Turkay B., (2019).

Personal privacy can only be accessed through the private key, every transaction information stored in the blockchain can be viewed through the public key and the content cannot be altered. To modify the information, more than 51% of all nodes in the network must vote in favor of the modification, which is nearly impossible according to Nakamoto (2008). The framework proposes a decentralized mechanism based on blockchain technology that enables individual travelers to interact with various stakeholders using a single wallet identifier to make payments Pérez-Sánchez, (2021).

Blockchain technology is based on a series of models to study consumer behavior in the online environment, such as the Theory of Reasoned Action, the Theory of Planned Behavior, and the Technology Acceptance Model by Raluca-Florentina T. (2022).

In the final discussion of this article, the results obtained from the viewpoints on blockchain technology proposed by leading researchers in the field are synthesized. This analysis reveals how various perspectives and findings contribute to a deeper understanding of the applications and challenges associated with blockchain.

3.1 Architecture and General Model of Blockchain in Tourism

Reference Architecture

The blockchain reference architecture consists of three different networks that combine and execute the entire blockchain application for users. The three different networks are the public network, the cloud network, and the enterprise network, as shown in Fig. 3. Each one has its own capabilities and functionalities for the proper functioning of decentralized applications.

Public network in this network, users, edge services, and peer cloud providers are connected or linked.

a: users In the public network, users manage the distribution and creation of the decentralized blockchain application and carry out operations with the help of the blockchain network.

b: developer Developers create various types of applications for clients or users with different functionalities. They develop the intelligence.

Blockchain is not just an application but a means to fundamentally transform how data is managed, enhance transparency, and ensure the security of transactions. It provides a robust foundation for creating decentralized solutions that can drive innovation and efficiency across various sectors, including tourism.

The figure presented, corresponding to the blockchain-based architecture, is derived from the study by Bodkhe et al. (2020) titled "Blockchain for Industry 4.0: A Comprehensive Review. This figure is essential for illustrating the structure and functioning of blockchain architecture, providing a fundamental visual foundation for understanding how different networks and components integrate in the implementation of decentralized applications.

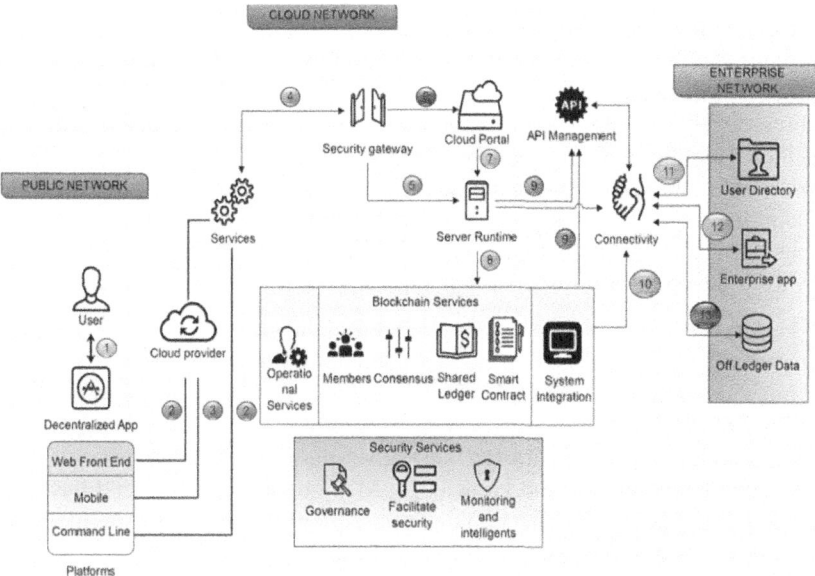

Fig. 3. Reference architecture./Source taken from Blockchain for Industry 4.0: A Comprehensive Review.

Below is an explanation of each component of the architecture:

User: The user is anyone interacting with the decentralized application through their mobile device or other means.

Decentralized Application (dApp): An application running on a decentralized network of blockchain nodes instead of a centralized server. Users access it through various interfaces such as mobile or command lines.

Cloud Provider: The entity offering cloud hosting services for the decentralized application and other architecture components.

Command Line: A user interface enabling direct interaction with the application using text commands.

Services: Different services provided by the architecture, such as storage, transaction processing, etc.

Security Gateway: A component ensuring the security of transactions between the decentralized application and other services like cloud services.

Runtime: The execution environment where decentralized applications and other services are run and managed.

Cloud Portal: A user interface providing access to cloud services and the decentralized application.

Blockchain Services: Services related to blockchain technology, such as member consensus, shared smart ledger contracts, etc.

System Integration: Connecting different systems and services to work efficiently together.

API Management: Tools and processes used to create, manage, and control APIs enabling communication between different architecture components.

Connectivity: It refers to the ability of different components of the architecture to connect with each other and with other external systems.

User Directory: It is a component that stores information about the application users, such as their identities and permissions.

Enterprise Application: It is the main application that uses the reference architecture, designed to meet the specific needs of a company or industry.

Off-Ledger Data: These are data that reside outside the blockchain ledger but are still important for the application, such as backup data, audit logs, etc.

This reference architecture provides a solid framework for developing a new model that integrates with the tourism business value chain model through the implementation of blockchain-based decentralized applications, ensuring security, efficiency, and integration with other systems.

There is a significant demand for innovative platforms in the tourism industry to integrate technology, money, and knowledge. Many companies like TUI have already started using blockchains for ticket booking transactions. Many companies like Expedia, Cheap Air, Webjet, and One-Shot Hotels have begun using bitcoins for their transactions.

Digital currencies seamlessly integrate with smart contracts that have the potential to develop highly disruptive technologies for the tourism industry. Typically, customers turn to online reviews of different tourism products to make a decision.

TUI and many other companies have already begun using blockchain to implement functionalities such as booking tickets and making payments. Companies like Expedia, Cheap Air, Webjet, and One-Shot Hotels utilize bitcoins for travel, that is, for booking tickets. Digital currencies seamlessly integrate with smart contracts that have the potential to develop highly disruptive technologies for the tourism industry. Umesh Bodkhe et al. (2020).

3.2 General Blockchain Model

The general blockchain model for the tourism industry represents an evolution in how travelers, businesses, and destinations interact. By leveraging distributed ledger technology, a more secure, transparent, and efficient approach can be offered to manage the tourism supply chain, service authenticity, booking and payment management, and the creation of personalized travel experiences (Fig. 4).

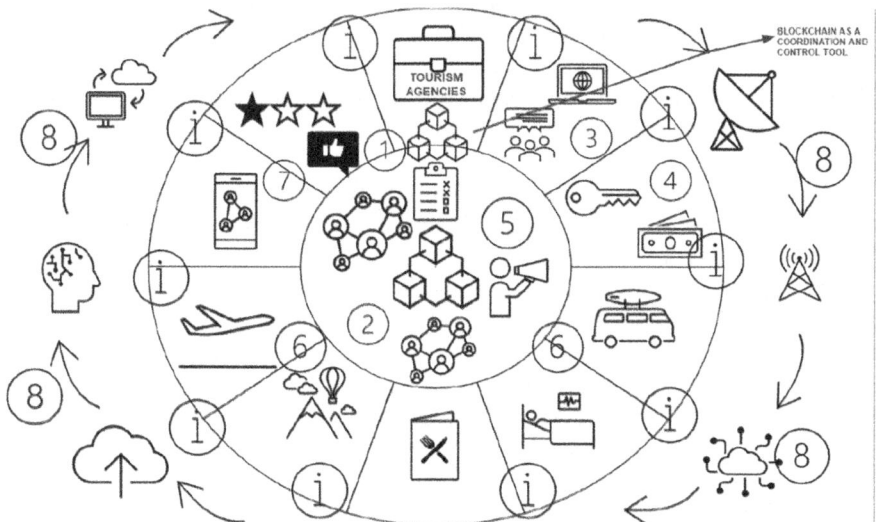

Fig. 4. Blockchain General Tourism Model/Source: Author's own elaboration

The following is a specified general model framework for the tourism industry on the blockchain platform applying the proposed general model: The model's proposed solutions are outlined below, explaining the key components.

1. Data registration and verification in tourism:

    Blockchain Platform: Implementing a specific blockchain platform for the tourism industry where key data related to destinations, hotels, activities, traveler reviews, etc., can be registered and verified. This ensures the authenticity of information and builds trust.

    Smart Contracts: Using smart contracts to establish agreements between different stakeholders in the tourism industry, such as hotels, travel agencies, airlines, and destinations. These contracts can automate bookings, payments, and ensure service quality.

2. Stakeholder management:

In the blockchain tourism industry, it involves coordination and transparency among all stakeholders, including travelers, service providers (hotels, airlines), travel agencies, and regulatory authorities. Blockchain facilitates the authenticity and security of transactions, enhances operational efficiency through smart contracts, and enables traceability of services and payments.

3. Development of tourism products and services:

Tokenization of tourist assets: Destinations can tokenize tourist assets such as properties, monuments, or tourist routes. Travelers can purchase tokens that grant them exclusive access to specific experiences at the destination. Token tourism markets: Establish secondary markets where travelers can buy and sell tourist tokens, facilitating travel planning flexibility and enabling investment in tourist destinations.

4. Financial management perspectives:

Crowdfunding: Utilize blockchain to transparently and decentralize fundraising for tourist infrastructure development. Investors can receive tokens representing their stake in the destination. Revenue management: Implement smart contracts to automatically distribute tourism-generated revenue among different stakeholders such as hotels, tour guides, and local authorities.

5. Destination marketing:

Segmentation based on immutable data: Utilize immutable blockchain data to segment travelers based on their preferences and behaviors, enabling highly targeted marketing campaigns. Token-based loyalty programs: Implement loyalty programs that reward travelers with tokens for their trips and activities at a destination, encouraging repeat visits and recommendations.

6. Customer experience management:

Traveler profiles on the blockchain: Travelers can create traveler profiles on the blockchain storing their preferences and reviews. This allows tourism companies to personalize offers and experiences.

7. Opinions and ratings collection:

Travelers' opinions are recorded immutably on the blockchain, guaranteeing the authenticity of reviews and enabling other travelers to make informed decisions. The introduction of blockchain in the tourism industry sets a precedent in the sector's evolution. In this final section of the research, we will address the core question: To what extent can blockchain technology be used to transform traditional business models and create new opportunities for social and sustainable innovation for the tourism industry in the Caldas region?

The tourism sector involves various stakeholders such as hotels, airlines, travel agents, tour operators, insurance companies, payment service providers, and others. According to Rana, R.L. et al. (2022), leveraging blockchain along with other technologies like ICT, AI, smartphones, and mobile devices can help overcome these complexities and enhance the quality of services offered to customers.

However, the extent of its utilization depends primarily on governments and banks. According to J. Kwok (2018), central banks and governments of major economies have been controversial regarding blockchain regulation, particularly cryptocurrencies. The regulation is mainly influenced by the inability to determine financial risks, liquidity risks, and market risks inherent in a transaction, coupled with challenges related to low internet penetration as noted by Palencia Silva (2023). While infrastructure processes have significantly improved, there is still much progress to be made.

8. Digital Collaboration: The cooperation of integrated communication networks with 5G network, cloud storage, big data, AI, Internet of Things through blockchain as a tool for coordination and control.

Embracing blockchain technology necessitates investment in new hardware and software for data collection. Mention must be made of cybersecurity, which instills confidence in the realm of digital solutions and ensures data integrity.

Currently, the Colombian government is introducing cybersecurity and blockchain courses to train personnel; we are in a learning phase. It can be concluded that the ongoing revolution is blockchain because these blockchains help enhance security. Unfortunately, we are still in experimental stages, as seen in the experiments being conducted by banks, such as Bank de Occident and Group Aval with their digital laboratory, aiming to digitize the Colombian peso in the coming years. This will be an adaptation process that will take some time as people become connected and ready to embrace digital models.

3.3 Conclusion

The analysis of the potential of Blockchain technology to transform traditional business models and generate opportunities for social and sustainable innovation in the tourism industry in Caldas region has been systematically addressed throughout this research. The implementation of blockchain in Colombian tourism not only requires technical knowledge but also a proactive approach to overcoming barriers and managing risks. From financial challenges to market risks in transactions, it is essential that the companies and professionals involved in this transformation understand and address these risks comprehensively.

The research conducted has been exploratory in nature, focusing on analyzing the transformative potential of Blockchain technology in traditional business models and its capacity to generate opportunities for social and sustainable innovation in the tourism industry of the Caldas region. This study has identified both the opportunities and challenges associated with the implementation of Blockchain in the tourism sector and has proposed key strategies for its successful adoption. Through an exploratory approach, a comprehensive view has been provided that supports the proposal of a general model for the integration of Blockchain in tourism, highlighting its relevance for the social and economic transformation of the sector in the region.

The first specific objective, focusing on conducting a literature review on the central concepts addressed in this study, was fully achieved through the comprehensive use of WOS and Scopus databases. Utilizing the R programming language with the bibliometrix package enabled an in-depth bibliometric analysis, offering a solid theoretical foundation to understand the current state of research at the intersection of Blockchain technology and tourism.

The second specific objective, aiming to assess the potential of Blockchain technology in fostering opportunities for social and sustainable innovation in tourism, was thoroughly addressed through the methodology employed. The combination of bibliometric analysis and visualization using Vosviewer helped in pinpointing trends, patterns, and key approaches, laying a strong groundwork to grasp the potential impact of Blockchain on the social and sustainable transformation of the tourism industry.

The third specific objective, which aimed to propose a general model for implementing Blockchain technology in reshaping traditional businesses within the tourism industry in Caldas, was successfully tackled through the methodology applied. Leveraging the R programming language and Vosviewer not only facilitated the identification of key elements of the proposed model but also strategically supported its feasibility and significance. Overall, these methodological tools established a solid and well-founded proposition for the successful integration of Blockchain technology in the tourism context of Caldas region.

The blockchain is revolutionizing the tourism sector, especially in terms of security. Aside from that, there are other significant implications of blockchain technology for tourism, such as enhancing the tourist experience, facilitating fast and hassle-free transactions, and providing financial inclusion for the unbanked. J. Kwok (2018); Li et al., (2019).

These findings suggest that integrating blockchain into tourism not only enhances operational efficiency but also creates new opportunities for developing safer and more personalized travel experiences. Moreover, this shift towards more sustainable and data-driven tourism aligns with the current demands of environmentally conscious travelers and increasingly stringent government regulations regarding sustainability.

A blockchain model for tourism as proposed in this work would support transparency and traceability, tourist empowerment, inclusion and accessibility, environmental sustainability, economic efficiency, collaborative governance, authenticity and cultural value, secure and transparent payments, combating exploitation and fraud, innovation, and the development of new business models.

References

Echeverri-Rubio, A., Flórez-Yepes, G.Y., Vargas-Marin, L.A.: Gobernanza para el desarrollo y la sostenibilidad de los destinos turísticos: una revisión de la literatura con ToS. Revista de Ingenierias Interfaces 5(1), 1–22 (2022)

Kwok, A.O.J., Koh, S.G.M.: Is blockchain technology a watershed for tourism development? Curr. Issues Tourism 22, 2447–2452 (2018). https://doi.org/10.1080/13683500.2018.1513460

Chen, G., Xu, B., Lu, M., Chen, N.: Explorando blockchain y sus aplicaciones potenciales. TIC y desarrollo turístico sostenible. 5(1), 1–10 (2018)

Crosby, M., Pattanayak, P., Verma, S., Kalyanaraman, V.: Blockchain technology: beyond bitcoin. Appl. Innov. (2016)

Kizildag, M., Dogru, T., Zhang, T., et al.: Blockchain: a paradigm shifts in business practices. Int. J. Contemp. Hospitality Manag. 32, 953–975 (2020). https://doi.org/10.1108/IJCHM-12-2018-0958

Kshetri, N.: Blockchain's roles in meeting key supply chain management objectives. Int. J. Inf. Manag. 39, 80–89 (2018). https://doi.org/10.1016/j.ijinfomgt.2017.12.005

Liu, Y., Lace, N., Chen, L.: Application of blockchain technology to smart sustainable tourism destination marketing. Int. Inst. Inform. Cybern. 2, 177–182 (2022). https://doi.org/10.54808/IMCIC2022.02.177

Trujillo, M., Marulanda, C.E.: Turismo transformador Gestión del conocimiento y tecnologías digitales en el turismo. Bogotá, Editorial Universidad Nacional de Colombia (2021)

Nakamoto, S.: Bitcoin: un sistema de efectivo electrónico peer-to-peer, pp. 1–9. Web of Science, Bitcoin.org (2008). https://bitcoin.org/bitcoin.pdf. Accessed 30 Apr 2023

Willie, P.A.: Todos los sectores de la industria hotelera y turística pueden verse influenciados por la innovación de la tecnología Blockchain? Emerald Publishing Limited, vol. 11, 112/120 (2019). https://doi.org/10.1108/WHATT-11-2018-0077

Palencia Silva, D.: Occiverso Capitulo 7: Criptoactivos [Linkedin] (2023). https://www.linkedin. com/posts/banco-de-occidente_occivernauta-criptomonedas-podcast-activity-712266712358 7325955-dMDr?utm_source=share&utm_medium=member_android

Pérez-Sánchez, M.Á., Barrientos-Báez, A.: Blockchain technology for winning consumer loyalty: Social norm analysis using structural equation modeling. Mathematics 9, 1–18 (2021)

Rana, R., Adamashvili, N., Tricase, C.: The impact of blockchain technology adoption on tourism industry: a systematic literature review. Sustainability 14(12), 7383 (2022). https://doi.org/10. 3390/su14127383

Raluca-Florentina, T.: The utility of blockchain technology in the electronic commerce of tourism services: an exploratory study on Romanian consumers. Sustainability (Switzerland) 14 (2022)

Saberi, S., Kouhizadeh, M., Sarkis, J., Shen, L.: Blockchain technology and its relationships to sustainable supply chain management. Int. J. Prod. Res. 57(7), 2117–2135 (2018). https://doi. org/10.1080/00207543.2018.1533261

Turkay, B., Dıncer, F.I., Dincer, M.Z.: An evaluation of new values in economy and their impacts on future transformation in tourism. Procedia Comput. Sci. 158, 1095–1102 (2019). https:// doi.org/10.1016/j.procs.2019.09.151

Tyan, I., Yagüe, M.I., Guevara-Plaza, A.: Blockchain technology's potential for sustainable tourism. In: Information and Communication Technologies in Tourism 2021: Proceedings of the ENTER 2021 eTourism Conference, 19–22 January 2021, pp. 17–29. Springer (2021)

Bodkhe, U., et al.: Blockchain para la industria 4.0 una revisón completa. ALGORITMOS PARA INTERNET DE COSAS MÉDICAS (2020). https://doi.org/10.1109/ACCESS.2020.2988579

Van Eck, N.J., Waltman, L.: Software survey: VOSviewer, a computer program for bibliometric mapping. Scientometrics 84(2), 523–538 (2014)

https://www.linkedin.com/posts/banco-de-occidente_occivernauta-criptomonedas-podcast-act ivity-7122667123587325955-dMDr?utm_source=share&utm_medium=member_android

Vulnerability Analysis of IoT Devices (IP Cameras) in Ocaña Norte de Santander

Luis A. Coronel-Rojas⬛, Yesenia Areniz-Arévalo⬛, Fabián R. Cuesta-Quintero⬛, and Dewar Rico-Bautista(✉) ⬛

Universidad Francisco de Paula Santander, Ocaña, Colombia

{lacoronelr,yareniza,fcuestaq,dwricob}@ufpso.edu.co

Abstract. The objective of this research is to identify and analyze IOT devices connected to the internet, case study IP cameras in the city of Ocaña Norte de Santander. For the development of the study and to avoid the violation of unauthorized IP cameras, a test laboratory was established in collaboration with the company Ctraker GPS specializes in the installation of such devices in the city. The research was divided into three phases: first, an exhaustive diagnosis of the situation of IP cameras in the region, using specialized search tools, allowing the characterization of each device according to technical aspects. Subsequently, an analysis of the security vulnerabilities identified in the IP cameras, considering their brands and versions, to understand the weak points of each device. Finally, recommendations were made to strengthen the security of IP cameras and reduce the risk to the privacy and security of individuals and organizations.

Keywords: IoT · IP Cameras · Shodan · Vulnerability

1 Introduction

Searching for connected devices on the Internet and accessing data related to their identity has become a reality using IOT search engines. There are different engines, including Shodan, Censys and Thingful. These engines are valuable tools that aim to solve the Internet of Things search problem [1], chose to work on research with Shodan, as it has been recognized as one of the most popular web crawlers available today, intended to crawl the Internet and index discovered services, with the ability to search for devices, available services and store the collected data, has a graphical user interface that provides a huge amount of data, using a variety of filters [2], that can be used in useful as well as dangerous ways.

Shodan, which was launched in 2009 by John Matherly [3, 4], this tool can identify different devices with IP addresses accessible among those IP cameras, which have their IP address assigned and are connected to a network and can be installed anywhere in the network scheme to which it belongs, the main function is to monitor and monitor places in which permanent control is required either entire cities, businesses or homes, through commands and different filters suitable for the search in shodan, the necessary information for the study is obtained [5, 6].

N. D. Duque-Méndez et al. (Eds.): CCC 2024, CCIS 2208, pp. 328–343, 2024.
https://doi.org/10.1007/978-3-031-75233-9_23

Similarly, the Shodan computer search engine has received much attention for its ability to identify and index components of industrial control systems connected to the Internet, this study investigates the functionality of the computer search engine that provides attackers with a powerful reconnaissance tool for attacking industrial control systems [7, 8], the Shodan search engine has become one of the favorite tools of attackers and penetration testers [9, 10].

Video surveillance, CCTV, and IP camera systems have become virtually ubiquitous and indispensable for many organizations. In this article, a systematic review of existing and new threats in video surveillance, CCTV and IP camera systems based on publicly available data was conducted [5, 8].

The Cyber Threats Report 2022, presented by SonicWall, states that Colombia is in the top 10 of the most attacked countries, having more than eleven million threats in 2021, ranking 7th in total volume of ransomware [10]. In the same way [11, 12], in their study, based on information gathered from 4 IP cameras, they were able to identify frequent vulnerabilities related to authentication of factory default passwords, weaknesses in password complexity, failures in access controls to the real-time video display interface, code injection, and firmware updates to correct vulnerabilities.

On the other hand, Semana Magazine describes how Shodan works and how it can be used to discover vulnerable systems and devices on the Internet [2]; In the same article, there is an excerpt on cybersecurity in Colombia, with the following question: "Are we prepared to face the risks?" where it mentions the importance of cybersecurity in our country and the use of tools such as Shodan to identify vulnerabilities in systems and devices connected to the Internet. It is important to keep in mind that the use of Shodan in Colombia, as elsewhere, must be done legally and ethically, respecting the privacy and security of IoT systems and devices that are being explored.

In Colombia, no law prohibits or prevents the installation of surveillance cameras, although of course, there is a law that must be taken into account and it is the Habeas Data Law (Law 1581 of 2012, Law of Treatment and Protection of data), likewise, the Colombian Constitutional Court, in the sentence 406 of November 22, 2022, expresses regarding the access of the National Police of Colombia to private security systems for "Prevention" actions as a reason described in the Citizen Security Law (Law 2197 of 2022), for such reason, the study covers the search of IP cameras, performing vulnerability discovery.

At the local level, no relevant studies have been conducted with tools for searching for devices connected to the Internet through an IP address, hence the importance of developing research in this region that is growing technologically. Ocaña is a city of approximately 130,000 inhabitants and has few companies, but it is advancing in the implementation of IoT devices and web services. Currently in the city, the installation of IP cameras has been increasing in homes, streets, clinics, businesses and/or schools, where they have seen the need to be protected and keep their properties monitored, likewise the City Council, opted for the installation of 104 IP cameras, of which, to date, are fully operational 30. These were installed in order to counter common crime in the city. [8].

2 Methodology

For the development of this study, we chose to work with projective research, this type of qualitative research is associated with the development of a model, plan, and proposal as a solution to a problem detected by the researcher, the vision of projective research is oriented to find practical solutions to a research purpose; proposals are generated that provide practical answers in specific contexts. The application of a methodology based on projective research or feasible project "consists of the research, elaboration, and development of a proposal for a viable operational model to solve problems, requirements or needs of organizations or social groups..." [13, 14].

Likewise, states that projective research is given as a solution to a practical problem or need, either of an institution or in a particular area of knowledge, based on a precise diagnosis of the needs of the moment, of the explanatory processes and the future trends [15, 16].

According to the above, the study works projective research, as it seeks to propose in the following research a security guide for the mitigation of vulnerabilities in IP cameras in Ocaña, Norte de Santander, this guide will contain good practices with recommendations to improve the security of IP cameras and reduce the risk of compromising the privacy and security of individuals and organizations that use them, by a technical expert.

3 Results

3.1 Device Identification

To obtain the results, we start with the identification of the IP cameras connected to the Internet, using the Shodan and Shodan wave search engines. It is important to point out that it is necessary to be registered in this platform-form to access to higher privileges and thus optimize the search process. Shodan is a tool that operates using filters; in our case, to focus on the area of Ocaña, located in Norte de Santander, a filter "camera country: 'CO' city: 'Ocaña'" was used. This resulted in the location of 13 cameras in Ocaña, see Fig. 1. Since the necessary authorizations were not available to carry out tests on these identified cameras, a laboratory environment was set up within the Ctracker GPS company. In this controlled environment, the relevant permissions and authorizations were obtained to carry out an exhaustive analysis of the IP cameras in question.

Within the assembly of the laboratory for the tests, a public address is enabled to search for IP cameras in the city of Ocaña, Norte de Santander, which are in the network with prior authorization for the investigation, the public address assigned is 190.90.**.**, remember that by security policies the addresses cannot be visible in this document, because they are IP addresses that are currently being used in the IP cameras installed in Ocaña Norte de Santander and are in operation.

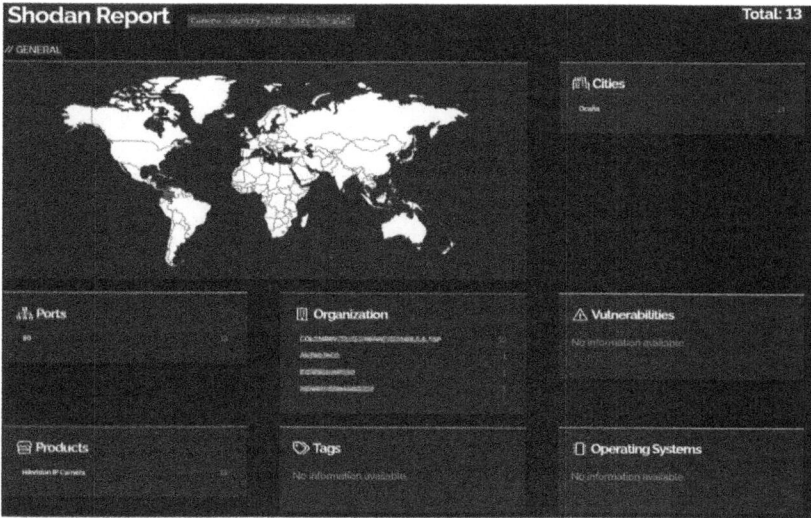

Fig. 1. Search IP cameras Shodan tool.

Once the public address was found with the different tools being used for the research, such as Shodan and Shodan wave, different analyses could be made. A search of the public address is made to find the exact location with the use of the internet, the public address is in the city of Ocaña, Norte de Santander, see Fig. 2.

Fig. 2. Location public IP address.

After a process of corroborating the authenticity of the public address, a thorough scan was performed on the machine in question to identify the IP addresses associated with that public address. During this analysis, multiple addresses assigned to servers were detected, as well as addresses linked to uninterruptible power supplies (UPS). In addition, a total of 23 IP addresses destined for IP cameras were identified. This procedure not only confirms the connection but also provides valuable information such as MAC addresses, manufacturer names, and the number of associated ports. This data adds a significant level of detail for future consideration, see Fig. 3.

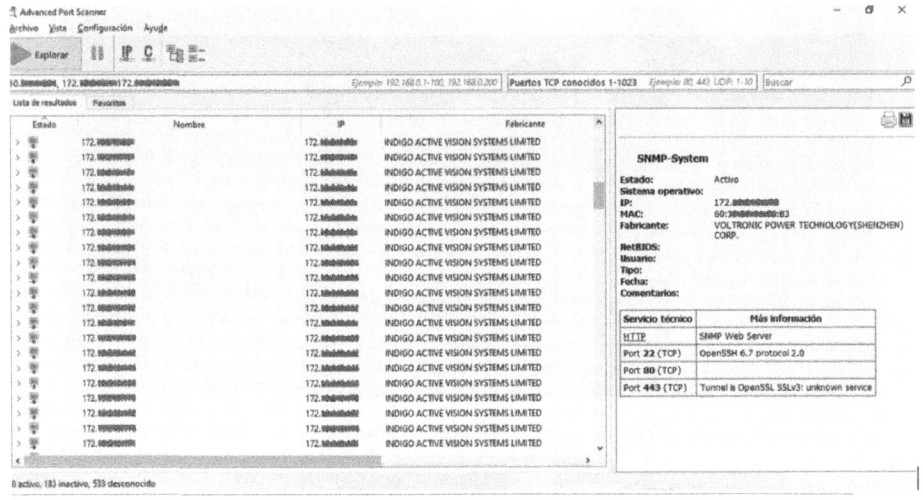

Fig. 3. IP cameras found.

Once the presence of Indigo Active Vision Systems Limited cameras, which the company has installed and configured, is confirmed, the first step is to search the Internet for the possible default passwords associated with these devices and try to access the operator's control panel, the default credentials for Indigo cameras are as follows: User = "Admin" and Password = "1234", see Fig. 4.

HIKVision	admin	12345
Honeywell	administrator	1234
Honeywell	admin	1234
IndigoVision BX/GX	Admin	1234
IndigoVision Ultra	<blank>	<blank>
Intellio	admin	admin

Fig. 4. Default passwords of IP cameras published on the Internet.

The default passwords are entered, and the authorization is obtained by identifying the exact location of each of the installed cameras. As shown in Fig. 5, there is a distribution of cameras by sectors named Branch 1, Branch 2, Branch 4, Branch 6, Branch 7, Branch 8, Branch 5, and Branch 3, by clicking on each group, it shows the exact location and number of cameras in the sector and described as CAM59, Tierra Santa, CAM 56, 26 de julio, among others.

Fig. 5. Cameras sector Branch 3.

In Fig. 6, the arrangement of the cameras in the two sectors is clearly shown. This representation provides a fundamental information base to start the process of analysis and search for possible vulnerabilities.

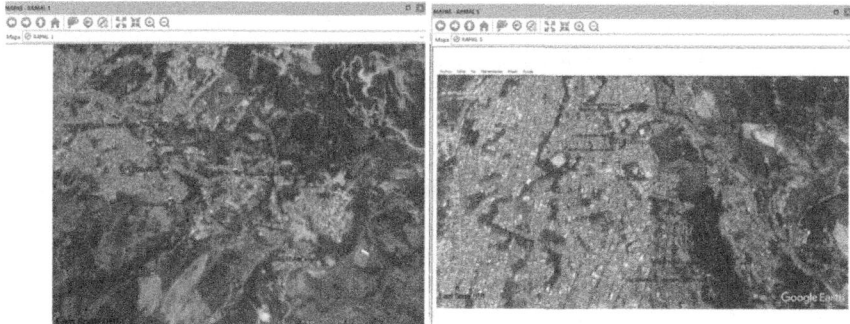

Fig. 6. Cameras sector Branch 1 and Branch 5.

Table 1 shows the distribution of IP addresses assigned to Sector Branch 1 and Branch 5, where 7 IP cameras are installed in each sector, for a total of 14 IP cameras.

Table 2 below details the characteristics of the cameras identified. It is important to note that, due to the nature of this ongoing research and its focus on a single company within a laboratory setting, only one brand of cameras has been examined. If this study were to be conducted on a broader spectrum, covering several organizations in Ocaña, Norte de Santander, a greater number of cameras with different specifications and different vulnerabilities could be discovered.

Table 1. Cameras assigned to Branch 1 and Branch 5.

Sector branch 1		Sector branch 5	
IP Address	Assigned Name	IP Address	Assigned Name
172.*.*.171	CAM 13 Entrada viacrucis cristo rey	172.*.*.40	CAM 52 26 de julio
172.*.*.176	CAM 36 Parque san Agustín	172.*.*.45	CAM 26 Ramal
172.*.*.172	CAM 73 la laguna	172.*.*.44	CAM 32 La Y Anasca
172.*.*.173	CAM 38 la rotina	172.*.*.35	CAM 32 Terminal
172.*.*.158	CAM 22 20 de julio	172.*.*.34	CAM 15 Promesa de dios
172.*.*.146	CAM 42 Canal	172.*.*.33	CAM 17 Acolsure
172.*.*.142	CAM 33 Coliseo Argelino	172.*.*.52	CAM 59 Entrada tierra santa

Table 2. Brand and version of IP cameras found.

Device found	Brand	version
Webcam	Indigo Active Vision Systems Limited	The Control Center v19.1 Ultra X PTZ HD Dome Camera

Once the search for IP cameras was completed using the Shodan and Shodanwave tools at the public address 190.*.*.*, a network of IP cameras distributed at various points in the city of Ocaña, in Norte de Santander, was identified. In total, 23 IP cameras were found installed and configured in the area.

During the scanning process, valuable information was obtained from each of these cameras, including the assigned IP address, the MAC address of the device, and the name of the manufacturer, in this case, Indigo Active Vision Systems limited version, the Control Center v19.1 Dome Camera HD Ultra X PTZ, also important information was collected about the technical service, including the protocol used, the assigned ports and a description of them.

3.2 Vulnerability Analysis

In addition, it was noted that the number of cameras associated with each sector varied according to specific needs. In this case, it was distributed in eight different sectors: branch 1, branch 2, branch 3, branch 4, branch 5, branch 6, branch 7 and branch 8. Each camera was identified by a name that followed a particular structure: the three initials "CAM," followed by a specific number and ending with the name of the corresponding location, examples of these names assigned "CAM 59 Tierra Santa" and "CAM 56 26 de Julio," among others.

To perform an analysis of the types of security vulnerabilities found in IP cameras according to their brands and versions, we proceed to work with Kali Linux to make a deeper scan, before performing the scan, from Kali Linux we prepare the environment to test the cameras, for this purpose we create two text files called USER and PASS, see Fig. 7.

USER: Contains information on possible user account names that a device may have to access.

PASS: Contains information on possible user passwords that a device may have to access.

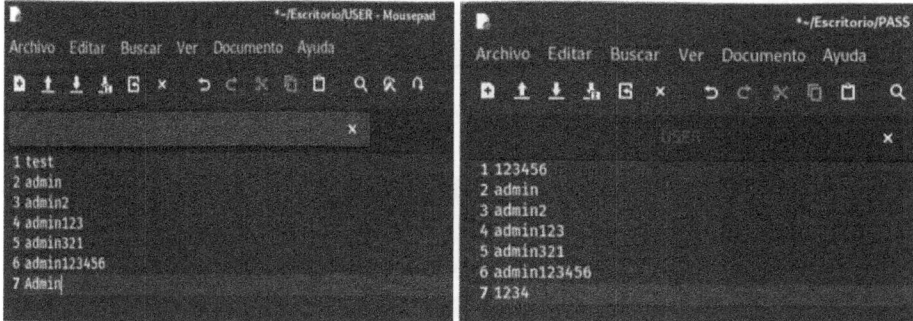

Fig. 7. Possible users and passwords for Indigo IP cameras.

In this test we work with 7 different users and passwords, the shodanwave tool will perform all the login attempts in the scanned cameras (brute force attack) until it obtains the device password, if its access credentials are in the combination of users and passwords mentioned above. Likewise, so that the Shodan wave tool can perform the scanning, it is necessary to link it with Shodan, this is done through the API Key, the API Key is obtained from the Shodan page. With the API Key see Fig. 8 and having read the text files USER and PASS we proceed to perform the scanning of cameras with the following command see Fig. 9.

Fig. 8. API Key page Shodan.

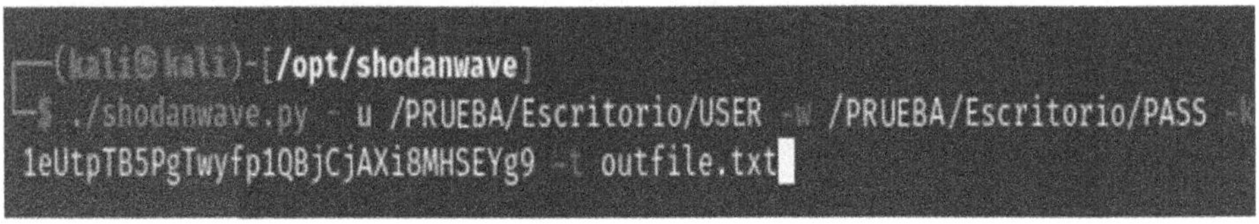

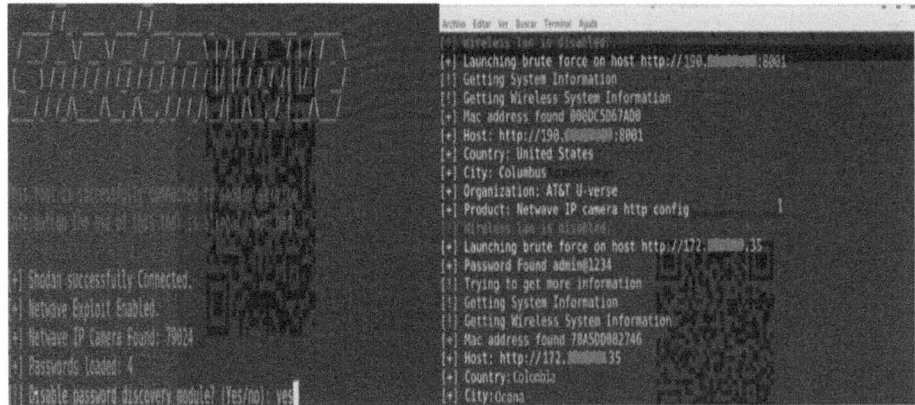

Fig. 9. IP camera scan command.

With the above command, shodan starts running, first, it asks if you want to disable the password discovery mode, in this case, it is important to give yes, so that the test is more effective Fig. 10.

Fig. 10. Shodanwave execution and password found by the shodanwave tool with the address 172.*.*.35.

Subsequently, we proceed to access the IP address of the camera through our browser and log in with the credentials obtained through shodanwave Fig. 11.

Fig. 11. Login with passwords Ip camera with address 172.*.*.35.

As shown in Fig. 11, you can access it with the credentials obtained Username = admin and Password = 1234, in this way we have control of the Indigo Vision Ultra center that provides us with the 23 cameras that are installed on the network, from here you can enter each camera and perform different activities such as manipulate and review the information according to its configuration see Fig. 12.

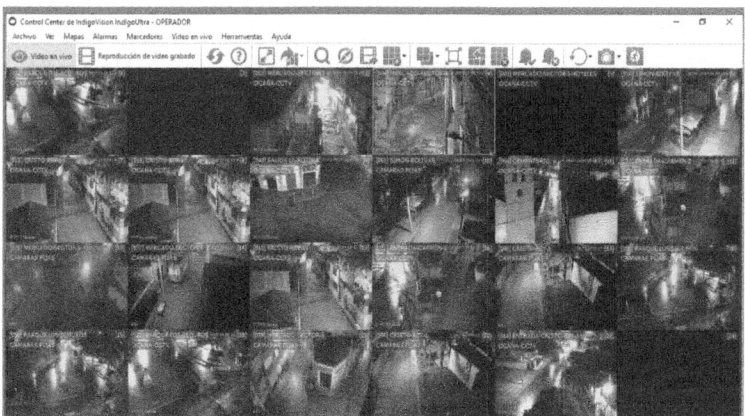

Fig. 12. Indigo Vision Ultra center control input.

To continue with the analysis and look for more options to have control of the installed IP cameras another test is performed with another IP address, in this case, once the other IP address, the following information was identified: Active Status, IP 172.*.*.52, Mac Address 00:90:**:**:**:65:97, manufacturer: indigo active Vision Systems limit, webcam type and active services Http Dahua webcam httpd, ports 80 TCP/port 554 TCP see Fig. 12. With this relevant information, you can get a search from the internet and investigate possible passwords, but as already with the shodan wave was identified, we proceed to load the browser and try to enter. Achieving access to the IP camera once in the configuration could make any type of modification see Fig. 13 can be seen that you have control to move the camera, this camera is mobile, in the cameras found there are mobile cameras and fixed cameras (Fig. 14).

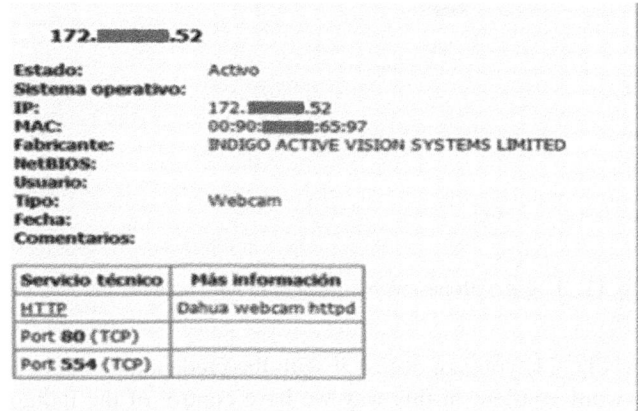

Fig. 13. Information found IP camera.

Fig. 14. Access to IP camera in the insurance sector.

3.3 Specific Findings

The analysis of the security vulnerabilities present in the IP cameras was carried out, taking into account the brand and version of the cameras, it is important to note that this research was developed in a single environment, and was carried out with the proper authorization of the company Ctracker GPS, therefore, it is relevant to note that the only cameras and brands that were included in the study were the following:

Brand: Indigo Active Vision Systems Limited
Version: Control Center v19.1
Camera model: Domo HD Ultra X PTZ

These were the specific devices that were subjected to analysis and evaluation in the context of the research. To obtain the results of this objective, a local laboratory was implemented using the Shodan wave tool from Shodan, so to start the analysis of the information from each IP camera, it was necessary to create a denoted text file "USER AND PASS". In the "user" field, the possible usernames were listed, while in "pass" the possible passwords used were included. In this case, after conducting an Internet search for the default passwords for the Indigo Active Vision Systems Limited IP cameras, it was discovered that the default credentials for the Indigo cameras are as follows: User = "Admin" and Password = "1234". This information is added to the "USER AND PASS" file.

After carrying out the password attack process, it was identified that several IP cameras installed and configured in each sector still maintained the default credentials. This allowed the researchers of this project to easily access the cameras, gaining full control and access to functions such as LIVE, Playback, and Setup, as well as the complete camera menu, which includes options such as Camera, IVS Setup, Network, Event, Storage, System, and Information.

Similarly, access was gained to the IP camera Control Center, and from the Control Center, direct access was obtained to the 23 Indigo IP cameras, which allowed any type of configuration or modification, it is important to note that, for this investigation, it was expressly stated in the authorization letter that no modification would be made to the configuration of the cameras.

A test was performed with a specific IP address 172.*.*.52 of an IP camera, which was assigned to the camera named CAM 59 ENTRADA TIERRA SANTA, located in sector branch 1, it was confirmed that this camera also had the default credentials, which allowed accessing and obtaining control and information, the results of this scan revealed that the camera was in active status, had the MAC address 00:90:**:**:**:65:97 and was manufactured by Indigo Active Vision Systems Limited. It was classified as a webcam-type camera, with the following services active: Dahua HTTP, Webcam HTTPD, port 80 TCP, and port 554 TCP.

An attempt was made to enter another IP address 172.**.**.131, this address assigned to the IP camera CAM 59 ENTRADA TIERRA SANTA, an attempt was made to enter with the default credentials, showing that this camera had changed the password, for this reason, an attempt was made to search for the possible password and it was not possible to find it according to the laboratory created in this research.

In terms of security, it was identified that some of the cameras used information encryption, which means that if an attacker were to enter, he could see what the camera was transmitting in real time but could not access the encrypted information stored on the hard disk. Indigo cameras currently have a high degree of protection against cybercriminals, especially when properly configured.

It was noted that these cameras have a system called "Cybervigilante" that provides constant monitoring to detect network anomalies, including unauthorized configuration attempts, recognition of network scans, access attempts from unauthorized clients, shell access attempts, use of unauthorized DNS, NTP, and SMTP servers, and the most common attack: denial of service. If a company uses this type of cybersecurity system, it is protecting its system from malicious cyber-attacks by automatically detecting threats and taking quick and decisive action.

According to the above, some of the vulnerabilities found in the cameras installed by the company were identified as shown in Table 3.

Table 3. Vulnerabilities found in IP cameras.

cameras	vulnerabilities
Indigo Active Vision Systems Limited	Default passwords Weak passwords Common passwords
Server	Assigned the first address

4   Discussion

Nozomi Networks Labs, who are dedicated to reducing cyber risk to the world's critical infrastructure and industrial organizations, conducted extensive security research on multiple IP camera and video surveillance systems [17]. This research led to the discovery of vulnerabilities in Axis and ThroughTek devices, a new vulnerability CVE-2022-30563, which affects the implementation of the WS-UsernameToken, the Open Network Video Interface Forum (ONVIF) authentication mechanism, in some IP cameras developed by Dahua, a major IP camera manufacturer, was discovered. Attackers could abuse this vulnerability to compromise network cameras by detecting a previous unencrypted ONVIF interaction and replaying the credentials in a new request to the camera [18, 19].

Similarly, in a graduate work developed at the Catholic University of Colombia, on the "Identification of information security risks in domestic surveillance cameras in IoT environments", the authors initially describe the problem, which is the insecurity in the country and the use and installation of security cameras as a measure to add another layer of protection through Shodan. A scan was performed, identifying the top cities in Colombia that use these security systems (Bogota, Medellin), then they were able to establish a connection with a camera (from the city of Medellin - Colombia), thus obtaining a live view and making some adjustments to the camera.

On the other hand, in the present investigation, in terms of security, it was identified that some of the cameras used encryption of the information, which means that if an attacker were to enter, he could see what the camera was transmitting in real time but could not access the encrypted information stored on the hard disk. Indigo cameras currently have a high degree of protection against cybercriminals, especially when properly configured [20–22].

It was noted that these cameras have a system called "Cybervigilante" that provides constant monitoring to detect network anomalies, including unauthorized configuration attempts, recognition of network scans, access attempts from unauthorized clients, shell access attempts, use of unauthorized DNS, NTP and SMTP servers, and the most common attack: denial of service. If a company uses this type of cybersecurity system, it is protecting its system from malicious cyber-attacks by automatically detecting threats and taking quick and decisive action [23].

5   Conclusions

The process of diagnosing the current situation of IP cameras through search tools and characterizing them according to their technical aspects is fundamental to understanding and effectively managing the security infrastructure based on IP cameras. Through this analysis, it was possible to identify the available cameras, their technical characteristics, and their operational status. This provided a comprehensive view of the IP camera network, facilitating informed decisions on maintenance, upgrades, and optimization. In addition, this process ensured optimal performance and robust security, which is essential in today's surveillance and monitoring applications.

Ultimately, diagnosing and characterizing IP cameras is a crucial step in managing security and technology infrastructure. When a proper configuration is not applied, considering all the important aspects, from the first change of passwords, it causes easy-to-detect vulnerabilities, just by looking for the manuals of the cameras where the default credentials are found, putting at risk the confidentiality, integrity, and availability of the captured data, when the default credentials are left. Any device connected to the internet and left with default passwords for accessing it, shows a worrying lack of security awareness in the configuration. This not only puts privacy and data security at risk but can also open the door to potential intrusions and security breaches. Immediate steps must be taken to change these default passwords and adopt robust security practices to protect network integrity and sensitive data.

The analysis also highlights the importance of keeping IP camera firmware up to date and following security best practices to mitigate potential risks. In addition, awareness of specific vulnerabilities associated with versions can help security professionals make more informed decisions when selecting and configuring IP cameras in their systems. Today, all IoT devices connected to the Internet are an integral part of our daily lives, from smart homes to critical infrastructures, so manufacturers, users, and companies must work together to establish robust security standards, promote cybersecurity education, and encourage the adoption of good security practices at all levels. This study highlights the continued need to investigate and address security vulnerabilities in IP cameras to ensure effective protection of information and infrastructure in modern surveillance and security environments.

References

1. Fagroud, F.Z., Ajallouda, L., Ben Lahmar, E.H., Toumi, H., Achtaich, K., El, F.S.: IOT search engines: exploratory data analysis. Procedia Comput. Sci. **175**, 572–577 (2020)
2. Chen, Y., Lian, X., Yu, D., Lv, S., Hao, S., Ma, Y.: Exploring Shodan from the perspective of industrial control systems. IEEE Access **8**, 75359–75369 (2020)
3. Hu, Z., Beuran, R., Tan, Y.: Automated penetration testing using deep reinforcement learning. In: 2020 IEEE European Symposium on Security and Privacy Workshops (EuroS&PW), pp. 2–10. IEEE (2020)
4. Vlajic, N., Zhou, D.: IoT as a land of opportunity for DDoS hackers. Computer (Long Beach Calif) **51**, 26–34 (2018)
5. Samtani, S., Yu, S., Zhu, H., Patton, M., Matherly, J., Chen, H.: Identifying SCADA systems and their vulnerabilities on the Internet of Things: a text-mining approach. IEEE Intell. Syst. **33**, 63–73 (2018)
6. Samtani, S., Yu, S., Zhu, H., Patton, M., Chen, H.: Identifying SCADA vulnerabilities using passive and active vulnerability assessment techniques. In: 2016 IEEE Conference on Intelligence and Security Informatics (ISI), pp. 25–30. IEEE (2016)
7. Beron, J., Benitez-Restrepo, H.D., Bovik, A.C.: Blind image quality assessment for super resolution via optimal feature selection. IEEE Access **8**, 143201–143218 (2020)
8. Gomez-Nieto, R., Ruiz-Muñoz, J.F., Beron, J., Franco, C.A.A., Benítez-Restrepo, H.D., Bovik, A.C.: Quality aware features for performance prediction and time reduction in video object tracking. IEEE Access **10**, 13290–13310 (2022)
9. Mosenia, A., Jha, N.K.: A comprehensive study of security of Internet-of-Things. IEEE Trans. Emerg. Top. Comput. **5**, 586–602 (2017)
10. Humayed, A., Lin, J., Li, F., Luo, B.: Cyber-physical systems security—a survey. IEEE Internet Things J. **4**, 1802–1831 (2017)
11. Yang, J., Zou, H., Jiang, H., Xie, L.: Device-free occupant activity sensing using WiFi-enabled IoT devices for smart homes. IEEE Internet Things J. **5**, 3991–4002 (2018)
12. Minoli, D., Sohraby, K., Occhiogrosso, B.: IoT considerations, requirements, and architectures for smart buildings—energy optimization and next-generation building management systems. IEEE Internet Things J. **4**, 269–283 (2017)
13. Rivas, L.: Elaboración de Tesis (Estructura y metodologia). Revista Trillas **1**, 1–384 (2016)
14. Salazar, F., Torres, M.: Métodos de recolección de datos para una investigación. Revista Indian J. Dental Res. **27**, 21 (2016)
15. Danel, O.: Metodologia de la investigación. Técnicas e instrumentos (2015)
16. Gómez, M.: Introducción a la metodología de la investigación científica, 1st edn. Argentina (2015)
17. Lopes, S.F., Silva, S., Mendes, J., Metrolho, J.C., Duque, D.: Development of a library for clients of ONVIF video cameras: challenges and solutions. In: 2013 IEEE International Conference on Industrial Technology (ICIT), pp. 1260–1266. IEEE (2013)
18. Abdalla, P.A., Varol, C.: Testing IoT security: the case study of an IP camera. In: 2020 8th International Symposium on Digital Forensics and Security (ISDFS), pp. 1–5. IEEE (2020)
19. Senst, T., Patzold, M., Evangelio, R.H., Eiselein, V., Keller, I., Sikora, T.: On building decentralized wide-area surveillance networks based on ONVIF. In: 2011 8th IEEE International Conference on Advanced Video and Signal Based Surveillance (AVSS), pp. 420–423. IEEE (2011)
20. Davy, D., Ashley, S., Davison, B., Ashcroft, A., McEwen, R.K., Moore, R.: A long-range camera based on an HD MCT array of 12μm pixels. In: Andresen, B.F., Fulop, G.F., Hanson, C.M., Norton, P.R. (eds), p. 90700G (2014)

21. Arroyo, S., Garcia, L., Safar, F., Oliva, D.: Urban dual mode video detection system based on fisheye and PTZ cameras. IEEE Lat. Am. Trans. **19**, 1537–1545 (2021)
22. Lin, C.-F., Chiao, H.-T., Sheu, R.-K., Chang, Y.-S., Yuan, S.-M.: A fault-tolerant ONVIF protocol extension for seamless surveillance video stream recording. Comput. Stand. Interfaces **55**, 55–72 (2018)
23. Lobo, J., Dewar, R.-B.: Implementación de la seguridad del protocolo de internet versión 6. Gerencia tecnológica informática **11**, 35–46 (2012)

Safeguarding Connected Autonomous Vehicles: A Cybersecurity Perspective

Cesar R. Beltrán-Hernández$^{(\boxtimes)}$ ⓘ, Rafael V. Páez-Méndez ⓘ, and Luisa F. Amaya ⓘ

Pontificia Universidad Javeriana, Bogotá, Colombia
cesarr_beltran@javeriana.edu.co

Abstract. This article analyzes the cybersecurity challenges of connected autonomous vehicles, emphasizing critical components such as sensors and communication technologies. The work proposes comprehensive security strategies to protect drivers, pedestrians, and the road environment, focusing on LTE and 5G reliance and concerns such as integrity, availability, and confidentiality attacks. Strategies like cryptography, redundancy, and network monitoring are proposed, alongside countermeasures for GPS, LiDAR, and network attacks. The results contribute significantly to the development of security standards for autonomous vehicles and offer practical guidelines for the automotive industry. Finally, it concludes by emphasizing the role of regulatory standards such as SAE J3016 and ISO/PAS 21448 in ensuring safety in autonomous driving. Through a thorough analysis of risks and countermeasures, this study aims to enhance safety in autonomous driving systems and expand knowledge in vehicle security.

Keywords: AV Attacks · AV safety · Cybersecurity in autonomous vehicles · In-vehicle network security · Intelligent Transportation Systems (ITSs)

1 Introduction

New technologies, including Machine Learning and Big Data, are revolutionizing multiple industries. One of the most affected is the automotive industry, especially with the rise of autonomous vehicles.

Nevertheless, the widespread adoption of the "autonomous vehicle" concept without considering the different levels of autonomy can pose security risks due to a lack of understanding of the vehicle's actual capabilities.

Given the rapid advancements in autonomous vehicle technology and their expanding capabilities, it is essential to guarantee the safety of both drivers and pedestrians, along with secure communications, systems, sensors, and data processed by these vehicles. These vehicles are exposed to various cyber-attacks that could trigger incidents related to both information security and road safety, potentially resulting in loss of human lives or harm to people's health. Authors in [1] and [2] highlight the serious consequences of security attacks on autonomous vehicles.

Various authors emphasize the importance of implementing methodologies to assess the weaknesses and impacts of possible attacks. Vitale et al. [3] proposes adopting established methodologies from the field of information and communication technologies.

N. D. Duque-Méndez et al. (Eds.): CCC 2024, CCIS 2208, pp. 344–359, 2024.
https://doi.org/10.1007/978-3-031-75233-9_24

Similarly, other authors in [4] and [5] present methodologies for managing cybersecurity risks in connected vehicles. However, it is crucial to clearly understand the vulnerabilities and different attack vectors to ensure comprehensive security measures.

This article breaks down the fundamental components of autonomous vehicles (AVs) and examines the main types of risks they pose, along with the countermeasures proposed by various authors based on a literature review. The study seeks to contribute to knowledge in this field by providing a comprehensive analysis of AVs and practical recommendations to enhance safety in autonomous driving systems.

2 Methodology

This study employed a literature review to explore cybersecurity in autonomous vehicles (AVs). A comprehensive search of academic databases and the snowball technique were employed to identify relevant studies. Rigorous criteria were applied to select and systematically analyze studies for vulnerabilities, attack types, and countermeasures. Attacks were categorized into three groups: GPS sensor security, LiDAR sensor security, and vehicular network attacks.

3 Autonomous Vehicle Technology

3.1 What is an Autonomous Vehicle?

An autonomous vehicle is defined as a vehicle that can drive independently by utilizing various data sources and analyzing the environment to control its operations, without the need for intervention [6].

Autonomous driving is considered a significant challenge because it must contemplate different such as a vehicle's knowledge or manipulation capability, compliance with traffic regulations, adaptation to variable road conditions, safe interaction with other vehicles and pedestrians, and the ability to make decisions in adverse situations.

3.2 Fundamental Pillars of Autonomous Driving

Autonomous driving is based on three pillars: perception, planning, and control [7].

Perception. The vehicle must have the ability to accurately and reliably perceive its surroundings, employing various sensors such as cameras, LiDAR and radars.
Planning. The vehicle's ability to design a safe and efficient route stands on the information collected through perception and adherence to driving regulations.
Control. Safe and precise management of the vehicle's movement is achieved through the implementation of a motion control system.

3.3 Levels of Autonomous Driving

The Society of Automotive Engineers (SAE) introduced the categorization of autonomous driving functions into six tiers of autonomy, numbered from 0 to 5 [8]. This classification depends on the level of reliance the vehicle has on the human driver, specifically, the extent of human participation in the driving process [9].

Level 0. At this level, vehicles are traditional and do not feature automation or autonomous driving systems, requiring a human driver to manage all car functions [10].
Level 1. The human driver is primarily responsible for most vehicle functions, with limited autonomous support, which may include features like parking assistance and dynamic cruise control to regulate speed and distance, among other functions [10].
Level 2. Vehicles offer partial automation, allowing drivers to delegate certain driving tasks. Drivers must remain attentive and ready to intervene, while the vehicle aids with acceleration, steering, braking, and speed control [10].
Level 3. Conditional automation is achieved only under optimal road conditions, enabling vehicles to operate autonomously. This enables drivers to relax, engage in other activities, and fully delegate driving tasks to the vehicle.[10].
Level 4. High automation allows vehicles to steer, accelerate, and brake autonomously, though this capability is limited to certain conditions or optimal road environments. Vehicles are equipped to observe road conditions and reacting to obstacles [10].
Level 5. Involves complete automation in vehicles, requiring no human input. At this degree of autonomy, vehicles employ Artificial Intelligence to process real-time data collected from sensors [10].

3.4 Security Principles

Autonomous vehicles are a crucial component of an interconnected network that includes various devices and other vehicles, becoming a key component of an extensive infrastructure where sensor and personal device information are shared. This interconnection increases the risk of cyberattacks on autonomous vehicles. An attacker could compromise not just the vehicle's security but also the integrity of its occupants, as well as that of other vehicles and pedestrians in the road environment. Therefore, it is essential to thoroughly tackle the cybersecurity of autonomous vehicles to reduce the likelihood of material damage, injuries, and even human losses.

When discussing cybersecurity in autonomous vehicles, it's crucial to take into account the three pillars of information security outlined in the ISO/IEC 27001:2022 standard: confidentiality, integrity, and availability. These pillars form the foundation upon which various techniques, guidelines, and strategies are developed to ensure effective risk management in autonomous vehicle security.

Confidentiality refers to data protection, ensuring that only authorized individuals can access information, thereby preventing unauthorized disclosures. Integrity ensures that data is accurate and consistent, preventing any unauthorized modification, destruction, or loss of information. Finally, availability focuses on keeping resources accessible to authorized users at all times [11].

4 Main Sensors in Autonomous Vehicles

To allow a vehicle to function autonomously, it must be outfitted with a range of sensors that enable it to accurately perceive its surroundings. This constitutes the vehicle's perception system. Among the most used sensors in the automotive industry are:

Internal Sensors. Cameras and audio sensors for monitoring the driver's state.
External Sensors. They include obstacle detection such as Radars, LiDAR, Cameras, Location sensors such as GPS/IMU, Audio sensors, rain, noise and suchlike.

The main sensors in autonomous vehicles focused on scene segmentation and object detection are: Radar, LiDAR, and Cameras.

4.1 Radar

A radar sensor uses electromagnetic waves to determine distances, altitudes, directions, and speeds of objects. This 'echo' collects extensive data, enabling the detection of objects beyond visible light or sound ranges. The distance calculation depends on factors like transmitted and received power, antenna gains, signal wavelength, target reflection capability, and system loss [12]. Simultaneous Localization and Mapping (SLAM) systems based on radar have advantages such as modeling the environment's structural features, cost-effectiveness, and performance in dusty or foggy conditions [13]. However, they may struggle with changing conditions and dynamic objects [14], and their mapping resolution may be less effective over long distances compared to LiDAR.

4.2 LiDAR

LiDAR (Light Detection and Ranging) emits thousands of laser signals with a specific frequency towards a target [15]. When these signals hit a target, they reflect back to the receiver, allowing the Time of Flight (ToF) to be calculated. ToF represents the time for the signal's round trip, enabling distance calculation to the target. The distance is determined by dividing the ToF by twice the speed of light, providing an accurate distance estimate [16]. The common location of the sensors is shown in Fig. 1.

LiDAR captures data with a 360° horizontal angle, creating a 3D point cloud. This data undergoes Point Cloud Semantic Segmentation (PCSS) using deep learning to identify and monitor objects around the vehicle, which is crucial for scene segmentation [16]. Compared to radars, LiDAR offers higher precision, resolution, and sensing speed, making it reliable even at night without external lighting. However, the cost is higher and the large volume of rapidly collected data requires high processing power.

4.3 Cameras

In autonomous vehicles, Complementary Metal-Oxide-Semiconductor (CMOS) sensors are commonly used. In CMOS sensors, photons convert into voltage in each pixel, with multiple transistors amplifying the signal for further processing and image creation [17]. These sensors offer high resolution at a low cost but struggle with depth estimation and efficiency in extreme weather conditions such as rain or snow.

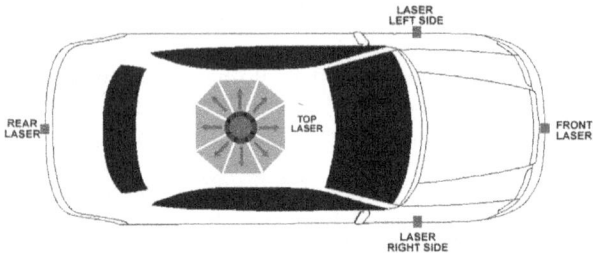

Fig. 1. Location of LiDAR sensors on a Waymo vehicle.

In conclusion, every sensor type has distinct advantages and limitations. A significant challenge in designing perception systems for autonomous vehicles is detecting and classifying objects in various scenarios. To address this, automotive companies often integrate multiple sensor types to enhance system reliability in diverse environments.

5   Fundamentals of Artificial Intelligence Applied to Autonomous Vehicles

After collecting data using the perception system of an autonomous vehicle, processing and interpreting the information is essential for generating control commands. This step is crucial for autonomous driving, where Machine Learning (ML) plays a key role. Machine learning is a subfield of artificial intelligence that develops algorithms enabling computers to learn and improve from experience without explicit programming [18].

5.1   Machine Learning Algorithms

The main types of machine learning algorithms are:

Supervised Learning. This approach uses a large labeled dataset for training, followed by evaluation on a separate test dataset. These algorithms are commonly employed for classification or regression tasks [18].
Unsupervised Learning. Unlike the previous one, this type of algorithm process unlabeled data sets. It is usually used for clustering or estimating data density [18].
Reinforcement Learning. Such an algorithm determines the optimal actions through rewards or penalties. It can be viewed as a fusion of the two previous types of algorithms. It is beneficial in scenarios where data is scarce or unavailable [18].

5.2   Machine Learning Applications in Autonomous Vehicles

Machine learning algorithms empower autonomous vehicles to sense and comprehend their environment by identifying and categorize objects like traffic signs, vehicles and pedestrians. These algorithms not only recognize objects but also inform the vehicle's decision-making process. Through a filtering and convolution process, images captured by cameras are prepared for neural network input by maintaining clarity while simplifying for processing. Each pixel's value serves as input for the network, with the network's input neuron count determined by the image's pixel count.

Scene Segmentation. Among the key applications of machine learning in self-driving cars is scene segmentation, iinvolving the identification and classification of objects within a scene. Object masks are essential for the algorithm to learn effectively. In the research field, various studies have utilized the Microsoft Common Objects in Context (COCO) dataset, which includes over 90 categories of common objects, of which 82 have more than 5000 labeled instances. Overall, the dataset contains 2.5 million labeled instances distributed across 328,000 images [19].

Lane Detection. The lane detection process proposed by Guerrero-Ibañez et al. begins with the lane image taken by the vehicle's front camera, which is transformed into grayscale to enhance edge detection. A Gaussian filter is employed to reduce noise and smooth the image. The Canny method is used for edge detection, reducing the amount of required learning data. Subsequently, a Region of Interest (ROI) is defined to focus on the lane lines, and a mask is applied to isolate this area. This method is based on computer vision and combines various techniques to achieve precise lane detection and is widely used in the automotive industry [20].

6 Infrastructure and Communication Technologies

6.1 Connected Autonomous Vehicles (CAVs)

An increasingly relevant concept is that of Connected Autonomous Vehicles (CAVs), where various technologies are integrated to facilitate connectivity between vehicles and offer a variety of services including road safety, remote monitoring, roadside support, driving optimization and system maintenance and malfunction prevention [21].

The main types of communication in autonomous vehicles are divided into several types, each with its specific function and determines the range of communication and interoperability with other systems.

V2V (Vehicle-to-Vehicle). This modality focuses on the direct connection between vehicles, functioning at short distances. Its purpose is to facilitate communication of driving plans between drivers, allowing for traffic analysis, optimization, and congestion reduction [22]. Additionally, it can significantly contribute to accident prevention by allowing the transmission of emergency situations, such as sudden braking or other incidents on the road.

V2I (Vehicle-to-Infrastructure). This communication type facilitates interactions between vehicles and the existing road network infrastructure, encompassing traffic lights, tolls, antennas, and other elements. Its primary function is to enhance traffic management at intersections and support highway operations. [23, 24].

V2X (Vehicle-to-Everything). This concept encompasses communication between vehicles (V2V), infrastructure (V2I), networks (V2N), and pedestrians (V2P), including elements such as headlights, traffic lights, antennas, mobile devices, and cloud servers. Recently, it has introduced the concept of Vehicle as a Service (VaaS) [25]. This comprehensive approach allows for more complete control in terms of road safety and traffic efficiency (Fig. 2).

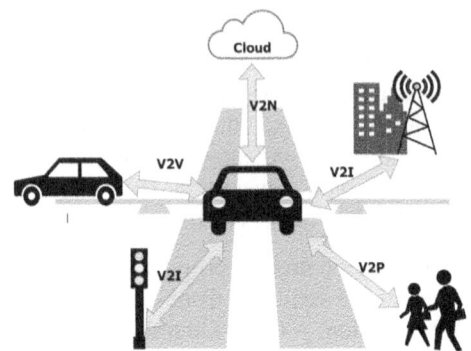

Fig. 2. Connected autonomous vehicle environment.

6.2 Communication Technologies in A.V.

To understand how autonomous vehicles exchange data with their environment, it is essential to explore the technologies, protocols, and common standards in this context, as well as the associated risks that an autonomous vehicle may be exposed to.

VANET (Vehicular Ad-Hoc Network). This is a highly utilized network in smart vehicles. Specifically designed for communication between different vehicles (V2V) and with infrastructure elements (V2I), VANET is characterized by its flexibility, as it does not require a fixed network infrastructure. Instead, the topology and connections are changing as vehicles move [26, 27]. Typically, VANETs use Dedicated Short Range Communications (DSRC) for their operations.

DSRC (Dedicated Short Range Communication). DSRC, widely used in VANETs for V2V and V2I communication, operates in the 5.9 GHz frequency band with a range of 100 to 300 m [28]. It has a 75 MHz spectrum allocation from the FCC in the US [29] and adheres to the IEEE 802.11p standard, created for wireless connectivity in vehicular networks. This protocol provides a reliable framework for safe and efficient communication, enabling interoperability between devices [30, 31].

C-V2X (Cellular Vehicle-to-Everything). This technology surpasses DSRC in maturity and facilitates seamless vehicle interconnection and a variety of entities, including other vehicles, pedestrians, infrastructures, and networks. It is based on LTE (4G) and 5G cellular technology [32]. C-V2X employs two communication modes. In the first mode, it conducts direct communications with vehicles, infrastructures, and pedestrians on the road, independently of cellular networks, using the PC5 interface. In the second mode, it uses the conventional cellular network to inform vehicles about traffic conditions and the state of the road in their area. The interface used is (Uu) [33].

7 Security in Autonomous Vehicles

Autonomous driving presents significant challenges in infrastructure and security, guaranteed in its 3 pillars: confidentiality, integrity, and availability.

The success in integrating vehicles with the environment largely depends on the strength of the implemented security measures. Given the complexity of the system, the autonomous vehicle is considered a critical asset, which implies identifying, evaluating, and mitigating the potential security risks related to its operation. It is crucial to identify the different components that constitute Autonomous Vehicles (AVs), such as sensors, software, hardware, systems, and communication. By knowing the vulnerabilities inherent in each of these elements and the potential threats, a solid starting point is established for the comprehensive protection of AVs.

7.1 Main Security Challenges

Integrity Attacks. These attacks could involve identity theft or data forgery, which could result in unauthorized vehicle control. For example, the Sybil attack, where an attacker impersonates legitimate entities by using counterfeit identities. In this type of attack, solid cryptographic schemes can be implemented as a preventive measure [34].

Availability Attacks. Threats to security, like denial-of-service (DoS) attacks can disrupt the operation of autonomous vehicles and overload systems. Mitigation involves access control through packet filtering [35]. Malware injection, firmware modification, and jamming of communication channels are other methods used to compromise availability. The vulnerability of autonomous vehicles is due to their wireless connectivity that makes them targets for man-in-the-middle attacks. These attacks can intercept communication signals and take control of vehicles or infrastructure system.

Confidentiality Attacks. Autonomous vehicles, handling vast data volumes, face threats compromising data confidentiality and privacy. Attacks may target data interception or storage system breaches. Essential measures include data encryption and powerful privacy management to safeguard information. Data exposure, a primary risk, stems from the extensive data generated by vehicle sensors. This data is accessible to authorities and shared among vehicles and infrastructures to enhance decision-making.

7.2 Security Strategies for Autonomous Vehicles

The security of autonomous vehicles requires the implementation of robust measures aimed at preventing and reducing potential threats. This involves incorporating rigorous cybersecurity protocols to safeguard the vehicle's computer systems against possible malicious attacks. Below are some of the most common and important controls:

Cryptography. Cryptography-based controls are an efficient strategy to counter potential threats to integrity. It is crucial to make sure that data is encrypted with strong algorithms during storage, and if necessary, secure data transmission to the command center must be ensured through an encrypted communication protocol [36].

Redundancy. Refers to duplicating critical devices and functionalities of an AV system from sensors to navigation systems. This measure allows the system to be more fault-tolerant. If one component fails, the system's functionality can be supported by redundant components, thus avoiding catastrophic failure. This aspect should be considered from the initial stages of design.

Digital Twins. Refers to a digital model of a real-world entity or system. This representation can be created using computer programs or integrated software models that interact and synchronize with their tangible equivalent. S. Almeaibed in [37] proposed a framework based on digital twins of vehicles to facilitate gathering and processing the vast amount of data an AV must operate with and thus improve decision-making in real systems. This helps reduce risk from both perspectives, cybersecurity, and accidents.

K-Anonymization. Is a crucial privacy protection method for handling the substantial volume of personal data collected by autonomous vehicle systems. It aims to prevent re-identification risks by making it difficult to distinguish individuals in datasets. This technique compared to traditional anonymization ensures that each record in the dataset is essentially indistinguishable from a minimum ok k-1 additional records, thus strengthening privacy protection in the context of autonomous vehicles [38].

Security by Design. Maintaining software and firmware security from the early stages of development involves implementing secure coding practices, applying controls to mitigate threats, and using methods like the principle of least privilege, static and dynamic testing, and Secure Development Methodologies.

Security Updates in Hardware and Software. A systematic process of hardware and software updates for autonomous vehicle systems must be implemented, ensuring their periodic application. This guarantees that each vehicle has the latest security measures. It is essential that vehicle maintenance not only includes routine mechanical checks but also makes this specific security update phase mandatory.

Penetration Testing. Conducting periodic penetration tests to detect potential vulnerabilities in the system and correct them before they can be exploited. These tests should be aligned with a properly structured risk analysis and management plan.

Network Monitoring. In connected autonomous vehicles, real-time monitoring through Machine Learning (ML) is crucial for predicting, detecting, and mitigating security threats. In [39], a proposed approach utilizes a dynamic Bayesian network model to evaluate real-time risks. This allows vehicles to analyze environmental risk factors, deduce, and quantify potential risks.

7.3   Main Types of Attacks and Countermeasures in Autonomous Vehicles

This section reviews the main attacks on Autonomous Vehicles (AVs) and the countermeasures proposed by authors to mitigate these risks, classifying them into three groups: GPS sensor security, LiDAR sensor security and attacks on vehicular networks.

GPS Sensor Security Attacks

Identity Spoofing Attacks. Attackers can generate synthetic signals to mimic legitimate GPS signals, which can lead to incorrect calculation of a vehicle's location and potentially cause a traffic accident [40]. To counteract this attack, Chu et al. [41] suggest a method that uses a specialized server generates hash-based authentication codes from GPS navigation data, enabling receivers to simultaneously authenticate the server and validate GPS signals.

Jamming Attacks. Attackers emit signals that coincide with those of GPS to saturate the original ones, preventing the receiver from establishing a connection and determining the precise location [40]. Against this type of attack, it is highly recommended to use intrusion detection systems, as well as redundancy in critical systems [42].

Replay Attacks. This attack entails intercepting GPS signals and subsequently broadcasting them at a different time or location. Such actions may mislead the GPS receiver and result in inaccurate positioning or timing [40]. Ucar et al. [43] propose several strategies to mitigate these attacks, such as using clocks synchronized with GPS or packet sequence numbers, along with the use of cryptographic keys, digital certificates, and signatures to defend against data packet spoofing attacks.

LiDAR Sensor Security Attacks

Relay Signal Attacks. Attackers can divert autonomous vehicles from their route by relaying LiDAR signals. This method, similar to a replay attack, creates a false image of the environment for the target vehicle. The relay is performed from a location different from the vehicle's real one, using only two transceivers [44]. In this context, using short pulse periods and redundancy of probing and wavelength in the LiDAR system is recommended to hinder the synchronization of false signals and detect anomalies [45].

Spoofing Attacks. They involve creating false objects by transmitting signals that mimic the LiDAR scanner frequency [15]. As a countermeasure, it is recommended to use short pulse periods and redundancy of probing and wavelength in the LiDAR system [45]. Additionally, the same authors suggest implementing an anomaly detection system in LiDAR data, which embeds a low-interference binary watermark to detect alterations during transmission. This watermark facilitates the detection of alterations, while real-time verification identifies discrepancies.

Electromagnetic Interference (EMI). It can manipulate LiDAR sensor data, generating false positives that can cause traffic accidents [46]. Bhupathiraju et al. [46] recommend implementing a ground point-based detection system, which analyzes the average intensity of LiDAR point clouds to identify potential interference.

Attacks on Vehicular Networks

Sybil Attack. A malicious actor within a vehicular network can create multiple false identities to affect the operation of applications [34]. An effective control to timely detect this type of attack is intrusion detection systems (IDS). Yang and colleagues [47] suggest an intelligent intrusion detection system based on machine learning designed to recognize various types of attacks within autonomous vehicle networks.

Denial of Service (DoS). By using a vehicular botnet, the communication capacity of a vehicle or group of vehicles can be saturated. Countermeasures include anomaly detection systems and secret key agreements in platoon-based systems to encrypt vehicle communications [48].

Radio Interference. It involves disrupting communications and affecting the IEEE 802.11p standard, which employs a control channel (CCH) in addition to various service channels (SCH). During this type of attack, the aggressor can generate interference by blocking a specific channel or rapidly switching between all channels to block them all [48], Ali Hamieh et al. [49] present a solution that uses the correlation between errors and correct reception times to identify potential interferences in vehicular networks.

Eavesdropping Attack. Both passive and active attacks can target communication lines, enabling adversaries to track and manipulate vehicle locations [50]. Proposed solutions include secure communication schemes and reliable anomaly detection systems to minimize false positives [51]. A secret key agreement system in platoon-based systems encrypts vehicle communications, ensuring confidentiality of driving patterns and sensitive information, protecting against eavesdropping and enhancing overall network security [52].

Software and Hardware Vulnerability Attacks. They pose a significant threat to autonomous vehicles, exploiting vulnerabilities in software or hardware for unauthorized access or unwanted actions, including fault injections and wireless attacks on connected vehicles [53, 54]. To counter this, various security measures are proposed. Woo et al. [54] suggest a protocol using certificate-based authentication and a long-term symmetric key in Authenticated Key Exchange Protocol 2 (AKEP2). Additionally, code obfuscation techniques and whitelist-based firewalls are proposed to regulate access to vehicle components, highlighting the importance of proactive security measures [53].

8   Regulations Governing Autonomous Vehicles

The SAE J3016, developed by the Society of Automotive Engineers (SAE), is the main regulation for autonomous vehicles, defining levels of autonomous driving and offering a standardized framework for manufacturers [9, 37]. The ISO/PAS 21448:2022 standard also plays a crucial role in specifying safety standards for autonomous vehicle systems and driver assistance functions. Globally, local government regulations vary: the National Highway Traffic Safety Administration (NHTSA) based in the United States, approved cars without steering wheel or pedals; across the European Union, Directive 2010/40/EU guides the deployment of intelligent transport systems. Regarding information security, international standards like ISO 27001 and NIST 800-53 are widely adopted, providing essential guidelines for ensuring data integrity and confidentiality.

9   Results

Table 1 summarizes the types of attacks previously analyzed and classified, along with some proposed solutions, and their impact on integrity (I), availability (A), and confidentiality (C).

This study highlights the importance of adopting multi-layered security approaches to strengthen the defenses of autonomous vehicles against these cyberattacks. The results underscore the need for integrating countermeasures such as cryptographic authentication and anomaly detection systems for GPS sensor security, advanced filtering techniques and sensor fusion for LiDAR sensor security, and secure communication protocols

Table 1. Summary of Attacks and Countermeasures

Attack Surface	I	A	C	Type of attack	Countermeasures
GPS Sensors	X	X	X	Spoofing	Cryptographic solutions [41]
		X		Jamming	IDS, Redundancy in critical systems [42]
	X			Reply	Synchronised clocks with GPS, packet sequence numbers, cryptographic solutions [43]
LiDAR Sensors	X	X		Relay signal attacks	Short pulse periods, redundancy [45]
	X	X		Spoofing signal	Short pulse periods, redundancy [45]
	X	X	X	EMI	Detection based on ground points [46]
Networks Vehicular	X	X	X	Sybil	IDS, Cryptographic solutions [47]
	X	X	X	DoS	IDS, Cryptographic solutions [48]
	X	X	X	Radio Interference	Errors and reception time correlation [49]
	X	X	X	Eavesdropping	IDS, Cryptographic solutions [51, 52]
Software/hardware	X	X	X	Fault Injections, Wireless Attacks	IDS, Cryptographic solutions, Code obfuscation, ACLs [53, 54]

and intrusion detection systems for vehicular network security. These measures should be prioritized from the design phase by manufacturers and developers, while policy-makers should establish stringent standards and guidelines to guarantee the security and reliability of autonomous vehicles on public roads.

10   Conclusions and Future Work

This study highlights the broad range of cyberattacks to which autonomous vehicles are vulnerable. While offering significant advantages, like increased road safety and improved accessibility for individuals with reduced mobility, they also face considerable challenges related to security and regulatory compliance. Addressing these challenges requires viewing autonomous vehicles and their ecosystem as interconnected systems that require robust frameworks for analyzing and mitigating security risks.

The findings underscore the need for strict regulatory oversight, not only for AV manufacturers but also for quality control authorities, to ensure adherence to safety standards throughout the vehicle's lifecycle. Local regulations should mandate periodic updates to firmware and software, as these measures are essential for reducing the risk of cyber-attacks on AV systems. Additionally, automotive companies must prioritize data privacy and confidentiality, in alignment with regulations like the General Data Protection Regulation (GDPR), to safeguard user information.

As future work, the goal is to expand the research and develop a security threat model specifically focused on autonomous vehicles. Additionally, developing best practice guidelines will be crucial for the industry to effectively implement security measures.

References

1. Chowdhury, A., Karmakar, G., Kamruzzaman, J., Jolfaei, A., Das, R.: Attacks on self-driving cars and their countermeasures: a survey. IEEE Access **8**, 207308–207342 (2020). https://doi.org/10.1109/ACCESS.2020.3037705
2. Algarni, A.M., Thayananthan, V.: Autonomous vehicles with a 6G-based intelligent cybersecurity model. IEEE Access **11**, 15284–15296 (2023). https://doi.org/10.1109/ACCESS.2023.3244883
3. Vitale, C., et al.: CARAMEL: results on a secure architecture for connected and autonomous vehicles detecting GPS spoofing attacks. EURASIP J. Wirel. Commun. Netw. **2021**, 1–28 (2021). https://doi.org/10.1186/S13638-021-01971-X
4. Nguyen, H.N., Shaikh, S.A., Kutsal, E., Stylianou, A., Potter, R., Sors, T.: Addressing automotive cybersecurity risks with an ARM Morello capability-enhanced prototype. Authorea Preprints (2024). https://doi.org/10.22541/AU.171401582.21766451/V1
5. Cobos, L.-P., Ruddle, A.R., Sabaliauskaite, G.: Cybersecurity assurance challenges for future connected and automated vehicles. In: Proceedings of the 31st European Safety and Reliability Conference (ESREL 2021), pp. 2038–2045 (2021). https://doi.org/10.3850/978-981-18-2016-8
6. Collingwood, L.: Privacy implications and liability issues of autonomous vehicles. Inf. Commun. Technol. Law **26**, 32–45 (2017). https://doi.org/10.1080/13600834.2017.1269871
7. Pilar Arnanz, S.: Servicios de vehículo conectado y conducción autónoma en un Twizy. Repositorio Universidad de Valladolid (2021)
8. Kalda, K., Sell, R., Soe, R.M.: Use case of Autonomous Vehicle shuttle and passenger acceptance analysis. Proc. Est. Acad. Sci. **70**, 429–435 (2021)
9. SAE International: J3016_202104: Taxonomy and Definitions for Terms Related to Driving Automation Systems for On-Road Motor Vehicles - SAE International. https://www.sae.org/standards/content/j3016_202104/
10. Khan, W.Z., Khurram Khan, M., Arshad, Q.U.A., Malik, H., Almuhtadi, J.: Digital labels: influencing consumers trust and raising cybersecurity awareness for adopting autonomous vehicles. In: Digest of Technical Papers - IEEE International Conference on Consumer Electronics, 2021-January (2021). https://doi.org/10.1109/ICCE50685.2021.9427684
11. Maleh, Y., Sahid, A., Ezzati, A., Belaissaoui, M.: A capability maturity framework for IT security governance in organizations. Adv. Intell. Syst. Comput. **735**, 221–233 (2018). https://doi.org/10.1007/978-3-319-76354-5_20/TABLES/3
12. Liang, M., Chaoying, H., Xiaoyu, X.: Target dynamic radar echo simulation based on sensor. Procedia Comput. Sci. **174**, 706–711 (2020). https://doi.org/10.1016/J.PROCS.2020.06.146

13. Lohar, S., Zhu, L., Young, S., Graf, P., Blanton, M.: Sensing technology survey for obstacle detection in vegetation. Future Transp. **1**, 672–685 (2021). https://doi.org/10.3390/FUTURE TRANSP1030036

14. Ort, T., Gilitschenski, I., Rus, D.: Autonomous navigation in inclement weather based on a localizing ground penetrating radar. IEEE Robot Autom Lett. **5**, 3267–3274 (2020). https://doi.org/10.1109/LRA.2020.2976310

15. Cao, Y., et al.: Adversarial sensor attack on LiDAR-based perception in autonomous driving. In: Proceedings of the ACM Conference on Computer and Communications Security, pp. 2267–2281 (2019). https://doi.org/10.1145/3319535.3339815

16. Kim, S., Ha, J., Jo, K.: Semantic point cloud-based adaptive multiple object detection and tracking for autonomous vehicles. IEEE Access **9**, 157550–157562 (2021). https://doi.org/10.1109/ACCESS.2021.3130257

17. Taraba, M., Adamec, J., Danko, M., Drgona, P.: Utilization of modern sensors in autonomous vehicles. In: 12th International Conference ELEKTRO 2018, 2018 ELEKTRO Conference Proceedings, pp. 1–5 (2018). https://doi.org/10.1109/ELEKTRO.2018.8398279

18. Morovat, K., Panda, B.: A survey of artificial intelligence in cybersecurity. In: Proceedings of the 2020 International Conference on Computational Science and Computational Intelligence, CSCI 2020, pp. 109–115 (2020). https://doi.org/10.1109/CSCI51800.2020.00026

19. Lin, T.-Y., et al.: Microsoft COCO: common objects in context. Lecture Notes in Computer Science (including subseries Lecture Notes in Artificial Intelligence and Lecture Notes in Bioinformatics). LNCS, vol. 8693, pp. 740–755 (2014). https://doi.org/10.1007/978-3-319-10602-1_48/COVER

20. Guerrero-Ibáñez, J., Contreras-Castillo, J., Santana-Mancilla, P.: Modelo basado en visión por computadora para detección de carriles viales para la autonomía de vehículos. Transformación Digital: Avances y paradigmas tecnológicos 272–279 (2021)

21. Hakak, S., et al.: Autonomous vehicles in 5G and beyond: a survey. Veh. Commun. **39**, 100551 (2023). https://doi.org/10.1016/J.VEHCOM.2022.100551

22. Lahdya, S., Mazri, T.: Data security challenges in self-driving car. Int. Arch. Photogrammetry Remote Sens. Spat. Inf. Sci. XLVIII-4-W3-2022, 61–66 (2022)

23. Bučko, B., Michálek, M., Papierniková, K., Zábovská, K.: Smart mobility and aspects of vehicle-to-infrastructure: a data viewpoint. Appl. Sci. **11**, 10514 (2021). https://doi.org/10.3390/APP112210514

24. Wu, M., Jin, L., Amin, S., Jaillet, P.: Signaling game-based misbehavior inspection in V2I-enabled highway operations. In: Proceedings of the IEEE Conference on Decision and Control, vol. 2018-December, pp. 2728–2734 (2018). https://doi.org/10.1109/CDC.2018.8619109

25. Liu, S., Liu, L., Tang, J., Yu, B., Wang, Y., Shi, W.: Edge computing for autonomous driving: opportunities and challenges. Proc. IEEE **107**, 1697–1716 (2019). https://doi.org/10.1109/JPROC.2019.2915983

26. Ashokkumar, K., Sam, B., Arshadprabhu, R.: Britto: cloud based intelligent transport system. Procedia Comput. Sci. **50**, 58–63 (2015). https://doi.org/10.1016/J.PROCS.2015.04.061

27. Qamar, M., Fouzia, Khan, S., Aqsa, Mehmood, A., Raeena: MANet vs VANet - the applications & challenges. Lahore Garrison Univ. Res. J. Comput. Sci. Inf. Technol. **3**, 34–38 (2019)

28. Kabbur, M., Arul Kumar, V.: Detection and prevention of DoS attacks in VANET with RSU's cooperative message temporal signature. Int. J. Recent Technol. Eng. **8**, 6371–6377 (2019). https://doi.org/10.35940/IJRTE.B2210.078219

29. Guan, W., He, J., Ma, C., Tang, Z., Li, Y.: Adaptive message rate control of infrastructured DSRC vehicle networks for coexisting road safety and non-safety applications (2012). https://doi.org/10.1155/2012/134238

30. Ansari, S., Boutaleb, T., Sinanovic, S., Gamio, C., Krikidis, I.: On the design and deployment of multitier heterogeneous and adaptive vehicular networks. In: 2018 11th International Symposium on Communication Systems, Networks and Digital Signal Processing, CSNDSP 2018 (2018). https://doi.org/10.1109/CSNDSP.2018.8471807
31. Tahmasbi-Sarvestani, A., Nourkhiz Mahjoub, H., Fallah, Y.P., Moradi-Pari, E., Abuchaar, O.: Implementation and evaluation of a cooperative vehicle-to-pedestrian safety application. IEEE Intell. Transp. Syst. Mag. 9, 62–75 (2017)
32. Kim, K., Kim, J.S., Jeong, S., Park, J.H., Kim, H.K.: Cybersecurity for autonomous vehicles: review of attacks and defense. Comput. Secur. 103, 102150 (2021). https://doi.org/10.1016/J.COSE.2020.102150
33. Miao, L., Virtusio, J.J., Hua, K.-L.: PC5-based cellular-V2X evolution and deployment. Sensors 21(3), 843 (2021). https://doi.org/10.3390/s21030843
34. Mejri, M.N., Ben-Othman, J., Hamdi, M.: Survey on VANET security challenges and possible cryptographic solutions. Veh. Commun. 1, 53–66 (2014)
35. Deepa Thilak, K., Amuthan, A.: DoS attack on VANET routing and possible defending solutions-a survey. In: 2016 International Conference on Information Communication and Embedded Systems, ICICES 2016 (2016). https://doi.org/10.1109/ICICES.2016.7518892
36. Madan, B.B., Banik, M., Bein, D.: Securing unmanned autonomous systems from cyber threats. Sage J. 16, 119–136 (2016). https://doi.org/10.1177/1548512916628335
37. Almeaibed, S., Al-Rubaye, S., Tsourdos, A., Avdelidis, N.P.: Digital twin analysis to promote safety and security in autonomous vehicles. IEEE Commun Stand. Mag. 5, 40–46 (2021). https://doi.org/10.1109/MCOMSTD.011.2100004
38. Trujillo-Rasua, R., Domingo-Ferrer, J.: On the privacy offered by (k, δ)-anonymity. Inf. Syst. 38, 491–494 (2013). https://doi.org/10.1016/J.IS.2012.12.003
39. Wang, D., Fu, W., Song, Q., Zhou, J.: Potential risk assessment for safe driving of autonomous vehicles under occluded vision. Sci. Rep. 12(1), 1–14 (2022). https://doi.org/10.1038/s41598-022-08810-z
40. Giannaros, A., et al.: Autonomous vehicles: sophisticated attacks, safety issues, challenges, open topics, blockchain, and future directions. J. Cybersecurity Priv. 3(3), 493–543 (2023). https://doi.org/10.3390/jcp3030025
41. Chu, Y.H., Keoh, S.L., Seow, C.K., Cao, Q., Wen, K., Tan, S.Y.: GPS signal authentication using a chameleon hash keychain. In: IFIP Advanced Information and Communication Technologies, IFIPAICT, vol. 636, pp. 209–226 (2022). https://doi.org/10.1007/978-3-030-93511-5_10/COVER
42. Wang, Y., Masoud, N., Khojandi, A.: Real-time sensor anomaly detection and recovery in connected automated vehicle sensors. IEEE Trans. Intell. Transp. Syst. 22, 1411–1421 (2021). https://doi.org/10.1109/TITS.2020.2970295
43. Ucar, S., Ergen, S.C., Ozkasap, O.: IEEE 802.11p and visible light hybrid communication based secure autonomous platoon. IEEE Trans. Veh. Technol. 67(9), 8667–8681 (2018). https://doi.org/10.1109/TVT.2018.2840846
44. Petit, J., Stottelaar, B., Feiri, M.: Remote Attacks on Automated Vehicles Sensors: Experiments on Camera and LiDAR. Black Hat Europe (2015)
45. Changalvala, R., Malik, H.: LiDAR data integrity verification for autonomous vehicle. IEEE Access 7, 138018–138031 (2019). https://doi.org/10.1109/ACCESS.2019.2943207
46. Bhupathiraju, S.H.V., Sheldon, J., Bauer, L.A., Bindschaedler, V., Sugawara, T., Rampazzi, S.: EMI-LiDAR: uncovering vulnerabilities of LiDAR sensors in autonomous driving setting using electromagnetic interference. In: WiSec 2023 - Proceedings of the 16th ACM Conference on Security and Privacy in Wireless and Mobile Networks, pp. 329–340 (2023)
47. Yang, L., Moubayed, A., Hamieh, I., Shami, A.: Tree-based intelligent intrusion detection system in internet of vehicles. In: Proceedings - IEEE Global Communications Conference, GLOBECOM (2019). https://doi.org/10.1109/GLOBECOM38437.2019.9013892

48. Amoozadeh, M., et al.: Security vulnerabilities of connected vehicle streams and their impact on cooperative driving. IEEE Commun. Mag. **53**, 126–132 (2015)

49. Hamieh, A., Ben-Othman, J., Mokdad, L.: Detection of radio interference attacks in VANET. In: GLOBECOM - IEEE Global Telecommunications Conference (2009). https://doi.org/10.1109/GLOCOM.2009.5425381

50. Solnør, P., Volden, Ø., Gryte, K., Petrovic, S., Fossen, T.I.: Hijacking of unmanned surface vehicles: a demonstration of attacks and countermeasures in the field. J. Field Robot. **39**, 631–649 (2022). https://doi.org/10.1002/ROB.22068

51. Meyer, P., Hackel, T., Korf, F., Schmidt, T.C.: Network anomaly detection in cars based on time-sensitive ingress control. In: IEEE Vehicular Technology Conference, vol. 2020-November (2020). https://doi.org/10.1109/VTC2020-FALL49728.2020.9348746

52. Li, K., et al.: Design and implementation of secret key agreement for platoon-based vehicular cyber-physical systems. ACM Trans. Cyber-Phys. Syst. **4**(2), 1–20 (2019). https://doi.org/10.1145/3365996

53. Lee, Y., Woo, S., Lee, J., Song, Y., Moon, H., Lee, D.H.: Enhanced android app-repackaging attack on in-vehicle network. Wireless Commun. Mob. Comput. **2019**, 1–13 (2019). https://doi.org/10.1155/2019/5650245

54. Woo, S., Jo, H.J., Lee, D.H.: A practical wireless attack on the connected car and security protocol for in-vehicle CAN. IEEE Trans. Intell. Transp. Syst. **16**, 993–1006 (2015). https://doi.org/10.1109/TITS.2014.2351612

Author Index

SPRINGER NATURE

GPSR Compliance

The European Union's (EU) General Product Safety Regulation (GPSR) is a set of rules that requires consumer products to be safe and our obligations to ensure this.

If you have any concerns about our products, you can contact us on ProductSafety@springernature.com

In case Publisher is established outside the EU, the EU authorized representative is:

Springer Nature Customer Service Center GmbH
Europaplatz 3
69115 Heidelberg, Germany

The manufacturer's authorised representative in the EU is Springer
Nature Customer Service Centre GmbH, Europaplatz 3, 69115 Heidelberg,
Germany. If you have any concerns regarding our products, please
contact ProductSafety@springernature.com

Printed and bound by CPI Group (UK) Ltd, Croydon, CR0 4YY

24/04/2026
02096358-0012